SHOCKS IN ASTROPHYSICS

SHOCKS IN ASTROPHYSICS

Proceedings of an International Conference held at UMIST, Manchester, England from January 9–12, 1995

Edited by

T. J. MILLAR
UMIST, Manchester, United Kingdom

and

A. C. RAGA
UNAM, Mexico City, Mexico

Reprinted from *Astrophysics and Space Science*
Volume 233, Nos. 1–2, 1995

KLUWER ACADEMIC PUBLISHERS
DORDRECHT / BOSTON / LONDON

A C.I.P. Catalogue record for this book is available from the Library of Congress.

ISBN-13: 978-94-010-7225-0 e-ISBN-13: 978-94-009-1624-1
DOI: 10.1007/978-94-009-1624-1

Published by Kluwer Academic Publishers,
P.O. Box 17, 3300 AA Dordrecht, The Netherlands.

Kluwer Academic Publishers incorporates
the publishing programmes of
D. Reidel, Martinus Nijhoff, Dr W. Junk and MTP Press.

Sold and distributed in the U.S.A. and Canada
by Kluwer Academic Publishers,
101 Philip Drive, Norwell, MA 02061, U.S.A.

In all other countries, sold and distributed
by Kluwer Academic Publishers Group,
P.O. Box 322, 3300 AH Dordrecht, The Netherlands.

Printed on acid-free paper

TABLE OF CONTENTS

PUBLISHER'S NOTE

From January 1 1996, major changes in the editorial procedures of *Astrophysics and Space Science* will be implemented. From that date, authors should submit manuscripts directly to the Editorial Office in Dordrecht at the address given below and NOT to the Editors. Manuscripts submitted directly to the Editors may experience delays.

Astrophysics and Space Science
Journals Editorial Office
Kluwer Academic Publishers
P.O. Box 17
3300 AA Dordrecht
The Netherlands

Preface

This volume contains the proceedings of an international conference on *Shocks in Astrophysics* held at UMIST, Manchester, England from January 9–12, 1995. The study of interstellar and circumstellar gas dynamics has a long and distinguished history in Manchester and has been almost entirely concentrated in the school founded by Franz Kahn in the Astronomy Department, University of Manchester. In January 1993, one of us (AR) was appointed to the faculty of the Astrophysics Group in the Department of Mathematics at UMIST and astrophysical gas dynamics became a major interest of the Group. The subject of this conference was chosen partly for the topicality of the subject matter and partly to help synthesise this expertise with the expertise in interstellar chemistry already present in the Group. The first fruits of this synthesis are contained in this volume. As it happened, this conference celebrated, not so much the beginnings of a long and fruitful collaboration, but rather gave many of Alex's friends the chance to say a fond farewell as he departed UMIST at the end of January 1995 to take up a chair at UNAM, Mexico City.

The core of this volume consists of twelve review articles, marked (R) in the list of contents, incorporating observational and theoretical studies of shock waves in a variety of situations from Herbig-Haro objects to Supernova Remnants to Active Galactic Nuclei. We have also included the contributed (C) and poster (P) papers. The efficacy of including the latter is much debated, particularly by book reviewers, but we felt the speed of publication promised by Kluwer meant that even short papers could play an important role in showing the breadth of the subject and in giving participants the opportunity to present exciting new results to the wider community. Finally, we decided to omit the discussions from the published volume. Partly this was because there could be no recorded discussions of poster papers—which is often where the comments which are most debatable and contain the most insight are made—partly because all authors were encouraged to incorporate relevant discussion into their articles and, most importantly, because the participants were encouraged to 'show their ignorance' during the discussion periods. A careful reading of the articles in this volume, however, will show that this field of research continues to be very active and central to many of the major issues in astronomy and astrophysics. The availability of ever-increasing computer power means that major advances, and perhaps even consensus, in our understanding of subjects as diverse as molecular outflows in star-forming regions and the relationship between black holes and active galactic nuclei can be expected in the next 5–10 years.

The conference was supported financially by the Collaborative Computational Project No. 7 (on *The Analysis of Astronomical Spectra*, by the Astronomy Group in the Department of Physics and Astronomy at the University of Manchester, and by the Astrophysics Group in the Department of Mathematics at UMIST. We are grateful to them.

This volume marks the end of a 30 year association between the Astrophysics Group and the Department of Mathematics at UMIST. On August 1st 1995, the Group transfers to the Department of Physics at UMIST. We expect that the tradition of Manchester Conferences on Astrophysics will continue in our new home.

T. J. Millar
Department of Physics
UMIST
Manchester

A. C. Raga
Instituto de Astronomía
UNAM
Mexico City

CIRCUMSTELLAR SHOCKS

J. MEABURN
Department of Physics and Astronomy, University of Manchester
Oxford Rd., Manchester, M13 9PL, UK

Abstract. The presence of shocks in a wide variety of circumstellar phenomena will be illustrated and discussed.

Key words: Shocks – Circumstellar Medium

1. Introduction

Shocks are shown to play a part in the formation of a wide variety of circumstellar phenomena. Here, evidence for shocks will be presented i) in the giant halo of the planetary nebula (PN) NGC 6543, ii) within the episodic, bi-polar jets from the PN Fleming 1, iii) in the nebulosity around the Luminous Blue Variables (LBVs) P Cygni and Eta Carinae and iv) around the newly formed low mass stars (YSOs) near the Trapezium stars in the Orion nebula (M 42).

2. The Giant Halo of NGC 6543

The faint, giant, outer halo of the PN NGC 6543 is most likely a consequence of the earliest ejection of mass from the progenitor star during its Red Giant phase. This view has been substantiated by measurements of its bulk motions which have indicated an expansion velocity of ≤ 10 km s^{-1} (Bryce et al 1992). Consequently, it seemed surprising that this halo not only is exceptionally bright in the [O III]5007 Å emission line but also has an electron temperature of $T_e = 14\ 700 \pm 850$ K from the [O III]5007 Å/4363 Å intensity ratio (Middlemass et al 1989). A measurement of $T_e = 8860 \pm 1000$ K from the widths of the Hα and [N II]6584 Å line profiles from precisely the same region (Meaburn et al 1991) adds to the puzzle. The latter method for measuring T_e is often unreliable because of significantly different turbulent motions in the two lines. However, it can be applied with accuracy in such a kinematically inert halo. The model in Fig. 1 which involves additional heating of the radiatively ionized gas of the halo by low Mach shocks appears to resolve many of these apparent contradictions. Here, an ablated wind percolates out from the nebular core at 20 km s^{-1} to form a non-radiative bow shock around the radiatively ionized surface of a halo globule. Additional heat is given to the immediate post-shock gas where the [O III]5007 Å line is emitted,

Astrophysics and Space Science **233**: 1–10, 1995.

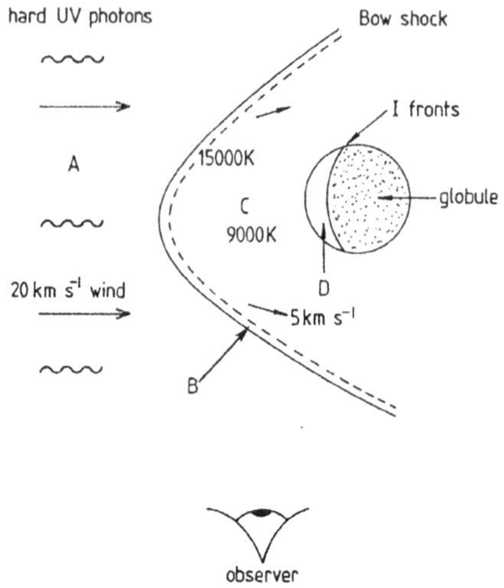

Fig. 1. A schematic model is shown of the interaction of the mass – loaded wind with a clump in the giant halo of NGC 6543. Faint and diffuse Hα & [O III]5007 Å emission will be seen from region A. Bright [O III]5007 Å emission will be predominant from the high temperature region B immediately behind the bow – shock whereas Hα will predominate from the cooled region C.

to raise the temperature of this region, whereas substantial cooling occurrs for the more diffuse Hα emission region further from the shock. Both the high [O III]5007 Å brightness and the temperature discrepancy are therefore explained.

3. The Bi-polar, Episodic Jets of Fleming 1

Lopez, Roth & Tapia (1993) discovered two diametrically opposite strings of lowly ionized knots extending 1′ on either side of the bright core of PN Fleming 1 (see Fig 2).

They proposed their ejection by a bi-polar, rotating, episodic jet from the nebular core. Observations of the kinematics of these strings of knots (Lopez, Meaburn & Palmer 1993) with the Manchester echelle spectrometer (MES – Meaburn et al. 1984) clearly confirmed this model for the radial velocity

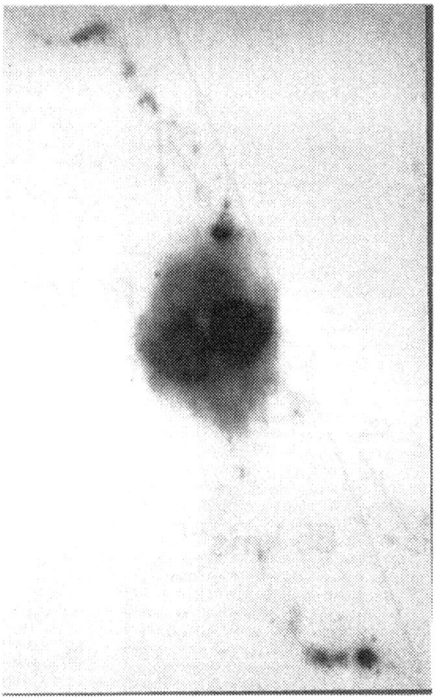

Fig. 2. An Hα + [N II]6584 Å image of the diametrically opposite strings of ionized knots which extend over 2$'$ and are moving out from the bright central regions of the PN Fleming 1.

differences of both strings systematically increase away from the nebular core. The north string is red-shifted whereas the south string is blue-shifted. The radial velocity difference from tip to tip of the north and south strings is 148 km s^{-1} which is consistent with an episodic jet of velocity 85 km s^{-1} rotating nearly perpendicularly to the plane of the sky at $\approx 7''$ yr^{-1}. A summary of the rotating jet model is shown schematically in Fig. 3.

A deeper analysis of the data (Palmer, Lopez & Meaburn 1995) has shown that the line emission from each knot is most likely that from a bow - shock driven into the ambient gas (the relics of the Red Giant wind from the precursor star) by an ejected globule moving supersonically.

An even more dramatic example of this episodic, rotating jet phenomenon may be at hand. Lopez et al (1995) have found similar diametrically opposite pairs of knots up to 7$'$ on either side of the otherwise innocuous PN, KjPn8. Morphologically they are similar to the prototypical strings in Fleming 1 but far more significant because of their huge angular extent out into the bi-polar halo of this PN. Kinematical confirmation of their nature is awaited.

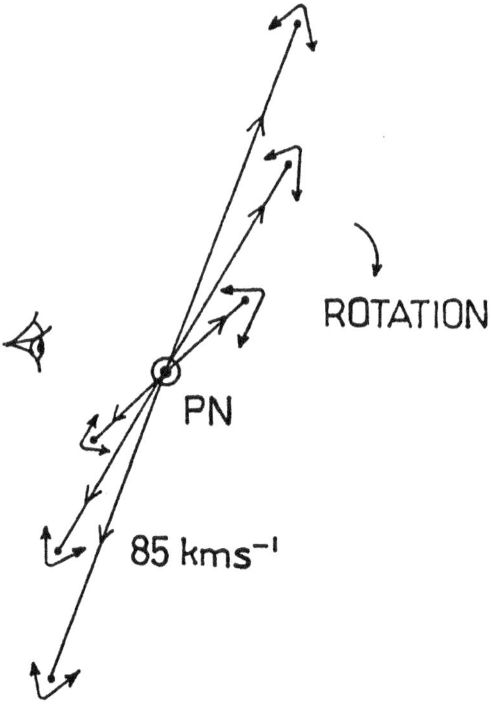

Fig. 3. Bullets of material are ejected at 85 km s^{-1}, episodically and in counter directions, from a rotating emitter at the centre of Fleming 1. The bright [N II]6584 Å emission seen in Fig. 2 originates from behind the bow – shocks which are formed as these bullets plough into the relics of the slow moving Red Giant wind of the precursor star.

Within this model the Fleming 1 strings of knots must be related to the high speed, lowly ionized knots found in the bright central regions of PNe e.g. the diametrically opposite pairs of high speed knots in NGC 6543 which were originally identified by Miranda & Solf (1992) and imaged with the HST by Harrington & Borkowsky (1995) and the similarly high-speed knots named 'fliers' by Balick et al (1994).

It seems that during the whole period of the post AGB out-bursts of PNe a mechanism for the periodic, bi-polar ejection of highly collimated gas (or bullets) is active. Soker & Livio (1992) invoke a low mass, close binary star within the primary star's envelope as the collimating mechanism. Magneto-hydrodynamical mechanisms should surely also be explored.

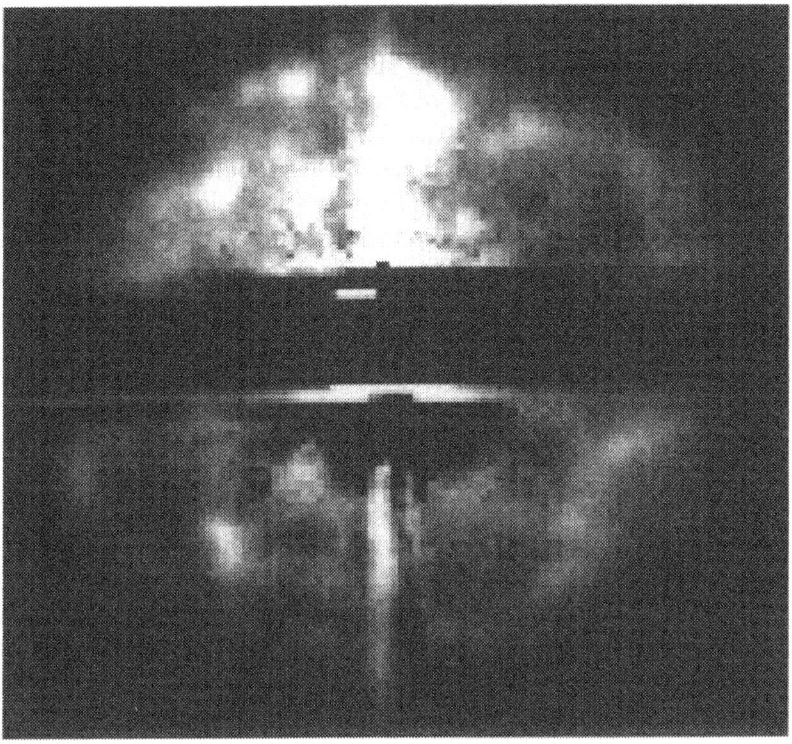

Fig. 4. A continuum – subtracted [N II]6584 Å image of the 22″ diameter inner shell around the LBV star P Cygni. The horizontal dark band is the occulting mask which eliminates the image of the star. The bright stripes are diffraction effects.

4. The Shocked Environment Surrounding Luminous Blue Variables

P Cygni, along with η Carinae, is one of a group of stars now known as luminous blue variables (LBVs) (Davidson, Moffat & Lamers 1989). Most, if not all, have nitrogen enriched circumstellar nebulae. The existence of that around P Cygni was first demonstrated by Johnson et al. (1992) and imaged in detail by Barlow, Drew, Meaburn & Massey (1994) in the light of the [N II]6584 Å emission line again using MES but with an occulting slit excluding the contaminating continuum from the massively bright (V = 4.8) star. An inner 22″ diameter shell (shown in Fig. 4) and two outer shells of 1′ and 1.5′ diameter respectively were found.

The inner [N II]6584 Å emitting nebula was shown to be expanding radially at 140 km s^{-1}. Curiously, the [Ni II]7412 Å over the same inner nebula revealed a knottier structure and a somewhat lower expansion velocity of 110 km s^{-1}. The [Ni II] emission lines had previously been shown by Johnson et al (1992) to be the brightest in this nebulosity.

Barlow et al (1994) demonstrate clearly that the [N II]6584 Å emission from the inner nebula is a consequence of the 206 km s^{-1} wind of P Cygni being ionized by passage through bow shocks around clumps in an expanding shell. This leaves the [Ni II]7412 Å observations to be explained. Lucy (1995) has now shown conclusively that the [Ni II]7412 Å emission originates in a fluorescence mechanism. Here the Ni gas in the globules within the expanding inner nebula is excited by the intense near uv (non – ionizing) radiation from P Cygni. The lower [Ni II]7412 Å expansion velocity, as well as the enhanced [Ni II]7412 Å knottiness, are both convincingly explained. The [N II]6584 Å profiles originate behind wind – formed bow – shocks around globules in the expanding, ejected, inner shell shown in Fig. 4. These localised flows add to the apparant expansion velocity of this shell. The [Ni II]7412 Å profiles originate within the same globules and therefore reflect the expansion velocity of the inner shell directly. Surely this fluorescently excited [Ni II] emission in this nebula will provide an important tool for investigating the conditions within the globular gas.

The nebulosity around the LBV star with the highest luminosity ($10^{6.6}$ $L_\odot$) (Davidson et al. 1986) η Car, is summarised in Fig. 5.

There is the dusty, bi-polar Homunculus nebula expanding at 650 km s^{-1} and scattering the Hα line from η Car itself (Meaburn et al. 1993a). Surrounding this is a nitrogen enriched shell expanding irregulary at $\approx$ 1000 km s^{-1} and an outer shell with similar motions (Meaburn et al 1987). With such high velocities the [N II]6584 Å etc. emission being observed is most likely the consequence of shocks being driven into the denser ejecta as they collide with previous, slower moving, envelopes. The post – shock velocities and temperatures are likely to be far too large for the emission of the visible nebular forbidden lines. However, if the speed differential between the shells involved in the collision are sufficiently small the creation of radiative bow – shocks around globules in either or both of these shells is a conceivable source of the [N II]6584 Å etc. emissions.

The 'jet' marked in Fig. 5 is of particular interest. This is shown to be composed of a series of bullets emerging from η Car at 1400 km s^{-1} and culminating in the knot NN. Contour maps of the position – velocity arrays of [N II]6584 Å profiles along slits 1 & 2 in Fig. 5, again obtained MES are shown in Figs. 6 a & b (Meaburn et al. 1993b).

The narrowness of the 'jet' [N II]6584 Å profiles compared with their displacement in radial velocity from the systemic radial velocity is notable. The [N II]6584 Å emission could arise from bow – shocks as bullets pass through a previous ejection only if again the speed differential between the bullets and these ejecta is small enough (say $\leq$ 150 km s^{-1}) to permit radiative bow – shocks to form. Otherwise, radiative shocks driven into the denser bullets themselves must be the source of the [N II]6584 Å emission.

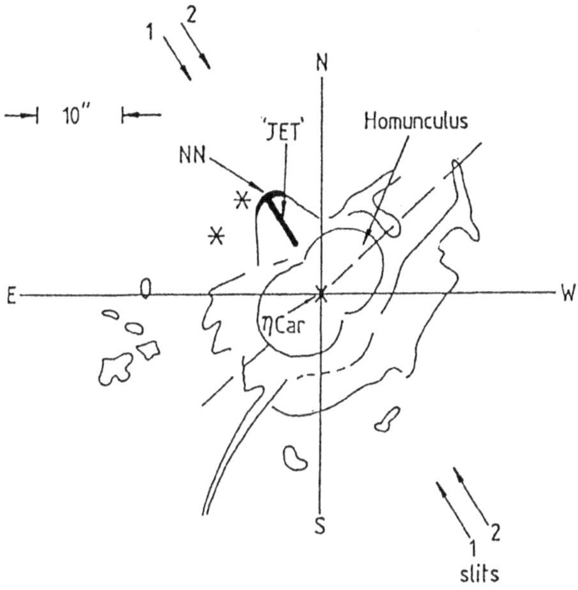

Fig. 5. A sketch of the nebulosity around η Carinae. The knot NN and apparent jet are identified. The major axis of the Homunculus nebulosity is shown as a dashed line and the slit positions 1 & 2 for the profiles in Fig. 6 are marked.

As with the PN Fleming 1 the origin of episodic, collimated ejecta in η Carinae is as yet not convincingly explained. The common feature of the otherwise very different situations encountered in Fleming 1 and η Car is that the collimated ejections have occurred during eruptive mass loss. The presence of low mass binary stars within the ejected envelopes close to the primary stars may be the key to the creation of these episodic jets.

5. The localised, high – speed flows near the M42 Trapezium stars

High-resolution, spatially resolved long-slit spectra of the [OIII]5007Å line in the vicinity of the Trapezium group of stars in the Orion Nebula have now been obtained for 17 slit positions with the MES in combination with the Isaac Newton, William Herschel and Anglo-Australian telescopes.

In a series of papers (Meaburn 1988, Meaburn et al. 1993, Massey & Meaburn 1993 & 1995) the discovery of variety of new high-speed kinemat-

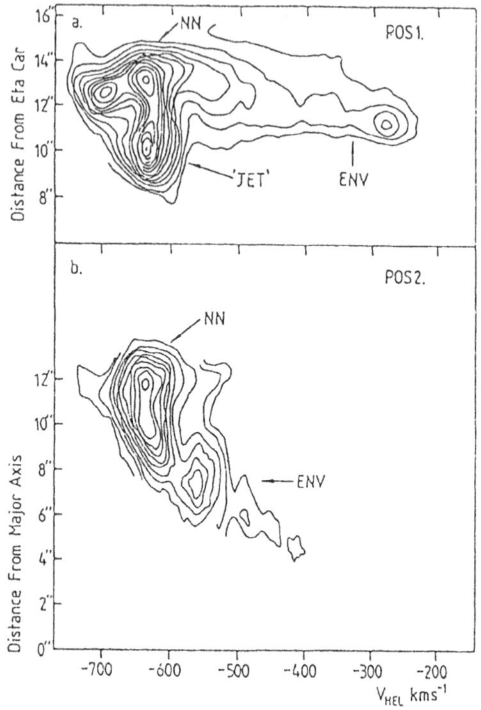

Fig. 6. Contour maps of the position – velocity arrays of [N II]6584 Å profiles along slit positions 1 & 2 in Fig. 5 are shown in **a** & **b** respectively. Slit position 1 is aligned with the η Carinae 'jet'.

ical phenomena are reported on a wide range of spatial scales . In particular, nearly all of the compact ($\leq 2''$ across), highly-ionized knots (Laques & Vidal 1979, Garay, Moran & Reid 1987, Odell, Wen & Hu 1993) covered by these observations in the core of M42 are associated with un-resolved ≈ 100 km s^{-1} mono- or bi-polar flows. Extended, high-speed features connect several compact sources. Jets from embedded YSOs as well as the particle winds from the Trapezium stars which form bow shocks around the globules, with radiatively ionized front surfaces, which surround these YSOs are both likely candidates for the origin of the high-speed, localised flows. These localised, high-speed, flows even if behind bow-shocks, are being irradiated by the Lyman photons from the Trapezium stars and are not related to the similarly localised, but purely shocked M42-HH objects (Canto et al. 1980).

A coherent structure is now found in two orthogonal position-velocity arrays of [OIII]5007Å profiles. This feature has a $1'$ radius and is composed of velocity 'spikes' and 'loops' whose extents in radial velocity diminish with increasing apparent distance from θ^1 Ori C. Its presence is considered to be strong evidence for the predominance of wind-driven ablated flows behind the bow – shocks around globules in the creation of the high-speed phenomena.

A 15″ wide, high-speed filament, bright in the [OIII]5007Å line and with a radial velocity of -90 km s^{-1} with respect to the systemic radial velocity, is traced for a length of 40″. This feature is possibly the surface, now exposed to the Lyman continuum photons of the Trapezium stars, of a high-speed, 'bullet' (or finger) in the underlying molecular cloud and driven by the burst of massive star formation associated with the Kleinmann-Low region. The infra-red observations of Allen & Burton (1993) have revealed the presence of such 'bullets' creating shocked cavity walls which fan out of the KL region. An alternative view is that these elongated features could be 'fingers' of material with shocked surfaces. Whatever their detailed nature, these 'bullets' or 'fingers' appear to have the shocked M42-HH objects at their tips which are seen at visible wavelengths through the front surface of the molecular cloud but shielded by the ionization front from the Lyman photons of the Trapezium stars.

A explanation of these various kinematical phenomena is presented schematically in Fig. 7.

References

Allen, D.A. and Burton, M.G.: 1993, *Nature,* **363** 54

Balick, B., Perinotto, M., Maccioni, A., Terzian, Y. and Hajian, A.: 1994, *Astrophys. J.,* **424** 816.

Barlow, M.J., Drew, J.E., Meaburn, J. and Massey, R.: 1994 *Mon. Not. R. astr. Soc.,* **268** L29 .

Bryce, M., Meaburn, J., Walsh, J.R. and Clegg, R.E.S.: 1992, *Mon. Not. R. astr. Soc.,* **254** 457.

Canto, J., Goudis, C.D., Johnson, P.G. and Meaburn, J.: 1980, *Astron. Astrophys.,* 85 128.

Davidson, K., Dufour, R.J., Walborn, N.R. and Gull, T.R.: 1986 *Astrophys. J.,* **305** 867.

Davidson, K., Moffat, A.F.J. and Lamers, H.J.G.L.M.: 1989, in Physics of Luminous Blue Variables, Kluwer, Dordrecht.

Garay, G., Moran, J.M. and Reid, M.J.: 1984 *Astrophys. J.,* **314** 535.

Harrington, J.P. and Borkowski, K.J.: 1995, NASA Press Release STScI – PRC95 – 01.

Johnson, D.R.H., Barlow, M.J., Drew, J.E. and Brinks, E.: 1992 *Mon. Not. R. astr. Soc.,* **255** 261.

Laques, P and Vidal, J.L.: 1979, *Astron. Astrophys.,* **73** 97.

Lopez, J.A., Roth, M. and Tapia, M: 1993, *Astron. Astrophys.,* **267** 194.

Lopez, J.A., Meaburn, J. and Palmer J.W.: 1994 *Astrophys. J.,* **415** L135.

Lopez, J.A., Vazquez, R. and Rodriguez, L.F.: 1995 *Astrophys. J.,* in press.

Lucy, L.B.: 1995 *Astron. Astrophys.,* in press.

Massey, R.M, and Meaburn, J.: 1993 *Mon. Not. R. astr. Soc.,* **262** L48.

Massey, R.M. and Meaburn, J.: 1995 *Mon. Not. R. astr. Soc.,* **273** 615.

Meaburn, J.: 1988 *Mon. Not. R. astr. Soc.,* **233** 79.

Meaburn, J., Blundell, B., Carling, R., Gregory, D.F., Keir, D. and Wynne, C.: 1984 *Mon. Not. R. astr. Soc.,* **219** 463.

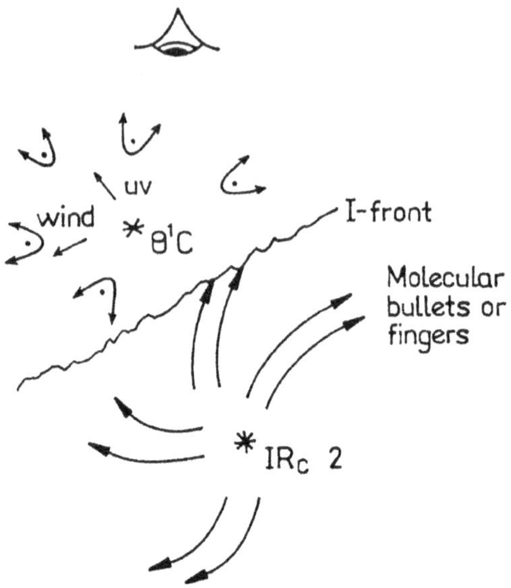

Fig. 7. An explanation of the complex, high – speed kinematical phenomena in the core of M42 is presented. Bow – shocks are formed by the wind from the Trapezium stars in the evaporating gas from the radiatively ionized surfaces of the cocoons around YSOs. Highly localised flows are observed whose radial velocities diminish with angular distance from θ^1C Orionis. Where a 'bullet' or 'finger' from IRc2 emerges into the ionizing radiation an extensive high – speed flow is observed. The HH - like objects at the tips of the 'bullets' or 'fingers' are not depicted.

Meaburn, J., Wolstencroft, R. and Walsh, J.R.: 1987 *Astron. Astrophys.*, **181** 333.
Meaburn, J., Nicholson, R., Bryce, M., Dyson, J.E. and Walsh, J.R: 1991 *Mon. Not. R. astr. Soc.*, **252** 535.
Meaburn, J., Walsh, J.R. and Wolstencroft, R.D.: 1993a *Astron. Astrophys.*, **268** 283.
Meaburn, J., Gehring, G., Walsh, J.R., Palmer, J.W., Lopez, J.A., Bryce, M. and Raga, A.C.: 1993b *Astron. Astrophys.*,**276** L21.
Meaburn, J., Massey, R.M., Raga, A.C. and Clayton, C.A.: 1993c *Mon. Not. R. astr. Soc.*, **260** 625.
Middlemass, D., Clegg, R.E.S. and Walsh, J.R.: 1989 *Mon. Not. R. astr. Soc.*, **239** 1.
Miranda, L.F. and Solf, J: 1992 *Astron. Astrophys.*, **260** 397.
O'Dell, C.R., Wen, Z. and Hu, X.: 1993 *Astrophys. J.*, **410** 696.
Palmer, J.W, Lopez, J.A. and Meaburn, J.: 1995 *Astron. Astrophys.*, in press.
Soker, N. and Livio, M.: 1992 *Astrophys. J.*, **339** 268.

THE OBSERVATIONAL STUDY OF HERBIG-HARO SHOCK WAVES

KARL-HEINZ BÖHM
Astronomy Department, University of Washington
Seattle, WA 98195

Abstract. We review the basic shock properties and the origin and the geometry of Herbig-Haro (H-H) shock waves. We first discuss different aspects of "normal" H-H objects which are connected with working surfaces (including internal working surfaces) of jets from young stellar objects. The emphasis is on unsolved problems of the H-H shock waves and not on the problems of the jet.

We study the line flux ratios of high excitation H-H objects (high velocity shocks) and low excitation HH objects (low velocity shocks) and carry out a comparison with theoretical predictions in both cases. We emphasize an unexplained deficit of higher ions (especially OIII and SIII, but also various other ions) in high excitation objects. This lets the line flux ratios of HH objects appear as if their shock velocities are almost never above 100 km s^{-1}, while other shock diagnostics (position-velocity diagrams, integrated line profiles, distributions of fluxes along the axis of the bow shock, etc.) definitely indicate higher shock velocities.

Some aspects of the spectrum interpretation of the very low velocity shocks (like HH7) are explained quite well by the theory. A basic unsolved problem is, however, the explanation of the CI lines whose flux is up to a factor 10 times stronger than predicted for any model. Obviously we are very far from correctly predicting the ionization of C in shock models.

In the last chapter we discuss, as one example of a very unusual HH-object, HH255 (Burnham's nebula). Detailed line fluxes in the immediate environment of T Tauri (the source of HH255) have shown that HH255 has a shock wave spectrum and is definitely an HH object. In the very narrow region between 3" and 4" S of T Tauri we find a sharp peak of the velocity dispersion, the centroid velocity, and N_e. In the same region there is an almost discontinous increase in ionization. Between 4" and 10" S (corresponding to 600-1600 a.u.) of T Tauri (the source of HH255) the ionization remains high but the centroid velocity is zero (with respect to T Tauri) and the velocity dispersion is very small. This result is completely surprising for a shock wave which according to the flux ratios must have ~90 km s^{-1-1} shock velocity. Why should a cooling region of a shock have a centroid velocity of ~0 km s^{-1} over a large range of distance from the stellar source? At present the geometry of the HH255 is enigmatic.

Key words: Shock Waves – Herbig-Haro Objects

1. Introduction

Herbig-Haro objects have been discovered and their (at that time) enigmatic optical spectra have been first described by Herbig (1951). The first quantitative spectrophotometry was carried out in 1956 (Böhm 1956). It took almost another 20 years before it was finally recognized that the optical spectra are formed in the recombination region of shock waves (Schwartz 1975).

Today Herbig-Haro (HH) objects are easily recognized as shock waves by the great strength of emission lines of neutral and single ionized parti-

Astrophysics and Space Science **233**: 11–25, 1995.

cles with low ionization energy ($\chi_{\rm ion} < 13.6$ eV). Some typical criteria for their identification are the flux of [OI] 6300, 6363 and [SII] 6716, 6731 (both comparable to Hβ or stronger in HH objects) and especially in low velocity shocks a strong [NI] 5200 and/or a strong [CI] 9850. Important supplementary information to the optical data is derived from IR (rotation-vibration) lines of H_2 and the UV fluorescent lines of H_2 (e.g. Elias 1980; Schwartz 1983; Schwartz, Cohen and Williams 1987; Hartigan, Curiel and Raymond 1989; Böhm, Scott and Solf 1991), see also the theoretical interpretation by Wolfire and Königl (1991). Additional helpful information is derived from the UV spectra in general (e.g. Ortolani and d'Odorica 1980; Böhm et al. 1981, 1987, 1993). Excellent detailed descriptions of observational results on Herbig-Haro objects are available in recent review articles. We mention specifically the outstanding articles by Reipurth (1989, 1991).

It is now also well understood that most (but not all) Herbig-Haro shocks are generated by the bipolar jets from young stellar objects and especially by their working surfaces.

In this review paper we shall mostly present basic observational information about Herbig-Haro shock objects relevant to the shock theory. Specifically we shall survey the following problems.

1. We first discuss properties of the shock which seem local and (in a crude approximation) can be qualitatively discussed in terms of a plane shock model.

2. We study the geometry and origin of the shocks in typical Herbig-Haro objects. We place especial emphasis of the (usually relative bright) HH shocks which are due to working surfaces of jets.

3. Finally, we shall consider Burnham's nebula (HH255) as an example of a highly unusual HH object. The spectrum of this object is clearly generated by a shock wave (like other HH object spectra). But in this case both the origin of the shock wave and, the self consistent shock wave structure pose difficult unsolved problems.

In general we shall emphasize in this paper the unsolved problems of HH shocks. We feel that there are a number of highly interesting, enigmatic results concerning these shocks. Normally these are not strongly emphasized because the shocks are mostly used as indicators of the bipolar jets and their working surfaces. The details of the shock waves are considered of lesser interest. However, the present conference is a good opportunity to discuss the problems of these shocks.

2. The Local Properties of HH Shock Waves

We shall first discuss the line flux ratios of HH shocks. Comparison will be made with standard emission line predictions for plane shocks (e.g. Hartigan, Raymond and Hartmann 1987; Binette, Dopita and Tuohy 1985; Shull and McKee 1979; Raymond 1979). If there is strong disagreement between observation and theoretical prediction for plane shocks we will also make a comparison with the line flux prediction for simple bow shock models.

Some basic properties of the optical spectra of a typical high excitation (fast shock) HH object HH1 and the very low excitation object HH7 (slow shock) are surveyed in Table I. The different ions which have been observed (distinguishing between permitted and forbidden lines) are indicated. We have also indicated the strongest line flux of the ion (in the usual units F (Hβ) = 100) provided the flux is larger than 15. Lines observed in HH1 but not in the low excitation object HH7 are indicated by the symbol –.

TABLE I

Comparison of Line Statistics of HH1 and HH7

	HH1 (for 3720 Å$\leq \lambda \leq$ 10830A)			HH7 (for 3900 $\leq \lambda \leq$ 9850)		
Ions	Perm. line(s)	For. line(s)	Strongest line flux >0.15 F Hβ	Perm. line(s)	For. line(s)	Strongest line flux >0.15 Hβ
H	X		X (311)	X		X (435)
He I	X		X (131)	–		
He II	X			–		
C I		X			X	X (1060)
N I		X	X (16)		X	X (268)
N II		X	X (151)		X	X (34)
O I	X	X	X (159)	–	X	X (577)
O II		X	X (135)		–	
O III		X	X (57)		–	
Ne III			X (15)		–	
Na I	X			–		
Mg I	X			X	X	
S II		X	X (105)		X	X (868)
S III		X			–	
Cl II		X			–	
Ar III		X			–	
Ar IV		X		–	–	
Ca II	X	X	X (45)	X	X	X (56)
Cr II		X			X	
Fe II		X	X (31)		X	X (52)
Fe III		X			–	
Ni II		X	X (16)		X	X (22)

If we use a plane shock wave model for the spectral diagnostics of a localized region in an HH object we can in principle determine the shock velocity, the pre-shock density and the pre-shock magnetic field. For typical HH values of shock velocities and densities, the influence of a magnetic field to up to 40 μG on the line ratios, is small (e.g. Hartigan et al. 1987; Noriega-Crespo et al. 1990). In high excitation HH objects the appearance of high ionization lines is (not surprisingly) most strongly influenced by the shock velocity (and much less by the pre-shock density). We shall consider now the [OIII] and the [SIII] lines which are highly sensitive to shock wave velocity. One surprising result is that the easily accessible high excitation HH objects (i.e., the high velocity shock waves) show a very limited range for the flux ratio of [OIII] 5007/Hβ compared to plane shock predictions. For the eight most easily observed (bright) HH objects the [OII] 5007/Hβ lies in the range 0.18 to 1.09 while the theoretical prediction is 0.07(!) for $V_s = 90$ km s^{-1} and 2.19(!) for $V_s = 100$ km s^{-1}. The results are shown in greater detail in Table II. If taken literally the results "show" that all these shock waves have a velocity of about 92-95 km s^{-1}. One might argue that bright HH objects correspond more to a bow shock than to a plane shock. But even for bow shocks (Hartigan et al. 1987) the theoretical ratio [OIII] 5007/Hβ changes from 0.43 at 100 km s^{-1} to 1.78 at $V_s = 150$ kms^{-1} so the interpretation of the observed ratio will lead us to the statement the V_s may reach $\sim$120 km s^{-1}, but no velocities above this value are "observed". This is astonishing because even at this value the thickness of the cooling region is only 26 a.u. (at $V_s = 140$ km s^{-1}, thickness of cooling region 29 a.u.) so this is a radiative shock which should be easily observed. (Quoted numbers apply to a pre-shock particle density $N = 10^2$ cm^{-3}.) Consequently the [OIII] 5007/Hβ flux ratio of the 8 best observed high excitation HH objects seems to indicate a strict upper limit of the shock velocity of 95 km s^{-1} (for plane shocks) or 120 km s^{-1} (for bow shocks). As we shall discuss below, other indicators of shock velocities are usually considerably higher so that the weakness of the [OIII] lines definitely appears as an enigma. The next question is, is this behavior attributable to the ionization and line formation of OIII or does it apply to all high ions. We have also studied the ratio of [SIII]/Hβ. ([SIII] is the other two-fold ionized particle whose line photometry is fairly reliable.) [SIII] results are only available for 2 HH objects. As the lower part of Table II shows, these ions indicate a shock velocity which is even lower than the one derived for [OIII]. It is clear that high ionization lines show a drastic deficit and even higher ionization lines are completely missing leading to the impression that (as far as line ratios go) practically all HH shocks have a shock velocity $V_s \lesssim$ 100 km s^{-1}. What are the highest ions present in HH objects? Solf, Böhm and Raga (1988) detected a very faint line of [AIV] (1F$_{H\beta}$) at 4740.3 Å. It requires 40.74 eV to ionize AIII. But this is only slightly different from obtaining OIII through the ionization of OII (35.12 eV). Obtaining NeIII (which is also observed in a number of high excitation HH objects) requires 40.96 eV (NeII). The only ion with (slightly) higher ionization energy is CIV whose UV line at 1550 Å is seen. (CIII has 47.89 eV ionization energy.) There are no higher ionization stages known in HH shock waves. One example is the missing NV line. The theoretical predictions for NV 1240 (the ionization energy for NIV is 77.42 eV) are 1.45 $F_{H\beta}$ at $V_s = 140$ km s^{-1} and 7.18 $F_{H\beta}$ at $V_s = 160$ km s^{-1} (a shock velocity which is certainly present according to other criteria).

TABLE II

Comparison of Observed Line Ratios [OIII/Hβ] and [SIII]/Hβ with Theory

Object	Observed [OIII] 5007/Hβ flux ratio	Plane Shock wave model	Theoretical [OIII] 5007/Hβ flux ratio	Bow Shock model	Theoretical [OIII] 5007/Hβ flux ratio
HH1	0.57	90 km s^{-1}	0.07	50 km s^{-1}	0
HH2H	0.63	100 km s^{-1}	2.19	100 km s^{-1}	0.43
HH2A	1.09			150 km s^{-1}	1.78
HH3	0.44				
HH24A	0.22				
HH32A	0.85				
HH40	0.60				
HH255	0.18				
			[SIII] (9069 + 9532)/Hβ		
Object	[SIII]/Hβ flux	Plane Shock Model	[SIII]/Hβ flux ratio	Bow Shock model	[SIII]/Hβ flux ratio
HH1	0.082	90 km s^{-1}	0.16		
HH255	0.097	100 km s^{-1}	0.235	100 km s^{-1}	0.23
		110 km s^{-1}	0.58	150 km s^{-1}	0.39

Now we look at the other extreme, the very low excitation objects for which we choose HH7 as an example. Although the theoretical predictions for very low shock velocities (20–30 km s^{-1}) seem to be somewhat questionable the agreement of observed (stronger) line flux ratios with the theoretical predictions is quite good (with one exception, see below). Table III shows that the "observed" V_s lies between 20 and 30 km s^{-1} ([OI]/Hβ and [NII]/Hβ) or somewhat below 20 km s^{-1} ([SII] (4068 + 4076)/Hβ, [SII] (6716 + 6731)/Hβ and [CaII] (7291 + 7324)/Hβ). The only line ratio which causes serious problems is [CI] (9849)/Hβ which is more than a factor 10 higher (!) than any prediction. It may be worth mentioning that the CI line also causes serious problems in high excitation HH shocks (Solf, Böhm and Raga 1988). In this case it is the spatial dependence which is very different from the theoretical prediction. These facts lead us to suspect that (for unknown reasons) the ionization of C is described very poorly in both plane as well as bow shock models of HH objects. Otherwise Table III also shows that the line ratios lead to a (more or less) definite value of V_s. However, even in low excitation objects the line profile apparently also leads to a much higher shock velocity. The FWZI (which should be equal to the bow shock velocity, Hartigan et al. 1987) is 150 km s^{-1} in HH7 (Solf and Böhm 1987).

TABLE III
Low Excitation Object HH7: Comparison Observation and Theory

Line ratio	Line flux ratio Observed	Model 30 kms^{-1}	20 kms^{-1}
[SII] (4068 + 4076)/Hβ	1.71 (1)	0.35	1.28
[OI] (6300 + 6363)/Hβ	7.70 (b)	4.05	14.68
NII] (6583)/Hβ	0.34 (b)	0.69	0.06
[SII] (6716 + 6731)/Hβ*	16.13 (1)	3.60	14.81
[CaII] (7291 + 7324)/Hβ	1.42 (1)	0.17	1.32
[CI] (9849)/Hβ	10.59 (e.l.)	1.04 (maximum for all models up to 100 km s^{-1})	0.57

*FWZI ([SII]) = 150 km s^{-1}

Recently we have begun to study another aspect of HH shocks. By determining the gas phase abundances of Fe we tried to find out whether the Fe is still mostly bound in dust grains (as in the original cloud material) or whether the HH shocks have been strong enough to destroy the grains and putting the metals (like Fe, Ca) back into the gas phase. We have studied HH1, HH7, HH11, HH43A and HH255 (Burnham's nebula). Among these HH1, HH43A, and HH255 are high excitation objects (high velocity shock waves) and HH7 and HH11 are low excitation objects (Table IV). We find that HH1, HH43A, H7 and H11 show normal solar abundance; only HH255 show an Fe depletion (with respect to S) of a factor ~3 (Beck-Winschatz, Böhm and Noriega-Crespo 1994).

3. The Geometrical Shape and the Origin of HH Shocks

In most cases HH shocks are formed by the jets from young stars; they are either parts of the working surfaces, or internal shocks of the jet. Of special interest are the working surfaces, because in many cases they led the brightest HH objects. We can either use the "idealized" working surface (shown

TABLE IV
Fe/S Gas Phase Abundance Ratio

Object	(Fe/S) ratio	Compare to
HH 1	3.1	Solar abundance
HH 7	2.5	ratio:
HH 11	3.05	Fe/S ~2.95
HH 43A	2.0	(Grevesse and
HH 255	1.03	Anders 1992)
(Burnham's nebula)		

e.g. by Dyson 1987). The most strongly emitting regions in this model are the jet shock (Mach disk, shocked jet gas) and the bow shock (shocked ISM). In a more realistic description one finds (dynamic and thermal) instabilities which change the appearance of the working surface considerable (Raga 1988; Blondin, Königl and Fryxell 1989; Stone and Norman 1993 a, b, 1994). However a detailed prediction of the spectrum is very difficult in this case. Going back to the idealized description the following is important. Hartigan (1989) has shown that the bow shock is brighter in the spectrum than the jet shock if the jet density is higher than the density of the ambient medium. For this reason the geometrical shape and the spectrum of the HH object can be predicted from a bow shock model (Hartmann and Raymond 1984; Choe, Böhm and Solf 1985; Raga and Böhm 1985, 1986, 1987; Hartigan et al. 1987; Indebetouw and Noriega-Crespo 1995). For the comparison with the bow shock prediction the following results have been used: 1) the shape of the bow shock (Raga 1986), 2) the (integrated) line profile (Hartigan et al. 1987), and 3) the position-velocity diagrams (from spatially resolved emission line profiles). We find that the position-velocity diagram gives an especially good diagnostic because it measures the bow shock velocity as well as the angle between the bow shock axis and the plane of the sky. In addition, it is fairly improbable (because of the complexity of the position-velocity diagram) that an agreement between observation and theoretical prediction would be purely accidental. The best known examples of good agreement between observation and theory are HH1 (motion in the plane of the sky) and HH32 (inclination of $\sim 20°$ between the axis of the bow shock and the line of sight). Position-velocity diagrams of HH1 are discussed by Choe et al. (1985), Böhm, Raga and Solf (1987), Raga et al. (1988), Indebetouw and Noriega-Crespo (1995), and of HH32 by Solf, Böhm and Raga (1986). The bow shock interpretation of HH32 is also interesting because it shows quite convincingly multiple bow shocks along the outflow (Solf et al. 1986). Recent results have shown that this type of result is quite common in a number of HH objects (Raga and Kofman 1992; Hartigan and Raymond 1993).

We now have available a number of shock velocity determinations which can be compared to each other. We make this comparison first for HH1 because the most data are available in this case. HH1 is, however, by far not the most enigmatic object.

Table V shows that for HH1 four out of five shock velocities (including even the Doppler effect of dust scattered light in front of HH1, see Solf and Böhm 1991; Noriega-Crespo, Calvet and Böhm 1991; Henney et al. 1994) agree with each other (150 - 200 km s^{-1}). Only, as emphasized above, the ratio of the line fluxes gives a very much lower shock velocity. We have excluded the space motion of HH1 from Table V ($\sim$380 km s^{-1} at the stagnation point, Herbig and Jones 1981, Eislöffel, Mundt and Böhm 1994) because the space motion does not necessarily have anything to do with the shock velocity. (The preshock gas may be moving with respect to the source.)

In all high excitation objects (studied so far) we find a situation similar to HH1, except that the discrepancy between the shock velocity determined from line flux ratios and the other shock velocity determinations is even larger. Taking HH32 as an example, we find that the line flux ratios give a shock velocity slightly below 120 km s^{-1} whereas the position velocity diagrams lead to a shock velocity of 300 km s^{-1} (Solf et al. 1986). The integrated line profiles gives even a velocity of 360 - 415 km s^{-1} (Hartigan et al. 1987).

In most of the low excitation objects the situation is similar. In Table III we have shown that the line flux ratios indicate a shock velocity of $\sim$20 km s^{-1}. On the other hand the FWZI of the [SII] line (which according to the theory should be identical to the shock velocity, Hartigan et al. 1987) is (for [SII] 6716, 6731) 150 km s^{-1} (!).

What about the working surfaces whose simple structure has completely dissolved and neither a bow shock nor a jet shock can be recognized? HH2 is a good example of this. In this case there is still a correlation between the space motion, the "excitation" (measured by Hα/[SII] flux ratio), and the velocity dispersion of each individual knot, very similar to the result for the bow shock like HH1 (Eislöffel et al. 1994, Fig. 11 and 12). There are several possible explanations for this. One interpretation is that in a completely "dissolved" HH object each knot may behave like an independent bow shock.

4. HH255 (Burnham's Nebula), a Very Unusual HH Shock Wave

There are quite a number of HH shocks which (according to our present understanding) do not come from the working surfaces of jets or interiors of

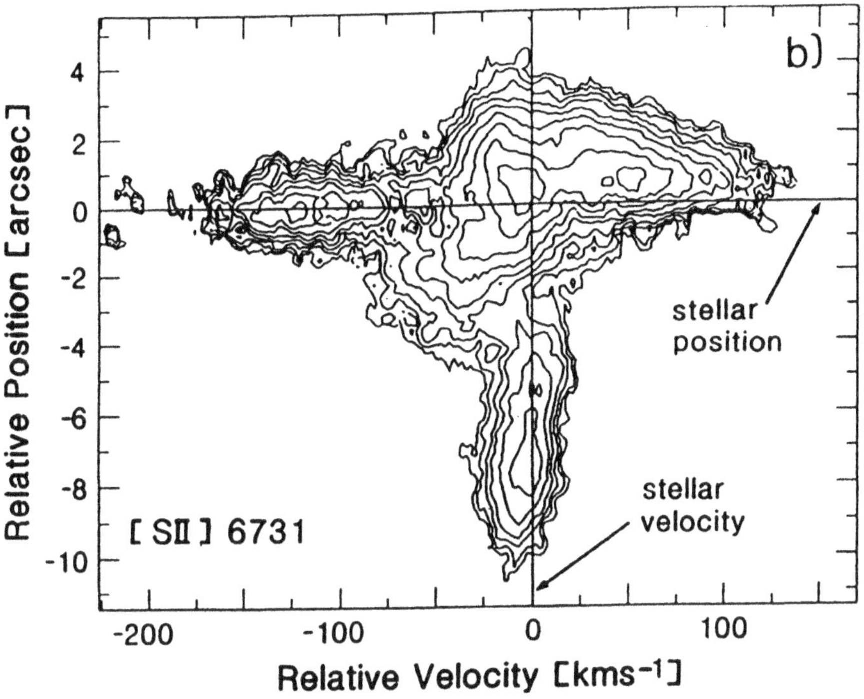

Fig. 1. Position-velocity diagram for the [SII] line 6731 in the immediate environment of T Tau (from Böhm and Solf 1994). The photospheric spectrum of T Tau has been eliminated. The slit position angle is 0°. The structure between 3."5 and 10" S of T Tau is the "proper" HH object. Note that the (centroid) radial velocity agrees very well with the stellar velocity in HH255 and that this object shows only a very small velocity dispersion.

the jets themselves. For instance, both Schwartz, Dopita and Cohen (1985) and Böhm and Solf (1990) have studied HH43 A, B, C and concluded (using different criteria) that HH43 is a "shocked cloudlet" (defined by Schwartz 1978) and not a bow shock. Such conclusions are interesting, but they are not too surprising. After all, the only spectroscopic criterion for HH object are (optically visibly) shocks in regions of star formation.

Here we want to introduce a much more uncommon HH object whose origin really seems to be an enigma, Burnham's nebula (HH255). Earlier spectroscopic investigation were made by Herbig (1950), Schwartz (1974), Bührke, Brugel and Mundt (1986) and others, and Burnham's nebula was tentatively identified as an HH-object.

Before we draw any conclusions about the peculiarity of HH255 we have to convince ourselves that the spectrum of this object corresponds completely to that of a real HH object. Solf et al. (1988) have made a spectrophotometric investigation of 81 optical lines. All the line fluxes are compatible with the

TABLE V
Comparison of Shock Velocity Determiniations for HH1

Determined from	Shock velocities	Data from
ratio of line fluxes		
(bow shock)	90 - 120 kms^{-1}	mostly from Solf et al. (1988)
line profiles (bow shock)	150 - 200 kms^{-1}	Hartigan et al. (1987)
from position-velocity diagram		Böhm et al. (1987)
(bow shock)	~185 kms^{-1}	Raga et al. (1988)
fluxes of different ions along		
the axis of bow shock	150 - 200 km	Noriega-Crespo et al. (1989)
Doppler effect of dust-scattered		Solf and Böhm (1991)
light in front of HH1	~200 kms^{-1}	Noriega-Crespo et al. (1991)

TABLE VI
Comparison of Strong Lines in HH 255 and HH1

	HH 255	HH1
[OIII]/Hβ	0.184	0.568
[OI]/Hβ	2.18	2.10
[NI]/Hβ	0.262	0.156
[SII]/Hβ	2.78	1.64

fluxes of an HH object of slightly lower excitation than HH1. In Table VI we show a comparison of the line ratios of very few of the strong lines in HH255 and HH1. The ratios [OI]/Hβ, [NO]/Hβ, and [SII]/Hβ are larger for HH255, showing a somewhat smaller shock velocity for HH255. [OIII]/Hβ is considerably smaller in HH255 than in HH1. However, the indicated difference in the shock velocity is also quite small because of the strong dependence of the [OIII] flux on the shocks velocity.

The enigmatic character of HH255 has become obvious through a detailed investigation of position-velocity (P-V) diagrams in the immediate environment of T Tauri (Böhm and Solf 1994). As an example, we show in Fig. 1 the (P-V) diagram of the [SII] 6731 line. The photospheric spectrum (continuum with absorption lines) has been subtracted out (see e.g. Solf and Böhm 1993). The position of the star is shown by a horizontal line, the stellar (photospheric) velocity by the vertical line. The P-V diagram refers to a position angle of 0°. The contour lines are logarithmic. Two successive lines correspond to an intensity ratio of $\sqrt{2}$. Near the star the P-V diagram north of T Tauri and south out to distance of 3."5 shows us essentially the formation of the complicated outflow from T Tauri (discussed by Böhm and Solf 1994). Starting at ~ 3."5 S of T Tauri and extending further to the S (to ~10") we see the main part of HH255. As discussed, this shows a spectrum indistinguishable from any other moderately high excitation HH shock (as long as one measures only the line fluxes). The position-velocity diagram shows an unexpected surprise in HH255. Compared to the outflow seen within 3."5 of the star, HH255 shows an extremely small centroid radial velocity with respect to the star and (for HH objects) an unusually small velocity dispersion. The centroid radial velocity is -3 km s^{-1} ([SII] 6731), -1 km s^{-1} ([OI] 6300), and -2 km s^{-1} ([NII] 6583). The FWHM is 25 km s^{-1} ([SII] 6731), 24 km s^{-1} ([OI] 6300), and 27 km s^{-1} ([NII] 6583). It is surprising that the recombination region of a shock wave (of at least 90 km s^{-1} according to the flux ratios) shows a very small radial velocity and a small velocity dispersion. What kind of structure does this shock wave have? It could definitely be not "bow-shock like" because then the velocity dispersion (FWZI ~40 km s^{-1} in HH255) must be equal to the bow shock velocity (see above). Normally the shock velocity determined from line profiles is much larger than the shock velocity determined from line flux ratios (see above); here it is the other way round and that is very hard to understand. It is clear that this shock wave must have a completely different geometry. The additional enigma is: Why should the matter in this shock wave with high accuracy be at rest (centroid radial velocity ~0 km s^{-1}) with respect to the star everywhere at a distance range from ~560 a.u. to ~1600 a.u. For a position angle of 45° (instead of 0°) we find approximately the same behavior of the radial velocity and the velocity dispersion as a function of the distance from the star. This cannot be the cooling region of, say, a spherically symmetric shock. The thickness is much too large (the measured thickness is larger than 1000 a.u.; the predicted thickness even for a pre-shock density of 100 cm^{-3} is only 17 a.u.).

In order to get more insight into this unusual object we have studied the following physical properties: 1) electron density, 2) centroid velocity, 3) velocity dispersion and 4) ionization (all quantities measured for the range from 0"-10" from T Tauri. We find the following: Approximately 0"-3" S of T Tauri we find quite low ionization, decreasing outward or constant density, with velocity dispersion mostly decreasing outward. At a distance of 3."0 to 4."0 S of T Tauri something unusual happens. The ionization increases very rapidly (shown in Fig. 2) from very small values to the maximum value. The ionization increase is shown by the almost simultaneous rapid rise of the [OIII] (5007) intensity and the steep rise of the line intensity ratio [NII] (6583)/[OI] (6300). At the same place (3"-4") the velocity dispersion shows a maximum of ~80 km s^{-1} and declines very rapidly to ~25 km s^{-1} at 4" (Fig.

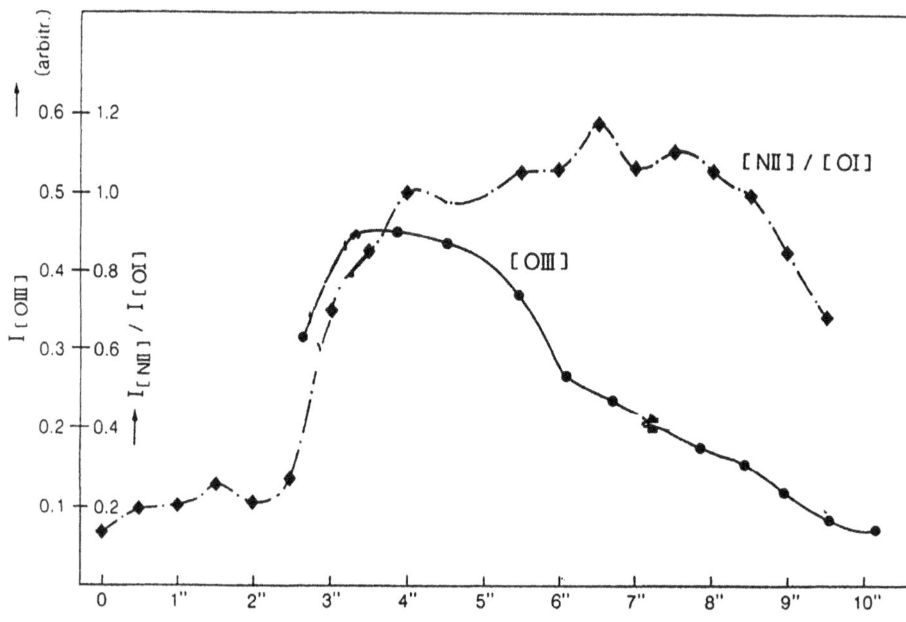

Fig. 2. The empirical indication of the ionization south of T Tau. The ionization is indicated by the flux in the [OIII] 5007 line and by the flux ratio of [NII] 6583/[OI] 6300. Note how abruptly the ionization increases at ~3".

3). A similar distribution is seen for the centroid velocities except that they go to 0 km s^{-1} (Böhm and Solf 1995, in preparation). The electron density stops declining and has a steep (intermediate) maximum also at 4" south of T Tauri.

In summary, we get the following picture: Between 3" and 4" south of T Tauri there is an almost discontinuous increase of the ionization, a rapid rise of the velocity dispersion and a rapid rise of the electron density. Further south the ionization declines very slowly (compared to the decline which one would get in plane shock wave, seen side on). The [OIII] number declines by a factor ~2 in 3", the [NII] declines by a factor 2 in about 6". The velocity dispersion, on the other hand declines from its maximum value (~80 km s^{-1}) to its minimum value (~25 km s^{-1}) in a very short distance, slightly more than 0."5. The extremely unusual properties of Burnham's nebula are recognizable in the following properties: The ionization looks as if there is a shock wave at ~3-4" S of T Tau, but the recombination region is unusually extended (almost 100 times larger than for a plane shock wave). Over a very short distance the centroid velocity becomes equal to zero and the velocity dispersion becomes very small, so we have a fairly extended range of the recombination region in which the shock wave velocity seems to be approx-

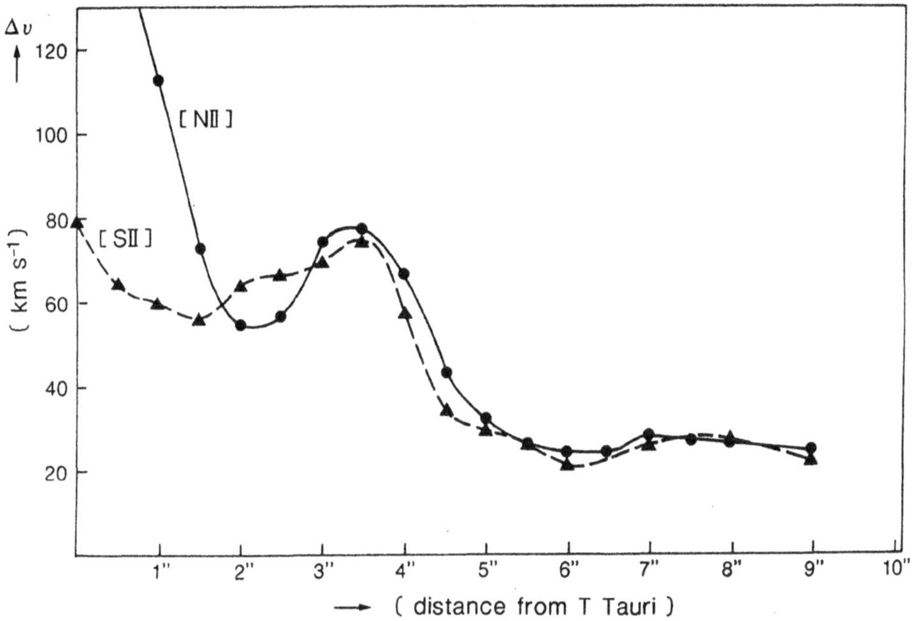

Fig. 3. The velocity dispersion of the [SII] 6731 and of the [NII] 6583 line. At about 3.5 south of T Tau the dispersion shows a steep intermediate maximum. This is very close to the place where the maximum of the ionization occurs. Beginning at ~4" the velocity dispersion reaches very low values (see text).

imately zero. It is extremely strange that the centroid velocity is identical to the stellar velocity over the essential range of the recombination region. Finally this behavior is not only visible for a position angle of 180° but is almost identical for a position angle 135°. It looks as if the region has developed a "standing" recombination region surrounding a part of the star (not moving with respect to the star). It is completely enigmatic why this should be the case.

References

Beck-Winchatz, B., Böhm, K.H. & Noriega-Crespo, A.: 1994, *Pub. Astr. Soc. Pac.* **106**, 1271

Binette, L., Dopita, M. A. & Tuohy, I. R.: 1985, *Astrophys. J.* **297**, 476

Blondin, J. M., Fryxell, B. A. & Königl, A.: 1990, *Astrophys. J.* **360**, 370

Blondin, J. M., Königl, A. & Fryxell, B. A.: 1989, *Astrophys. J.* **337**, L37

Böhm, K.H.: 1956, *Astrophys. J.* **123**, 379

Böhm, K.H., Böhm-Vitense, E. & Brugel, E. W.: 1981, *Astrophys. J.* **245**, L113

Böhm, K.H., Noriega-Crespo, A. & Solf, J.: 1993, *Astrophys. J.* **416**, 647

Böhm, K.H., Bührke, T., Raga, A., Brugel, E. W., Witt, A. N. & Mundt, R.: 1987, *Astrophys. J.* **316**, 349

Böhm. K.H., Raga, A. C. & Solf, J.: 1987, 'I.A.U. Symp. 122, Circumstellar Matter' in I. Appenzeller and C. Jordan, ed(s)., *Reidel*, Dordrecht, 187
Böhm, K.H., Scott, D. M. & Solf, J.: 1991, *Astrophys. J.* **371**, 248
Böhm. K.H. & Solf, J.: 1990, *Astrophys. J.* **348**, 297
Böhm, K.H. & Solf, J.: 1994, *Astrophys. J.* **430**, 277
Bührke, T., Brugel, E. W. & Mundt, R.: 1986, *Astron. Astrophys.* **164**, 83
Choe, S. U., Böhm. K.H. & Solf, J.: 1985, *Astrophys. J.* **288**, 338
Dyson, J.: 1987, 'I.A.U. Symp. 122, Circumstellar Matter' in I. Appenzeller and C. Jordan, ed(s)., *Reidel*, Dordrecht, 159
Eislöffel, J., Mundt, R. & Böhm, K.H.: 1994, *Astron. J.* **108**, 1042
Elias, J. H.: 1980, *Astrophys. J.* **241**, 728
Grevesse, N. and Anders, E.: 1991, 'Solar Interior and Atmosphere' in A. N. Cox, W. C. Livingstone and M. S. Matthews, ed(s)., *The University of Arizona Press*, Tucson, 227
Hartigan, P.: 1989, *Astrophys. J.* **339**, 987
Hartigan, P., Curiel, S. & Raymond, J.: 1989, *Astrophys. J.* **347**, L31
Hartigan, P. & Raymond, J.: 1993, *Astrophys. J.* **409**, 705
Hartigan, P., Raymond, J. & Hartmann, L.: 1987, *Astrophys. J.* **316**, 323
Hartmann, L. & Raymond, J.: 1984, *Astrophys. J.* **276**, 560
Henney, W. J., Raga, A. C. & Axon, D. J.: 1994, *Astrophys. J.* **427**, 305
Herbig, G. H.: 1950, *Astrophys. J.* **111**, 11
Herbig, G. H.: 1951, *Astrophys. J.* **113**, 697
Herbig, G. H. & Jones, B. F.: 1981, *Astron. J.* **86**, 1232
Indebetouw, R. & Noriega-Crespo, A.: 1995, *Astron. J.* **109**, 752
Noriega-Crespo, A., Böhm, K.H. & Raga, A. C.: 1989, *Astron. J.* **98**, 1388
Noriega-Crespo, A., Böhm, K.H. & Raga, A. C.: 1990, *Astron. J.* **99**, 1918
Noriega-Crespo, A., Calvet, N. & Böhm, K.H.: 1991, *Astrophys. J.* **379**, 676
Ortolani, S. & d'Odorico, S.: 1980, *Astron. Astrophys.* **83**, L8
Raga, A. C.: 1986, *Astron. J.* **92**, 637
Raga, A. C.: 1988, *Astrophys. J.* **335**, 820
Raga, A. C. & Böhm, K.H.: 1985, *Astrophys. J. Suppl.* **58**, 201
Raga, A. C. & Böhm, K.H.: 1986, *Astrophys. J.* **308**, 829
Raga, A. C. & Böhm, K.H.: 1987, *Astrophys. J.* **323**, 193
Raga, A. C. & Kofman, L.: 1992, *Astrophys. J.* **386**, 222
Raga, A. C., Mateo, M., Böhm, K.H. & Solf, J.: 1988, *Astron. J.* **95**, 1783
Raymond, J.: 1979, *Astrophys. J. Suppl.* **39**,
Reipurth, B.: 1989, 'ESO Workshop on Low Mass Star Formation and Pre-Main Sequence Objects' in B. Reipurth, ed(s)., *ESO*, Garching, 247
Reipurth, J.: 1991, 'Physics of Star Formation and Early Stellar Evolution Proc., Nato Advanced Study Institute' in C. L. Lada and N. D. Kyfalis, ed(s)., *Kluwer*, Dordrecht, 497
Shull, J. M. & McKee, C. F.: 1979, *Astrophys. J.* **227**, 131
Schwartz, R. D.: 1974, *Astrophys. J.* **191**, 419
Schwartz, R. D.: 1975, *Astrophys. J.* **195**, 631
Schwartz, R. D.: 1978, *Astrophys. J.* **223**, 884
Schwartz, R. D.: 1983, *Astrophys. J.* **268**, L87
Schwartz, R. D., Cohen, M. & Williams, P. M.: 1987, *Astrophys. J.* **322**, 403
Schwartz, R. D., Dopita, M. A. & Cohen, M.: 1985, *Astron. J.* **90**, 1820
Solf, J. & Böhm, K.H.: 1987, *Astron. J.* **93**, 1172
Solf, J. & Böhm, K.H.: 1991, *Astrophys. J.* **375**, 618
Solf, J. & Böhm, K.H.: 1993, *Astrophys. J.* **410**, L31
Solf, J., Böhm, K.H. & Raga, A. C.: 1986, *Astrophys. J.* **305**, 795
Solf, J., Böhm, K.H & Raga, A. C.: 1988, *Astrophys. J.* **334**, 229
Stone, J. M. & Norman, M. L.: 1993, *Astrophys. J.* **413**, 198
Stone, J. M. & Norman, M. L.: 1993, *Astrophys. J.* **413**, 210
Stone, J. M. & Norman, M. L.: 1994, *Astrophys. J.* **420**, 237

Wolfire, M. G. & Königl, A.: 1991, *Astrophys. J.* **383**, 205
Zinnecker, H., Mundt, A., Geballe, T. R. & Zealey, W. J.: 1989, *Astrophys. J.* **342**, 337

IR OBSERVATIONS OF SHOCK WAVES IN THE INTERSTELLAR MEDIUM

P. W. J. L. BRAND
Institute for Astronomy, University of Edinburgh
Royal Observatory Edinburgh, Edinburgh EH9 3HJ, U. K.

Abstract. A description of the development of the use of H_2 lines as diagnostics of shock excitation in molecular clouds is given. The use of both line ratios and of velocity profiles provides constraints on the models so tight that it may be possible to reject most or even all of the standard stationary shock models.

Key words: Molecular Hydrogen – Shock Waves – Outflows

1. Introduction

The subject started quite suddenly in 1976, when Gautier et al. observed infrared lines of excited molecular hydrogen in Orion (Gautier et al. 1976). This was followed by an avalanche of shock models (Hollenbach & Shull 1977, Kwan 1977, London et al. 1977), and by 1982 there was sufficient evidence to render the recently developed magnetically moderated C-shock model (Draine 1980, Draine, Roberge & Dalgarno 1983) the leading contender for the excitation mechanism of infrared lines of H_2 in molecular clouds (Chernoff, Hollenbach & McKee 1982, Draine & Roberge 1982).

It is perhaps ironic that the identification of *fluorescent* excitation had to wait until 1986 (Gatley et al. 1987, Sellgren 1986) although it had been observed in the 1976 wave of discovery (Treffers et al. 1976). Indeed the first serious explanation of IR excited H_2 was in terms of fluorescence (Black & Dalgarno 1976).

Hovering over the entire development of the shock excitation mechanism, and still as important today, has been the early observation (Nadeau & Geballe 1979) that the H_2 lines were extraordinarily wide in many cases. In the brightest part of the Orion outflow the range of velocities in a region only a few arcseconds across was observed to be greater than 140 km s^{-1} (Brand et al. 1989a). Explanations for such widths still do not seem entirely secure.

This paper will examine the current state of play, both for observations and for models. A thorough and detailed review of shock physics is given by Draine & McKee (1993).

Astrophysics and Space Science **233**: 27–38, 1995.

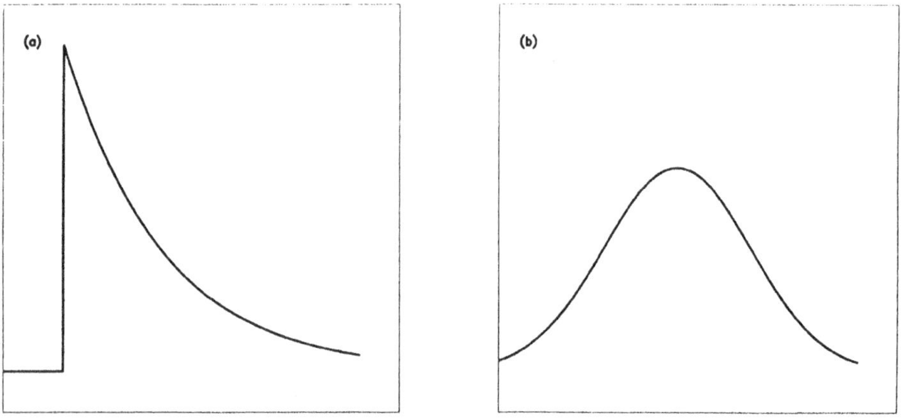

Fig. 1. The profile of temperature versus distance in (*a*) a J shock and (*b*) a C shock

2. Plane Models

The standard shock models are either pure hydrodynamic shocks, called J (or jump) shocks in Draine's original paper (1980), or magnetically moderated C ('continuous') shocks. The C shock is a two-fluid shock: neutrals, and magnetized ions (a small component by mass). The ionic magnetosonic speed is larger than that of the propagating disturbance and so the ions—which are locked to the magnetic field (the propagation is perpendicular to the field lines)—adjust their velocity by a smooth (magnetic) pressure wave, gradually accelerating to the downstream speed. The neutrals ahead of the disturbance experience the friction of these ions streaming through, and become heated and accelerated by the frictional drag. If the coupling by collisions is sufficient, the heating and acceleration so caused will bring the neutral fluid to its downstream accelerated and compressed state without the need for a hydrodynamic jump. There is still a change in neutral entropy, but in this case it is due to the randomizing collisions causing the friction.

Fig. 1 shows the temperature profile through J and C shocks, and the fact that these are of such a differing nature gives rise to one of the stronger diagnostic results.

2.1. J SHOCKS

Neglecting details such as the time required for the internal states of the molecule to achieve equilibrium, the gas temperature reaches its highest value just behind the hydrodynamic jump, and then rapidly cools to its final value. The cooling rate is set by the local constitution of the gas (the cooling function), and the final temperature is negligibly small in terms of the levels of H_2 giving rise to IR lines. It is generally assumed (see section 3.2) that the gas is sufficiently dense that it is in local thermodynamic equilibrium, i.e. that at every distance behind the thermodynamic jump there is a unique temperature which, via a Boltzmann distribution, simultaneously describes internal and translational states.

The pressure behind the jump is $F\rho_0 V_{\rm shock}^2$ where F increases from $2/(\gamma + 1)$ to unity; in the case of a diatomic gas, this is a 17% increase only. Thus the assumption of constant pressure behind the shock is reasonable. In the case of high magnetic fields, this means that the cooling becomes isochoric when the magnetic pressure dominates. Otherwise the cooling rate is the isobaric rate.

The time that the cooling gas spends in each region of the temperature range is inversely proportional to the cooling rate. Each IR line will be significantly excited only above a temperature which is proportional to the upper energy level of the line. Thus the strength of each line—and in particular the run of column densities of upper levels—is a map of the cooling function of the gas, and is independent of most other properties of the shock.

One property which *is* important is the fact that at a shock velocity of greater than approximately 23 km s^{-1} (Kwan 1977) the shock is so intense that the gas, cooling predominantly by dissociation, destroys all the molecules. This property of planar J shocks is obviously very embarrassing in a region with H_2 velocity profiles as wide as seen in Orion.

2.2. C SHOCKS

In contrast, C shocks have a very different temperature profile. The slow frictional heating phase which then achieves balance with cooling means that temperature rises to a peak value where $dT/dz = 0$ followed by a decline back to low temperatures. The simplest description of this mass of gas is of a slab at approximately the peak temperature, and with a range of velocities which, like the peak temperature itself, is shock velocity dependent.

C shocks are less likely to destroy the molecules, but even in this case there is a maximum value for shock speed (approximately 40 km s^{-1}, Draine et al. 1983) for typical ISM values.

How do these properties, of C and of J shocks, stack against the data?

3. Basic Data

The data are from spectra taken in the infrared between 1μm and 5μm, and are the relative line intensities, and their velocity profiles. The transitions in question are quadrupolar, and therefore it is straightforward to convert intensity to column density, from which relative population can be inferred.

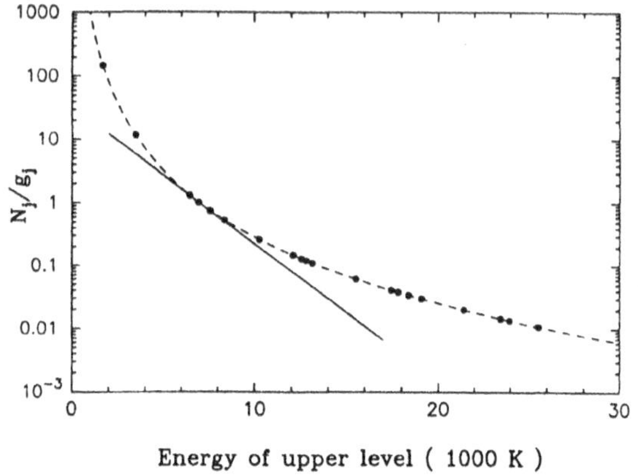

Fig. 2. Post-shock cooling: logarithm of column density in each energy level of H_2 versus energy (after Burton 1992). The straight line is the distribution of levels in a gas at 2000 K.

The effect of dust remains to obscure the result, but this can be turned to advantage as we shall see.

Fig. 2 (from Burton 1992) shows the distribution of column density with upper energy level expected from a plane J shock with a typical cooling function. Superposed is a line which represents the distribution expected from a slab of gas at the constant temperature of 2000K. It should be noted that while there are levels (at energies around 7000 Kelvins) whose column densities are well represented by this line, above and below this energy level the populations have a relative excess.

This relative excess is a functional of the cooling function in the case of a plane J shock, and is *not* predicted in the case of a C shock.

3.1. Excited hydrogen in Orion

Orion, in which one of the nearest and brightest regions of shocked molecular gas has been observed assiduously since 1976, yields some of the most complete data. Fig. 3 shows the column density distribution at the brightest part of the Orion outflow, the ordinate modified to show the excess over 2000 K emission (Brand et al. 1988). The continuous line is a best fit planar J shock, and the dashed lines are planar C shock predictions; these are consistent with the 'constant temperature' prediction referred to in section 2.2, with the lowest energy levels showing the characteristic cooling zone behaviour of a J shock (and therefore an excess) and the highest levels not fully thermalized at the lower density used in this type of model, and therefore falling below the horizontal line. This evidence seems clearly to rule out planar C shocks as contenders.

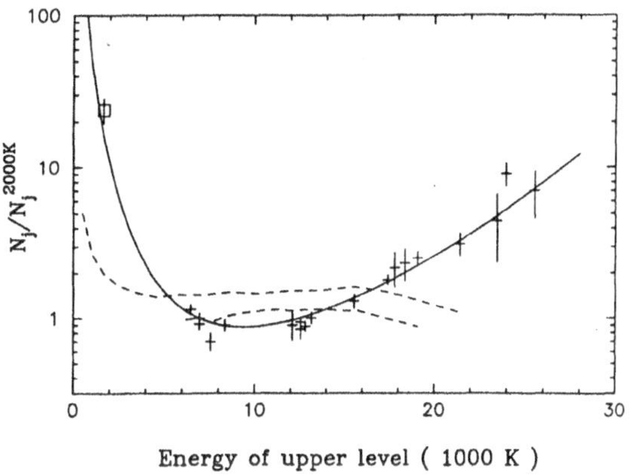

Fig. 3. Normalized column densities derived from the H_2 spectrum at peak 1 in the Orion outflow (Brand et al. 1988).

There is another puzzle posed by the data. A map over the brighter parts of the Orion outflow of the ratio of two of the more highly excited lines, 0-0 S(13) (17000 K) to 1-0 O(7) (8300 K), was used to test the variation of the shape of the curve in Fig. 3 (Brand et al., 1989b). Fig 4 shows the result, which implies that within the (10%) errors, the slope of the upper part of the curve did not vary within the flow. This would not seem unreasonable if a planar J shock were the culprit—merely demonstrating the invariance of the cooling function. But this requires the density dependence of all contributors to the cooling to be identical, and this is far from obvious (for instance, dissociational cooling by H_2 varies as density squared while line cooling by H_2 varies as density).

Conversely, if the curve is due to the superposition of two or more C shocks, the relative strengths and speeds of these shocks must be the same.

When we turn to the velocity resolved data (Fig. 5, Brand et al. 1989a), then the twin constraints on velocity in J shock models confront us. On the one hand, velocity changes greater than 23 km s^{-1} are forbidden as the molecules are destroyed, and on the other, the gas has completed much of its cooling by the time it is radiating the IR lines, and therefore will be very close (a few km s^{-1}) to its downstream velocity, implying a small observed velocity range. But Fig. 5—and indeed Nadeau & Geballe's 1979 results—give the lie to that!

Thus neither type of planar shock appears able to explain the observations in Orion.

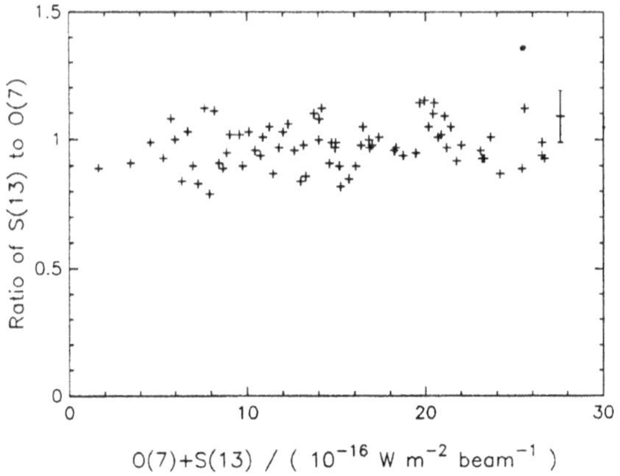

Fig. 4. The ratio of intensity in the 0-0 S(13) line to that in the 1-0 O(7) line of H_2 in the region of peak 1 in the Orion outflow (Brand et al. 1989b). The lack of variation with intensity in the map implies a lack of variation with position.

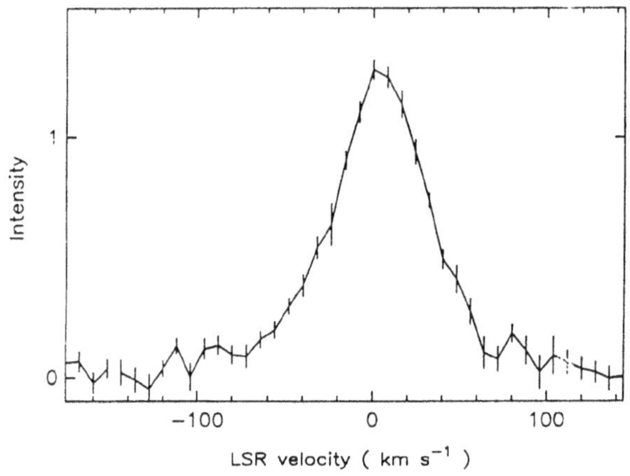

Fig. 5. The velocity profile of the 1-0 S(1) line of H_2 in a 5″ aperture at peak 1 in the Orion outflow (Brand et al. 1989a).

3.2. MODEL INDEPENDENT DATA

Before proceeding further, one should note that two inferences may be drawn even while our theory is in disarray. In each case the conclusion comes from the smooth distribution of column densities about a single curve in Fig. 3. The first is that the distribution of line wavelengths is quite different from the distribution of upper energy levels. Extinction due to dust will affect intensity and therefore derived column density according to wavelength (usually assumed to vary as $A_\lambda \propto \lambda^{-1.7}$, Mathis, 1990). If this correction has

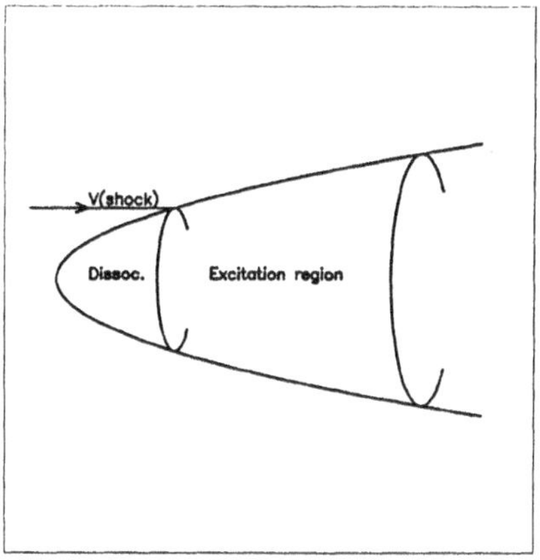

Fig. 6. A sketch of the main features of a bow shock propagating through molecular gas.

been wrongly applied, there will be an excess scatter in Fig. 3. This technique, which needs many lines, was applied to Peak 1 in Orion to derive a 2.2μm extinction of 0.9 (Brand et al. 1988).

The second model-free conclusion that can be drawn is from the upwards curve of column density with higher energy in Fig. 3. Since these levels have high critical densities, and since the curve shows no evidence of 'droop', this implies that even the highest observed levels are thermally populated, and so—barring a conspiracy in the interaction of excitation with the Einstein coefficients—if the collision cross sections of these levels are known, minimum densities can be inferred. Unfortunately, calculations for these cross sections are not yet to hand. The data are waiting!

4. Model Tests

4.1. PLANAR *versus* BOW

It is clear that planar shocks are out, at least for Orion. No other data seem free of this constraint. So we must turn to configurations that give a range of velocities. Bow shocks, the structure formed by the interaction of a blunt obstacle with a smooth flow, appear to have the necessary properties.

Such a structure is sketched in Fig. 6. The flow through the head of the shock is so fast that the molecules are dissociated. Here ionic emission is expected, but no contribution to the molecular line profiles. Further back in the structure the shock is oblique to the flow and is correspondingly weaker. Molecules will survive, and a telescope aperture that covers the entire flow will measure a profile that will have a range of line of sight velocities in it. If the unshocked gas is static with respect to the observer, there will be much of the emission with a line of sight velocity close to zero (for example at the tangent curves of the structure, and far down the flanks). It is interesting that the profiles observed thus far in Orion peak at near zero velocity.

A major problem persists with the model, however. It is clear that with the best will in the world it is not possible to get a velocity range in surviving molecules that is at the most up to twice the survival speed for molecules. Indeed, even without this constraint Brand et al. (1989a) showed that in J bow shocks it was impossible in a single structure to produce the broad range of velocities seen at Peak 1 in Orion.

4.2. C BOWS : "SHOCK ABSORBERS"

Smith et al. (1991a,b) proposed a solution to this pressing problem. The standard velocity limit for C shocks is derived from 'standard' conditions in the ISM. If the expectation that the Alfvén velocity is everywhere 1-3 km s^{-1}, and the magnetic field can be large, then the C shock structure can be arbitrarily lengthened to allow a gentle acceleration which does not destroy molecules to achieve very high post-shock molecular speeds. The simultaneous application of a bow structure then gives a spread of velocities which will match the observed profile in Orion. The range of shock speeds then gives a range of post shock excitations which when integrated together produce the characteristic upward curve of Fig. 3.

Furthermore, the distribution of column densities can simultaneously be brought into concordance with the Orion data. This seems promising.

There are several requirements, however. Two of the most pressing are that the bow shape is fairly closely prescribed, and that the entire bow structure must be in the entrance aperture of the spectrometer. This latter requirement is to ensure that the full velocity spread is observed, and that the range of shock speeds necessary to synthesise the N_j/g_j *vs.* T_j diagram (Fig. 3) is within the beam.

Hartigan et al. (1987) have demonstrated how the velocity profile of a bow shock structure behaves as a function of orientation and flow velocity, and a similar analysis (Smith et al. 1991) for C shocks confirms that so long as the bow shock shape is not far from parabolic, a C shock with a *high* magnetic field (the requirement is for an Alfvén speed of 30+ km s^{-1}) will give satisfactory results in Orion. A bonus is that the N_j/g_j *vs.* T_j curve is to first order independent of the flow velocity. This may be at the root of the otherwise puzzling constancy of the 0-0 S(13)/1-0 O(7) ratio referred to above.

The requirement that the entire bow be within the spectrometer beam is more troubling, however. The characteristic drag length of a C shock with such a high field is easily resolved at the distance to Orion, and such struc-

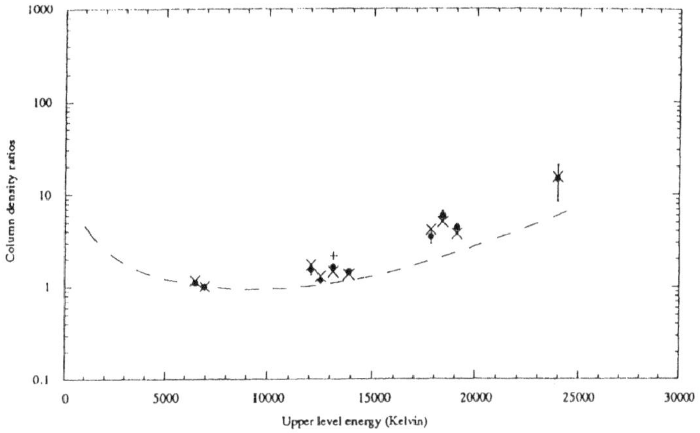

Fig. 7. A plot of occupancy versus energy level for H_2 observed in HH7 at a position on the axis of the bow shock, and some arc seconds downstream from the head.

tures should be readily observed. A clear-cut experiment on a clearly defined bow shock is the obvious next step.

5. Bow Shock Velocity Dissection

5.1. HH7

An object which appears to be an isolated bow shock emitting H_2 lines is HH7 in NGC 1333 (Hartigan et al. 1989, Carr 1993). The possibility now open of doing spectroscopy at relatively high spatial resolution means that it is possible in such an object to examine parts of the bow shock and see if the model stands up in detail. Fernandes et al. (1995) have started such an investigation. The slit of CGS4, the UK Infrared Telescope facility spectrometer, was placed along the axis of the bow, with a pixel size significantly less than the scale size of the bow structure. The spectrum resulting from an on-axis position some small distance down the bow shock is shown in Fig. 7.

The model fit, the curve, shown is for a C shock fit to *two* (front and back) oblique plane shocks representing the areas of the bow structure intersected by the spectrometer pixels. The crosses are the sum of the prediction of N_j/g_j, or relative occupancy, due to the shock and the corresponding occupancy due to a fluorescent contribution. The pattern of fluorescent relative occupancy is so characteristic that this is a highly significant fit. The exciting radiation is supposed to be Lyman alpha line emission from the highly excited head of the bow shock, and is not inconsistent with the expected flux.

This relies on the fact that the irradiated gas is excited by the shock, and has a wide velocity profile.

A J shock in any guise was found not to be able to fit these data.

This is an interesting step forward, and needs more work to render it secure. In particular, it will be interesting to obtain high signal to noise velocity profiles in positions corresponding to those for which these spectral data exist.

5.2. Orion bullets

The recent observation (Allen & Burton 1993) of 'bullets' in Orion has led to the interesting speculation that this outflow, or at least part of it, is explosive rather than steady. Several dozen 'bullets', identified by a tip emitting [FeII] lines and a wake of excited H_2 have been discovered or confirmed, with velocities up to 400 km s^{-1}. The traditional problems, on the one hand, of how a bullet is accelerated without being destroyed, and on the other, if these are not bullets but jets, what kind of multiple jet 'hedgehog' could produce them, remain imponderable. The current concern is to see if the shocked gas is explicable in terms of bow shocks.

Tedds et al. (*this volume*) describes a study of the velocity profiles of the [FeII] and H_2 in one of these objects. It is evident that downstream in the wake of a sideways moving bow shock, the gas in front should be moving forward, while that behind moves backward, with essentially zero crosstalk between the two profiles. That is *not* what is observed. Because of possible source confusion one cannot be dogmatic, but if this behaviour is the norm—and investigations are in progress in several of these wakes—it means that one or several of the model assumptions will have to go. In particular, such shocks may be unstable (cf. Tóth, *this volume* and Malone, Dyson & Hartquist (1994), Taylor & Raga, *this volume*). In that case we expect a significantly different level occupancy, although this has not been worked out in sufficient detail to make firm predictions.

6. Other Loose Ends

6.1. IC 443 supernova shocks

Richter et al. (*this volume*) present remarkable data on column densities in IC 443, a supernova remnant colliding with a molecular cloud (cf. also Moorhouse et 1992). In several positions, the relative level occupancy (i.e. N_j/g_j *vs.* T_j) is identical with that seen in Orion (Brand et al. 1988). It seems that whatever explains Orion's main outflow, also operates in IC 443 (although not in HH7). Are they J shocks, or are they super-high magnetic field C-shocks, or are they instabilities in the flow?

6.2. HH91A — DIFFERENT AGAIN?

Finally, the recent data (Gredel et al. 1992) on H_2 emission from HH91A tells a different story again! Although at first glance the distribution of relative occupancy looks like that of a constant temperature gas, close inspection reveals that the lowest levels are indicating the characteristic upward concavity of the shock plot. The flatness of the plot is also in part possibly due to the lack of levels in the diagram at the highest energies. Such a flat plot, however gybes with our prejudices about C shocks, so it is a surprise to find that Smith (1995) has an explanation for this data set based on a J shock model, albeit one with a magnetic field.

7. In Conclusion

The shocks described here are of vital importance in star formation, since they are probably the means whereby the star formation rate in dark clouds is regulated (Norman & Silk 1980, McKee 1989). But more than that, they are interesting in their own right, both as diagnostics of the local environment, and as examples of supposedly straightforward interstellar physics. It might be regarded as depressing to find ourselves in such dissarray over this simplest of challenges. I much prefer to take the view that since all the problems seem on the brink of solution, largely thanks to the exhilaratingly rapid progress in astronomical technology, we are in the middle of a rapid development of our understanding of outflows and their place in the scheme of interstellar things.

Acknowledgements

Thanks are due to colleagues who helped to form these ideas, particularly Michael Burton, Tom Geballe, Michael D. Smith, and also to the staff of the United Kingdom Infrared Telescope, a facility of the Particle Physics and Astronomy Research Council.

References

Allen, D. A., and Burton, M. G.: 1993, *Nature* , **363**,54

Black, J. H. and Dalgarno, A.: 1976, *Astrophys. J.*, **203**, 132

Brand, P.W.J.L., Moorhouse, A., Burton, M.G., Geballe, T.R., Bird, M., and Wade, R.: 1988, *Astrophys. J.*, **334** , L103

Brand, P.W.J.L., Toner, M.P., Geballe, T.R., Webster, A.S., Williams, P.M. and Burton, M.G.: 1989a, *Mon. Not. R. astr. Soc.*, **237**, 929

Brand, P.W.J.L., Toner, M.P., Geballe, T.R. and Webster, A.S.: 1989b, *Mon. Not. R. astr. Soc.*, **237**, 1009

Burton, M.G.: 1992, *Aust. J. Phys.*, **45**, 463
Carr, J.S.: 1993, *Astrophys. J.*, **406**, 553
Chernoff, D.F., Hollenbach, D.J. and McKee, C.F.: 1982, *Astrophys. J.*, **259**, L97
Draine, B.T.: 1980, *Astrophys. J.*, **241**, 1081
Draine, B.T. and McKee, C.F.: 1993, *Ann. Rev. Astron. Astrophys.*, **31**, 373
Draine, B.T. and Roberge, W.G.: 1982, *Astrophys. J.*, **259**, L91
Draine, B.T., Roberge, W.G. and Dalgarno, A.: 1983, *Astrophys. J.*, **264**, 485
Fernandes, A.J.L., Brand, P.W.J.L.: 1995, *Mon. Not. R. astr. Soc.*, in press
Gatley, I., Hasegawa, T., Suzuki, H., Garden, R. P., Brand, P. W. J. L., Lightfoot, J., Glencross, W., Okuda, H. and Nagata, T.: 1987, *Astrophys. J.*, **318**, L73
Gautier, T.N., Fink, U., Treffers, R.R., and Larson, H.P.: 1976, *Astrophys. J.*, **207**, L129
Gredel, R., Reipurth, B. and Heathcote, S.: 1992, *Astron. Astrophys.*, **266**, 439
Hartigan, P., Curiel, S. and Raymond, J.: 1989, *Astrophys. J.*, **347**, L31
Hartigan, P., Raymond, J., and Hartmann., L.: 1987, *Astrophys. J.*, **316**, 323
Hollenbach, D.J., and Shull, M.: 1977 *Astrophys. J.*, **216**, 419
Kwan, J.: 1977 *Astrophys. J.*, **216**, 713
London, R., McCray, R. and Chu S-I.: 1977. *Astrophys. J.*, **217**, 442
Mathis, J.S.: 1990, *Ann. Rev. Astron. Astrophys.*, **28**, 37
Malone, M.T., Dyson, J.E. and Hartquist, T.W.: 1994, *Astrophys. & Sp. Sci.*, **216**, 143
McKee, C.F., 1989, *Astrophys. J.*, **345**, 782
Moorhouse, A., Brand, P.W.J.L.,Geballe, T.R., and Burton, M.G.: 1992, *Mon. Not. R. astr. Soc.*, **253**, 232
Nadeau, D. and Geballe, T.R. : 1979: *Astrophys. J.*, **230**, L169
Norman, C. and Silk, J.: 1980, *Astrophys. J.*, **238**, 158
Sellgren, K.: 1986, *Astrophys. J.*, **305**, 399
Smith, M.D.: 1994, *Astron. Astrophys.*, **289**, 256
Smith, M.D.: 1995, *Astron. Astrophys.*, in press
Smith, M.D., Brand, P.W.J.L. and Moorhouse, A.: 1991a, *Mon. Not. R. astr. Soc.*, **248**, 451
Smith, M.D., Brand, P.W.J.L. and Moorhouse, A.: 1991b, *Mon. Not. R. astr. Soc.*, **248**, 730
Treffers, R. R., Fink, U., Larson, H. P., and Gautier, T.N.: 1976 *Astrophys. J.*, **209**, 793

OBSERVATIONS OF SHOCKED H_2 AND [FEII] LINE PROFILES IN ORION BULLET WAKES

JONATHAN A. TEDDS and PETER W.J.L. BRAND
Institute for Astronomy, University of Edinburgh, EH9 3HJ, U.K.
MICHAEL G. BURTON
School of Physics, University of New South Wales, Australia.
ANTONIO CHRYSOSTOMOU
Division of Physical Sciences, University of Hertfordshire, U.K.
and
AMADEU J.L. FERNANDES
Centro de Astrofisica, Universidade do Porto, Portugal.

Abstract. New observations of H_2 velocity profiles in the Orion bullet wakes are extremely difficult to reconcile with existing steady-state shock models. We have observed [FeII] 1.644μm velocity profiles of selected bullets and H_2 1-0 S(1) 2.122μm velocity profiles for a series of positions along and across the corresponding bow-shock wakes. Integrated [FeII] velocity profiles of the brightest bullets are consistent with theoretical bow shock predictions. Observations of broad, singly-peaked H_2 1-0 S(1) profiles in the most clearly resolved bullet wakes challenge our understanding of molecular shocks. It may be necessary to model the effects of instabilities and turbulence in the Orion bullet wakes in order to fit our observations.

1. Introduction

The nature of molecular shocks, which play an important role in the processes of momentum and energy transfer within star-forming molecular clouds (McKee 1989), is still uncertain (Draine & McKee 1993). The Orion molecular cloud is the brightest known source of shocked H_2 emission and as such has been the primary testbed for theoretical models. We have established that planar C(magnetised)-shock models cannot explain the H_2 line intensities in Orion or any other outflow, while planar J-shock models can (Brand et al. 1988). However, neither planar nor bow J-shock models can reproduce the line profiles observed at Peak 1 in Orion (Brand et al. 1989). It is just possible to fit both H_2 line profiles and ratios in this region using highly magnetised C-shock models (Smith et al. 1991). Measurements of H_2 excitation conditions which remain constant throughout the bright part of the Orion outflow, however, are not easily explained by any bow shock models (Brand et al. 1989, McKee 1989).

Recent near-IR imaging of Orion with 0.5″ spatial resolution in the emission lines of H_2 (2.122 μm) and [FeII] (1.644 μm) has clarified our view of the shocked molecular outflow (Allen & Burton 1993). Many new Herbig Haro-type objects were revealed, visible as [FeII] 'bullets', at the heads of wakes

Astrophysics and Space Science **233**: 39–44, 1995.

of H_2 emitting gas. The bullets (apparently originating within 5″ of IRc2) have been ejected over a wide opening angle and we have measured [FeII] line profile widths of up to 380 km s^{-1} (Tedds et al. 1994), in agreement with [OI] 6300Å linewidths (Axon & Taylor 1984) indicating an explosive origin in the core of Orion within the last 10^3 years. We cannot yet identify the cause of this event although an FU-Orionis event (Hartmann & Kenyon 1985) is speculated.

Fernandes and Brand (1995a) have demonstrated that molecular hydrogen emission from the bow shock-shaped HH7 is well explained as a C-shock in a bow configuration in which the 'cap' is completely dissociated, and H_2 fluorescence produced by Lyman alpha resonance pumping in the hot post-shocked layers (Black & van Dishoeck 1987) in the cap irradiate the downstream flow. The same type of shock + fluorescent spectrum as seen in HH7 has been seen in DR21 (Fernandes & Brand 1995b). It was expected that the bow shocks in Orion could be similarly excited.

2. Observations and Results

We have used the UKIRT near-IR spectrometer CGS4 in its echelle mode to obtain [FeII] line profiles of some of the brightest bullets together with H_2 line profiles in the most clearly defined bow shock wakes (Tedds et al., in preparation).

The two most important parameters in modelling integrated bow-shock line profiles are the incident shock velocity, v_s and the orientation, α. Observationally, we can directly measure v_s since it is equal to the full width near zero intensity of the integrated emission over an entire bow shock front independent of the shape of the bow-shock, orientation angle, preshock density, elemental abundances and reddening (Hartigan et al. 1989). Furthermore, by measuring the maximum and minimum radial velocities for the integrated line profile we obtain both v_s and α.

In the case of the bright bullet M42 HH1 (Axon & Taylor 1984), the integrated [FeII] profile corresponds to a shock velocity of $v_s = 380 \pm 10$ km s^{-1} and an orientation angle of $\alpha = 55 \pm 5°$ (Tedds et al. 1994). Comparison of the observed [FeII] profile shape with radiative bow-shock model predictions (Hartigan et al. 1989) shows close agreement for a $v_s = 400$ km s^{-1}, $\alpha = 60°$ model. The stronger of the two peaks lies near zero radial velocity as predicted in bullet models as opposed to models in which the molecular gas flows past an obstacle. However, both the [FeII] and H_2 profiles indicate an excess of this peak compared with model predictions. The H_2 wake itself is not clearly defined and coincides to some extent with the more confused Peak 1 region. This hampers a clear interpretation of the corresponding H_2 profiles (Tedds et al., in preparation).

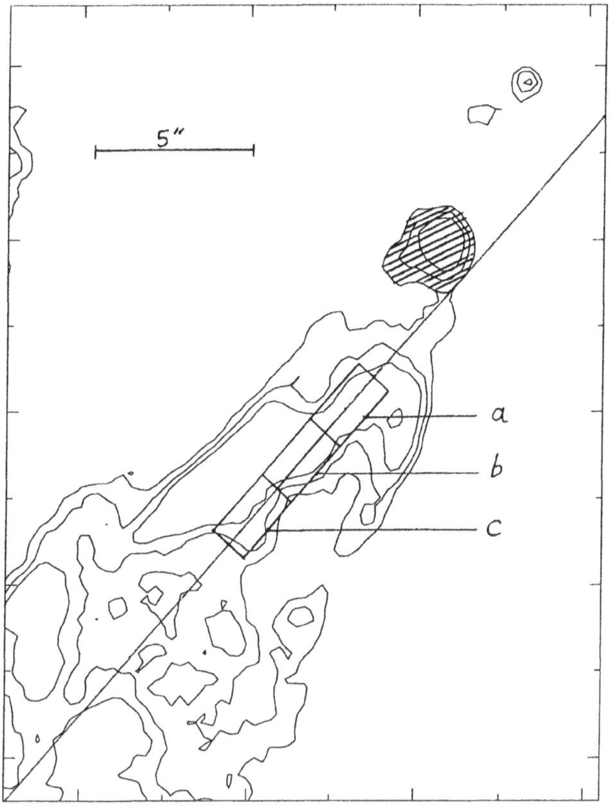

Fig. 1. H_2 1-0 S(1) 2.122 μm contour plot of second bullet wake (see section 2). The shaded area indicates the position of the [FeII] 1.644 μm bullet. Labelled boxes indicate the location of 3 H_2 line profile positions.

Observations of the integrated [FeII] 1.644 μm line profile of another bullet at [1950] 05^h 32^m 44.5^s, -05° 23′ 08" (Allen & Burton 1993, Fig 1.) oriented in or very close to the plane of sky, show a width of up to 120 km s^{-1}. We have observed H_2 1-0 S(1) 2.122 μm line profiles at positions along the corresponding, clearly resolved, bow-shock wake (Fig 1). At these positions, a line-of-sight should intersect two distinct and oppositely directed shock fronts on either side of the bow shock wake, giving a doubly-peaked line profile (Fig 2a). In moving down the wake, away from the bullet, the model velocity width decreases as the normal component of bow shock velocity of each gas component decreases (Fig 2b,c). Furthermore, the model shows that the ratio of individual component line width to the separation of the two components must be 0.5, independent of the speed of the shock structure. Figure 3, however, shows the observed profiles (resolution 15 km s^{-1}) at these

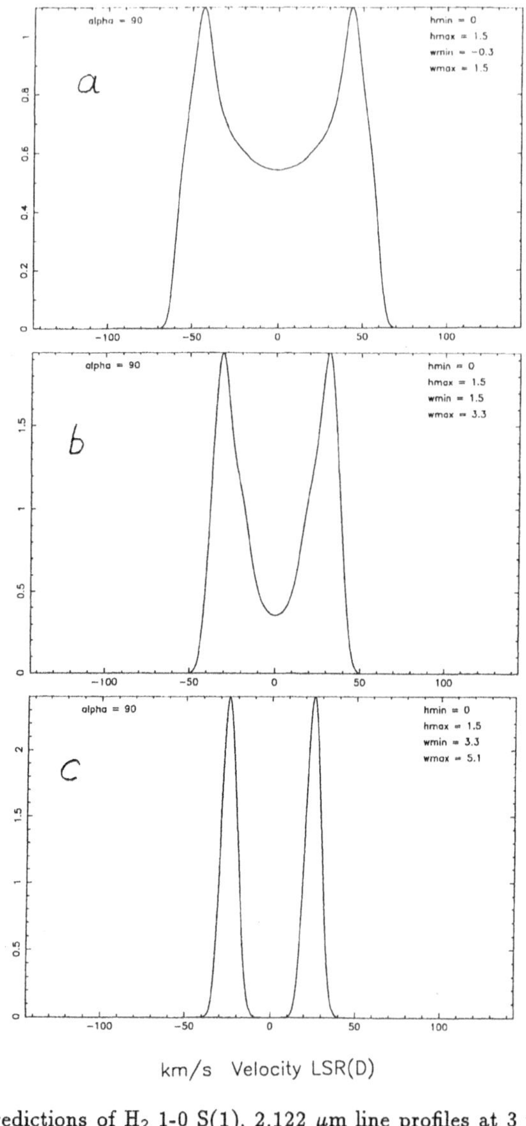

Fig. 2. Model predictions of H_2 1-0 S(1), 2.122 μm line profiles at 3 pixel positions (Fig 1) in a $v_s = 120$ kms^{-1} bow shock structure oriented in the plane-of-sky ($\alpha = 90°$). The model assumes post-shock relative velocity is negligible and that the shock moves into a stationary medium.

positions. The profiles are broad (≤ 80 km s^{-1}) but singly-peaked. Similar profiles are also seen in the plane-of-sky bullet wake M42 HH7 (Tedds et al., in preparation).

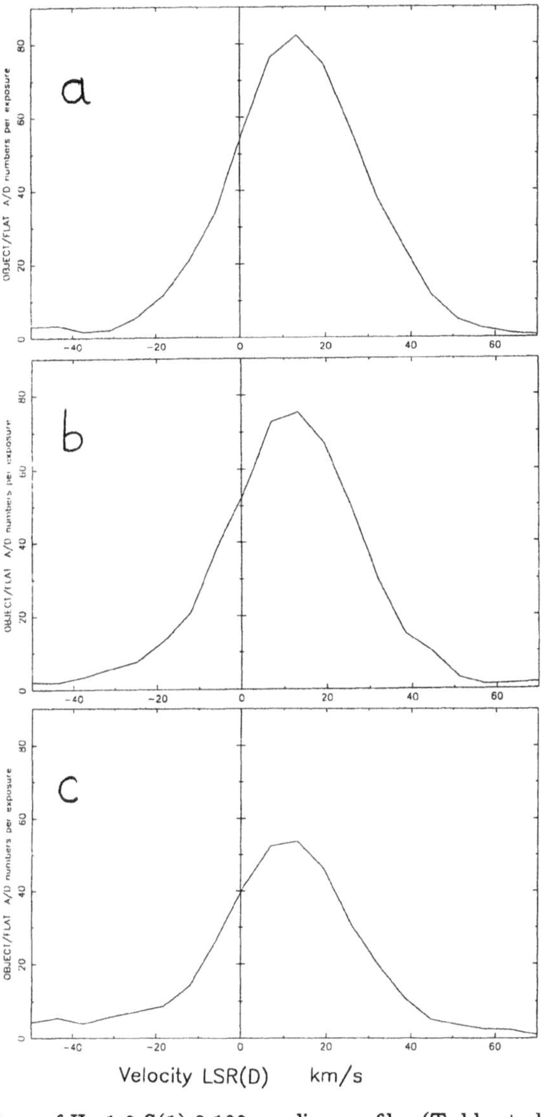

Fig. 3. Observations of H_2 1-0 S(1) 2.122 μm line profiles (Tedds et al., in preparation) at 3 wake positions (Fig.1)

3. Discussion

We have demonstrated that integrated [FeII] line profiles in Orion are consistent with theoretical bow-shock predictions for different values of v_s and α. The extended H_2 bullet wakes have allowed us to dissect individual molecular bow shock structures but the broad, singly-peaked profiles observed in the two most clearly resolved, plane-of-sky oriented wakes challenge our present understanding. Carr (1993) observed H_2 profiles in HH7 and detected double peaked structure consistent with the predictions for a radiating bow shock (Fig 2). However, it is very difficult to reconcile *any* steady-state bow shock model with these observations in Orion. The fluorescent component fitted to H_2 intensities in HH7 is very narrow in profile compared to shocked emission.

Alternatively, we may still not be resolving multiple shock fronts along the line-of-sight. For example, two or more overlapping bullet wakes could give rise to merged sets of doubly-peaked profiles resulting in the type of single Gaussian shaped profiles observed here. However, given the appearance of single bow shaped wakes, the velocity resolution of our observations and that we see this phenomenon in *two* different wakes, this explanation is expected to be excluded. New observational methods utilising adaptive optics techniques should help to resolve this issue and further constrain shock models by distinguishing the range of cooling lengths from 10^{13}cm for J shocks up to 10^{17}cm for some C shocks in the nearest sources.

If we cannot fit the profiles in Orion with steady state molecular shocks it may be necessary to model the effects of instabilities and turbulence. This will have important consequences. Not only will line profiles be broadened but level populations of shocked species will be altered and hence the observed column densities over a range of transitions. New observations of H_2 and [FeII] column densities in these bullet wakes will address this.

References

Allen, D. A. and Burton, M.G: 1993, *Nature* **363**, 54

Axon, D. A. and Taylor, K: 1984, *Mon. Not. R. astr. Soc.* **207**, 241

Black, J. & van Dishoeck, E.: 1987, *Astrophys. J.* **322**, 412

Brand, P. W. J. L., Moorhouse, A., Burton, M. G., Geballe, T. R., Bird, M. and Wade, R: 1988, *Astrophys. J. Lett.* **334**, L103

Brand, P. W. J. L., Toner, M. P., Geballe, T. R. and Webster, A. S: 1989, *Mon. Not. R. astr. Soc.* **237**, 1009

Carr, J.: 1993, *Astrophys. J.* **406**, 553

Draine, B. T. and McKee, C. F: 1993, *Ann. Rev. Astron. Astrophys.* **31**, 373

Fernandes, A. J. L. & Brand, P. W. J. L.: 1995a, *Mon. Not. R. Astron. Soc.* , in press

Fernandes, A. J. L. & Brand, P. W. J. L.: 1995b, *Mon. Not. R. Astron. Soc.* , submitted

Hartigan, P., Curiel, S. and Raymond, J.: 1989, *Astrophys. J. Lett.* **347**, L31

Hartmann, L. & Kenyon, S. J.: 1985, *Astrophys. J.* **299**, 462

McKee, C. F.: 1989, *Astrophys. J.* **345**, 782

Smith, M. D., Brand, P. W. J. L. & Moorhouse, A.: 1991, *Mon. Not. R. Astron. Soc.* **248**, 730

Tedds, J. A., Brand, P. W. J. L., Burton, M. G., Chrysostomou, A. & Fernandes, A. J. L.: 1994, *in Clouds, Cores and Low Mass Stars, ASP Conference Series, eds D.P. Clemens and R. Barvainis* **65**, 375

Tedds, J. A. & Brand, P. W. J. L.: 1995, *Mon. Not. R. Astron. Soc.* , in preparation.

H_2 FLUORESCENCE IN HH 7 AND DR 21

A. FERNANDES
Centro de Astrofísica da Universidade do Porto, Rua do Campo Alegre 823, 4150 Porto, Portugal

P. BRAND
Institute for Astronomy, Blackford Hill, Edinburgh EH9 3HJ, Scotland

and

M. BURTON
School of Physics, University of New South Wales, Sydney, NSW 2052, Australia

Abstract. We present evidence for Lyα pumping of the Lyman band system of molecular hydrogen in Herbig-Haro 7 and the bipolar outflow DR 21. For this study we have measured several vibrational-rotational emission lines of H_2 whose energy levels are widely spaced and ranging from 6000 ($v = 1$) to 25000 Kelvin ($v = 4$). We show that the near-infrared H_2 emission from the shocked gas in HH 7 can be well described by a bow C-type shock. The enhanced emission observed from the higher energy levels ($v > 3$) can be well modelled by employing the Lyα pumping mechanism.

In the DR 21 outflow the multi-line study showed that different physical conditions exist in the eastern and western emission lobes. The higher H_2 line ratios measured in the eastern lobe suggests a higher Lyα pump rate which may be locally produced in the fast bowshocks. The FUV radiation field emanating from the central HII regions may in addition be exciting the Lyman and Werner bands of H_2 in the molecular lobes.

We show that the observed H_2 emission can be interpreted in terms of a simple model consisting of a C-type bowshock, which produces the low excitation H_2 emission, and a FUV radiation field with enough Lyα line radiation to produce the high excitation H_2 emission through fluorescence.

Key words: ISM: Molecules – Shock Waves

1. Introduction

Supersonic winds produced in the early stages of the star formation process create energetic interactions with the immediate surrounding medium. The best known examples of such behaviour are the shocked molecular outflows in, for example, Orion (Allen and Burton 1993), DR 21 (Garden et al. 1986) and NGC 2071 (Burton et al. 1989). On a smaller scale, other examples include the optical line emission objects, or Herbig-Haro objects, that are commonly associated with the supersonic jets arising from young embedded stars (Reipurth et al. 1992). Generally, in most of these objects the observed molecular line radiation, and in particular the near-infrared lines of H_2, have been attributted to collisional excitation by shock waves. Early shock models were shown to be consistent with the observation that the 2-1 S(1) to 1-0 S(1) H_2 line ratio was small (~ 0.1) when compared to the value of 10 attained in a classical photo-dissociation (PDR) environment, where the H_2 molecule

Astrophysics and Space Science **233**: 45–50, 1995.

is radiatively excited (Black and van Dishoeck 1987). It was soon realised, however, that in dense ($n \geq 10^4$) PDRs the low-lying levels of H_2 may be collisionally quenched so that the 2-1 S(1) / 1-0 S(1) line ratio could attain a value that is similar to the shocked case (Burton et al. 1990). A clear discrimination between these two excitation processes is therefore of vital importance in allowing a detailed description of the underlying physics of each process and its relationship to the star formation mechanism.

The extremely broad line widths observed for the 1-0 S(1) line (FWHM $\sim$ 40km s^{-1}) (Zinnecker et al. 1989) and in the optical (Hα FWHM $\sim$ 206km s^{-1}) (Hartigan et al. 1987), suggested that the H_2 molecules could only survive in the wings of the bow where the effective shock velocity is lower than the velocity for H_2 dissociation (about 45km s^{-1} for MHD C-shocks). These bowshocks and in particular bow C-shocks, have been successful at reproducing many observed properties of the emitting gas (Smith and Brand 1990; Smith et al. 1991; Raymond et al. 1988).

Observations of an ultraviolet continuum and H_2 Lyα band emission from HH 43 (Schwartz 1983b; Schwartz et al. 1988) have revealed the importance of the radiative molecular excitation mechanism. It also appears that the H_2 infrared line ratios in HH 1/2 are consistent with both UV continuum pumping and collisional excitation in the post-shocked layer behind a dissociative shock (Harvey et al. 1986). An alternative mechanism for producing the vibrational-rotational lines of H_2 in which radiative excitation plays a role is Lyα pumping. This mechanism has been inferred from the ultraviolet spectra of HH 43 and HH 47 which shows a number of emission lines in the Lyman band of H_2 (Schwartz 1983b).

2. H_2 Excitation Processes in HH 7 and DR 21

Our recent investigation of the nature of shocks in HH 7 and the bipolar molecular outflow DR 21 suggest that shocks cannot be solely responsible for all the observed infrared H_2 line emission.

HH 7 is a low excitation Herbig-Haro object whose origin is presumably a supersonic wind from the nearby young star SVS13 (Herbig and Jones 1983). Its optical and infrared appearance resembles a bowshock (Garden et al. 1990; Solf and Böhm 1987) and its velocity structure appears consistent with a bow shaped C-shock (Carr 1993). For a typical bow, the critical speed for which dissociation takes place in C-shocks ($\sim$ 45km s^{-1}) is exceeded in the bow apex. Thus, H_2 can only survive in the wings of the bow surface and further downstream where the normal component of the shock velocity at the surface is lower than that critical speed. The resulting structure is of a tilted bow ($\theta \sim 40°$for HH 7) with a missing cap (see Fig. 1).

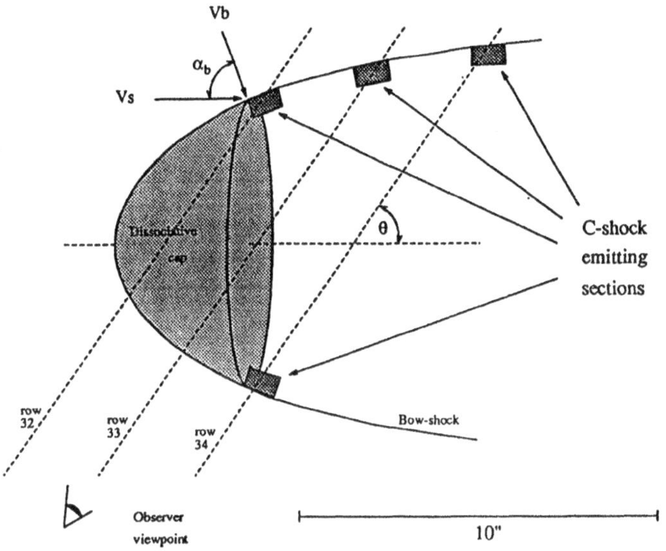

Fig. 1. Proposed structure of a bow C-shock

The dissociative shocks at the bow apex can be a source of strong UV and X-ray radiation. These high energy photons can travel downstream and excite quadrupole H_2 line emission in molecular condensations nearby by X-ray heating and FUV pumping (Wolfire and Königl 1991). In this situation, the H_2 molecules can absorb the non-ionizing ultraviolet photons ($\lambda > 912\text{Å}$) in the Lyman and Werner band systems. These absorptions are then followed by fluorescent emission to a vibrational bound level of the ground state 90% of the time while dissociation to the vibrational continuum occurs 10% of the time. The infrared H_2 line emission is then produced by the cascade down the rotational-vibrational energy ladder on the ground electronic state (Black and Dalgarno 1976).

In Fig. 2 we show the H_2 column density line ratios measured in HH 7. These line ratios (CDR) have been divided by those expected from a 2000 K slab of gas to emphasize the deviation of the data from the purely thermalised case (dashed curve). In addition, the data shows clearly the scatter amongst the column densities observed for each vibrational level, a feature that is precluded in any C-type shock model. We have shown (Fernandes and Brand 1995) that this scatter can be well modelled by allowing a Lyα pump to populate the higher energy levels of the ground electronic state of H_2. These excited states will then decay to lower levels leading to the mixing with the shocked population. The scatter is then produced by the enhanced population created by the action of the Lyα pump.

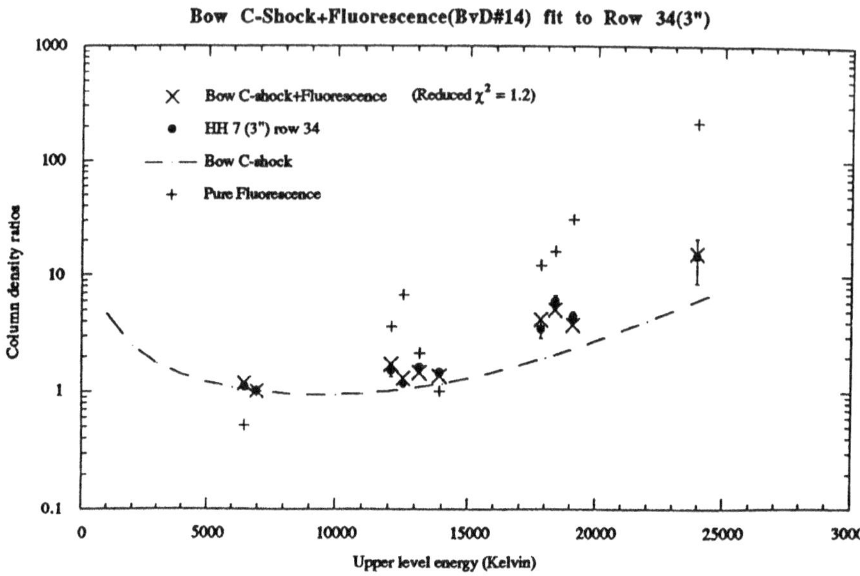

Fig. 2. H_2 column density ratios observed in HH 7

Near-infrared observations of the DR 21 bipolar outflow show also some evidence for fluorescent H_2 emission from the molecular lobes through the observed scatter amongst the points in the CDR diagram (Fernandes et al. 1995). Evidence for shocked emission from both emission lobes was provided by the detection of strong line emission at 2 μm from H_2 at temperatures of $\approx$ 2000 K and densities $\geq 10^5$ cm^{-3} (Fischer et al. 1980; Fischer et al. 1985), and the fact that the 1-0 S(1) to 2-1 S(1) line ratio was $\sim$ 10: the typical shock signature (Shull and Draine 1987). We have found a CDR signature at both emission lobes in DR 21 (Fernandes et al. 1995) similar to that found in HH 7. Although the spatial resolution of the DR 21 measurements did not permit to resolve individual bowshocks, it is conceivable to imagine several unresolved bowshocks within the beam contributing to the emission.

The high angular resolution observations of DR 21 presented by Garden et al. (Garden et al. 1990) shows that the H_2 line emission is intrinsically inhomogeneous and clumpy in both lobes down to scales of $\leq 10''$ (0.1 pc at a distance of 300 kpc). The 1-0 S(1) intensity structure which is observed within each lobe may correspond to variations of density within the outflowing gas; either pre-existing or the result of periodic outbursts in the acceleration mechanism. This complex structure suggests that the (molecular) clumps are probably being accelerated outward by the strong stellar (atomic) winds originating from the central DR 21 cloud core. The dense clumps of shocked gas may in fact represent a larger scale variation of the Herbig-

Haro phenomenon common to the high-velocity winds of less luminous pre-main sequence stars (Schwartz 1983a). A similar clumpy structure is observed for example in the NGC 2071 bipolar outflow source (Walther et al. 1991; Burton et al. 1989). It is interesting to note that the sizes of the clumps in these sources are of order $\sim 10''$; the same typical size of the Herbig-Haro 7 object. Thus, it is feasible that the origin of the shocked emission observed in these regions might arise from bow type shocks similar to those in HH 7.

3. Composite "Bow C-shock + Fluorescence" model

Extensive work has been done on modelling the infrared H_2 line emission from a C-shock bow structure. A very elaborate model is given by Smith et al. (Smith et al. 1991) and we use their analytic approximation to calculate the total number of molecules in each energy level that are excited in the shocks. The near-infrared H_2 line emission due to Lyα resonant pumping has been predicted using the PDR models of Burton, Hollenbach and Tielens (Burton et al. 1990) and Black and van Dishoeck (Black and van Dishoeck 1987). The "Bow C-shock + Fluorescence" model has been fully described in Fernandes (Fernandes 1993). From figs. 1 and 2 we can see the overall good fit to the data (crosses) of this composite model. This model has so far been succesfully applied to explain the near-infrared line emission from HH 7 and the bipolar outflow DR 21 (Fernandes and Brand 1995; Fernandes et al. 1995). Further testing of this model on other objects is now underway to test the conditions of validity and to establish the generality of the model.

References

Allen, D. and Burton, M.: 1993, *Nature* **363**, 54
Black, J. and Dalgarno, A.: 1976, *Astrophys. J.* **203**, 132
Black, J. and van Dishoeck, E.: 1987, *Astrophys. J.* **322**, 412,
Burton, M., Geballe, T., and Brand, P.: 1989, *Mon. Not. R. Astron. Soc.* **238**, 1513
Burton, M., Hollenbach, D., and Tielens, A.: 1990, *Astrophys. J.* **365**, 620,
Carr, J.: 1993, *Astrophys. J.* **406**, 553
Fernandes, A.: 1993, *Ph.D. thesis*, University of Edinburgh
Fernandes, A. and Brand, P.: 1995, *Mon. Not. R. Astron. Soc., accepted*
Fernandes, A., Brand, P., and Burton, M.: 1995, in preparation
Fischer, J., Righini-Cohen, G., and Simon, M.: 1980, *Astrophys. J., Lett.* **238**, L155
Fischer, J., Sanders, D., Simon, M., and Solomon, P.: 1985, *Astrophys. J.* **293**, 508
Garden, R., Geballe, T., Gatley, I., and Nadeau, D.: 1986, *Mon. Not. R. Astron. Soc.* **220**, 203
Garden, R., Russell, A., and Burton, M.: 1990, *Astrophys. J.* **354**, 232
Hartigan, P., Raymond, J., and Hartmann, L.: 1987, *Astrophys. J.* **316**, 323
Harvey, P., Joy, M., Lester, D., and Wilking, B.: 1986, *Astrophys. J.* **301**, 346
Herbig, G. and Jones, B.: 1983, *Astron. J.* **88**, 1040
Raymond, J., Hartigan, P., and Hartmann, L.: 1988, *Astrophys. J.* **326**, 323
Reipurth, B., Raga, A., and Heathcote, S.: 1992, *Astrophys. J.* **392**, 145

Schwartz, R.: 1983a, *Ann. Rev. Astron. Astrophys.* **21**, p. 209,
Schwartz, R.: 1983b, *Astrophys. J., Lett.* **268**, L37
Schwartz, R., Cohen, M., and Williams, P.: 1988, *Astrophys. J.* **333**, 1035
Shull, J. and Draine, B.: 1987, *Interstellar Processes*, D. Reidel Publishing Company:Dordrecht, p283,
Smith, M. and Brand, P.: 1990, *Mon. Not. R. Astron. Soc.* **245**, 108
Smith, M., Brand, P., and Moorhouse, A.: 1991, *Mon. Not. R. Astron. Soc.* **248**, 451
Solf, J. and Böhm, K.: 1987, *Astrophys. J.* **93**, 1172
Walther, D., Geballe, T., and Robson, E.: 1991, *Astrophys. J.* **377**, 246
Wolfire, C. and Königl, A.: 1991, *Astrophys. J.* **383**, 205
Zinnecker, H., Mundt, R., Geballe, T., and Zealey, W.: 1989, *Astrophys. J.* **342**, 337

SHOCKS IN THE L 1551-IRS 5 OUTFLOW OPTICAL AND NEAR-INFRARED IMAGING

CHRISTOPHER J. DAVIS and REINHARD MUNDT
Max-Planck-Institut für Astronomie, Königstuhl 17, D-69117 Heidelberg, Germany

JOCHEN EISLÖFFEL
Laboratoire d'Astrophysique, Observatoire de Grenoble, B.P. 53X, F-38041 Grenoble Cedex, France

and

THOMAS P. RAY
Dublin Institute for Advanced Studies, School of Cosmic Physics, 5 Merrion Square, Dublin 2, Ireland

Abstract. We present new images of the well-known molecular outflow and Herbig-Haro complex L 1551-IRS 5. Deep, high-resolution images of the central region of the flow in [SII] $\lambda\lambda$ 6716,6731 and Hα (6565 Å) are complemented by a mosaic of much of the CO outflow in H_2 v=1-0 S(1). While the optical data trace the intermediate-to-high excitation shocks in the flow ($v_{shock} > 30 - 50$ km s^{-1}), the near-IR data reveal the lower-excitation, molecular shocks ($v_{shock} \sim 10 - 50$ km s^{-1}). In particular, the H_2 data highlight the regions where the flow impacts and shocks ambient molecular gas.

Key words: Shock Waves – ISM: jets and outflows – ISM: kinematics and dynamics

1. Optical Imaging: A Cavity Within a Cavity?

The Hα image of Graham & Heyer (1990) of the outflow driven by L 1551-IRS 5 clearly shows the ovoidal morphology of the optical flow associated with the blue-shifted CO outflow. We interpret this structure in terms of a wind-swept cavity (see below, and Davis et al. 1995 for a more detail discussion) and refer to it as the "Hα cavity". A faint cavity is also observed in scattered light in our new [SII] and Hα images. Like the Hα cavity, this "scattered light cavity" fans out from the source of the flow towards the southwest. However, the opening angle of the scattered light cavity is clearly greater than that of the Hα cavity, so that the scattered light cavity "surrounds" the Hα cavity. The optical jet observed close to IRS 5 is directed approximately down the central axis of these "nested cavities", while the CO outflow occupies much of the scattered-light cavity area (e.g. Pound & Bally 1991).

The deep [SII]-continuum image shown in Fig. 1 clearly illustrates the complex structure of the shocks in the blue lobe of the IRS 5 outflow. They are highly structured, with many clumps and filaments occuring over a large range of length scales. Particularly prominent are many of the knots in HH 29, in HH 263, and in HH 264.

Astrophysics and Space Science **233**: 51–54, 1995.

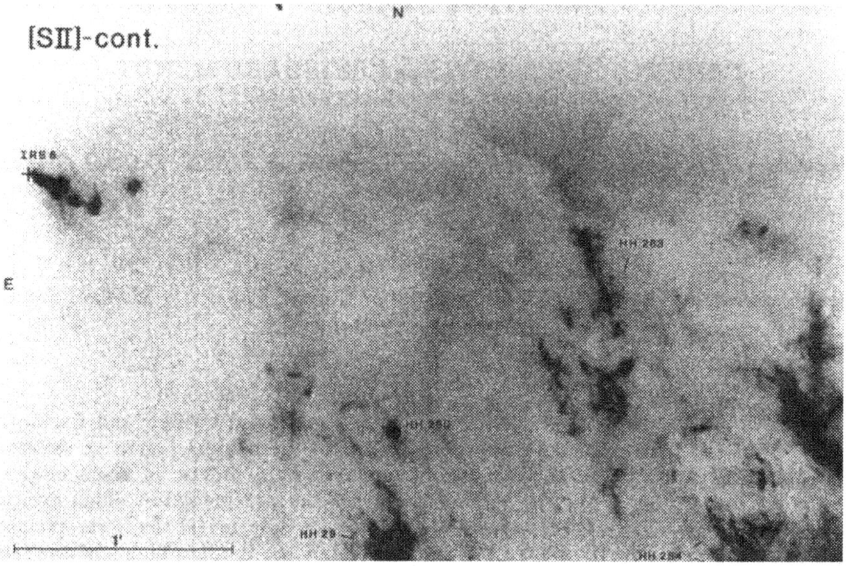

Fig. 1. [SII]–continuum image of the IRS 5 - HH 29 region.

2. H_2 Imaging

In Fig. 2 we show a H_2 + continuum mosaic of much of the blue-shifted CO outflow lobe (no H_2 emission or interesting features were observed in the red CO flow lobe – about 50% of this flow lobe was imaged – so this region is not shown). A (scaled) K-band mosaic of the same region has been used to establish which features are emission-line knots and which are stars or regions of scattered light. We found that, of the non-starlike sources, only the very diffuse, faint emission 2′- 3′ east of IRS 5, and of course the IRS 5 and L 1551 NE reflection nebulae are continuum sources: the chains of knots in HH 263 and HH 264, together with the knots, wisps and arcs in HH 260, HH 261 and HH 29, and the fainter filaments near HH 28, are all H_2 S(1) line-emission features.

Like the optical data, the H_2 emission traces a very clumpy flow. The data also illustrate, once again, the large opening angle of the flow. However, while the Hα emission traces much of the southern edge of the cavity (see Davis et al. 1995; and also Stocke et al. 1988; Graham & Heyer 1990), the H_2 is also prominent along the northwestern edge of the cavity (HH 263 to HH 264 and further west), where the flow is impacting the higher-density ambient gas in this region.

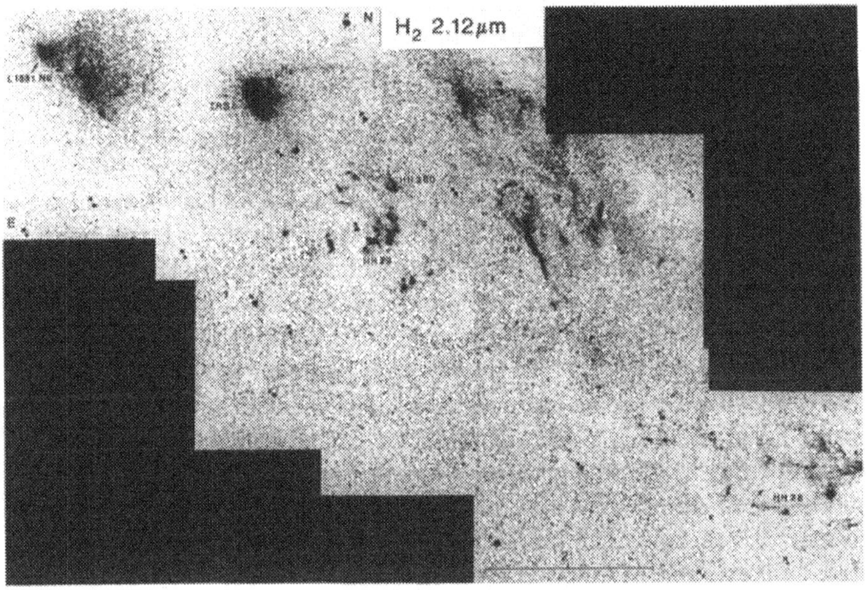

Fig. 2. H_2 v=1-0 S(1) (+continuum) image of the IRS 5 - HH 28 - HH 29 region, covering about 70% of the blue-shifted CO outflow lobe.

A comparison of the H_2 data with our [SII]-continuum image (Fig. 1) shows that in the brightest part of HH 29 and in some parts of HH 264, there is an overall spatial coincidence between these two types of emission. Note also the apparent bow shock morphology of HH 260, where the optical emission traces the head of the bow while the H_2 emission extends out into the wings. HH 29 may be a similar, though much larger and more unstable bow shock. There are, however, also some differences between the [SII] and H_2 data (e.g. north of HH 263 and northwest of HH 264), and since the detailed structure in Hα (see Davis et al. 1995) is also quite different from that in [SII], our imaging data obtained in the various shock tracers indicate a large range of shock velocities in the emitting regions. Therefore, a detailed explanation of our data probably requires a very inhomogenous mixture of atomic and molecular gas.

Finally, the clumpy nature of the CO outflow in L 1551 has recently been observed in CO J=2-1 strip-maps of the flow (Bachiller et al. 1994). In particular, a high-velocity, compact clump is observed close to HH 29. This clump may well be associated with the bright arc of molecular (H_2) shocks in this area; note that a similar correlation between bright H_2 features and peaks in CO outflow maps has been observed in some other outflows (e.g. Bally et al. 1993; Davis et al. 1994; Davis & Eislöffel 1995; see also Eislöffel & Davis, this volume). However, higher-resolution interferometer maps of this and other outflows are needed to establish a definite spatial coincidence between H_2

shocks and CO clumps in outflows from young stars.

3. Variability in the Outflow Opening Angle

Many of the knots in the Hα image of Graham & Heyer (1990; see also Stocke et al. 1988) appear to coincide with the southern edge of a large, wind-swept cavity. This cavity is itself enclosed by a much wider cavity that is observed in scattered light (see Davis et al. 1995 and also Stocke et al. 1988). We suggest that this "nested-cavity" structure may be the result of a variable opening angle. In such a model, the Hα shocks that coincide with the southern edge of the cavity illustrate the boundary of the present outflow, while the wider "scatter-light" cavity is a remnant from an early outflow phase for which the outflow opening angle was greater. The collimated, high-velocity CO outflow observed by Fridlund & White (1989) would then be part of the flow being currently driven down the inner cavity volume, while the more poorly collimated, hollow shell of lower-velocity CO may be "coasting" away from the source in the gap between the two cavities, i.e. the shell of low-velocity CO may be a result of the earlier, wider outflow phase.

Acknowledgements

TR and RM would like to acknowledge support from the German-Irish Research Fund. JE was supported by the Human Capital and Mobility programme of the European Union under grant ERBCHBGCT920205.

References

Bachiller, R., Tafalla, M., Cernicharo, J.: 1994, *preprint.*
Bally, J., Devine, D., Hereld, M., Rauscher, B.J.: 1993, *Astrophys. J.*, **418**, L75.
Davis, C.J., Dent, W.R.F., Matthews, H.E., Aspin, C., Lightfoot, J.F.: 1994, *Mon. Not. R. astr. Soc.*, **266**, 933.
Davis, C.J., Eislöffel, J.: 1995, *Astron. Astrophys.*, in press.
Davis, C.J., Mundt, R., Eislöffel, J., Ray, T.P.: 1995, *Astron. J.*, in press.
Fridlund, C.V.M., White, G.J.: 1989, *Astron. Astrophys.*, **223**, L13.
Graham, J.A., Heyer, M.H.: 1990, *PASP*, **102**, 972.
Pound, M. W., & Bally, J.: 1991, *Astrophys. J.*, **383**, 705.
Stocke, J. T., Hartigan, P. M., Strom, S. E., Strom, K. M., Anderson, E. R., Hartmann, L. W., & Kenyon, S. J.: 1988, *Astrophys. J. Suppl.*, **68**, 229.

PROPER MOTION MEASUREMENTS IN VARIOUS YOUNG STELLAR OBJECT JETS IN TAURUS

JOCHEN EISLÖFFEL
Laboratoire d'Astrophysique, Observatoire de Grenoble, B.P. 53X, F-38041 Grenoble Cedex, France

and

REINHARD MUNDT
Max-Planck-Institut für Astronomie, Königstuhl 17, D-69117 Heidelberg, Germany

Abstract. We present proper motion measurements for a number of knots in the jets and bow shocks of the outflows from DG Tau, DG Tau B, T Tau, CoKu Tau 1, FS Tau, and FS Tau B in the nearby Taurus-Auriga star forming region. From these measurements and the available radial velocity data we derive tangential velocities, spatial velocities, angles of the outflows with respect to the plane of the sky, and in a couple of cases the pattern motions of the knots relative to the flow speed.

Key words: Shock Waves – ISM: jets and outflows – ISM: kinematics and dynamics

1. Introduction

Recently, proper motion studies of several collimated outflows from young stars have been carried out on deep CCD images, e.g. HH 34 (Eislöffel & Mundt 1992; Heathcote & Reipurth 1992), HH 1/2 (Eislöffel, Mundt & Böhm 1994), HH 46/47 (Eislöffel & Mundt 1994). These measurements normally included many knots in each of the outflows and therefore a good sampling of their tangential velocity field could be acheived, opening up the possibility for a variety of tests of jet and bow shock models (see Eislöffel (1995) for some examples). While the objects investigated up to now mainly were situated in Orion, and therefore relatively far away, it seems especially advantageous to study objects in the nearby Taurus – Auriga star forming region, at a distance of only 150 pc. There, detailed measurements with small absolute measurement errors should be possible. We have carried out such a study for a number of outflows and in the following present some of our results. For a fuller account of this work the reader is referred to our forthcoming paper (Eislöffel & Mundt, in preparation).

2. Results and Discussion

Our measurements of the knots in the outflows from DG Tau, DG Tau B, T Tau, CoKu Tau 1, FS Tau, and FS Tau B are based on deep high-quality CCD images taken in the [SII]$\lambda\lambda$6716,6731 and Hα lines at the 3.5m-telescope

Astrophysics and Space Science **233**: 55–58, 1995.

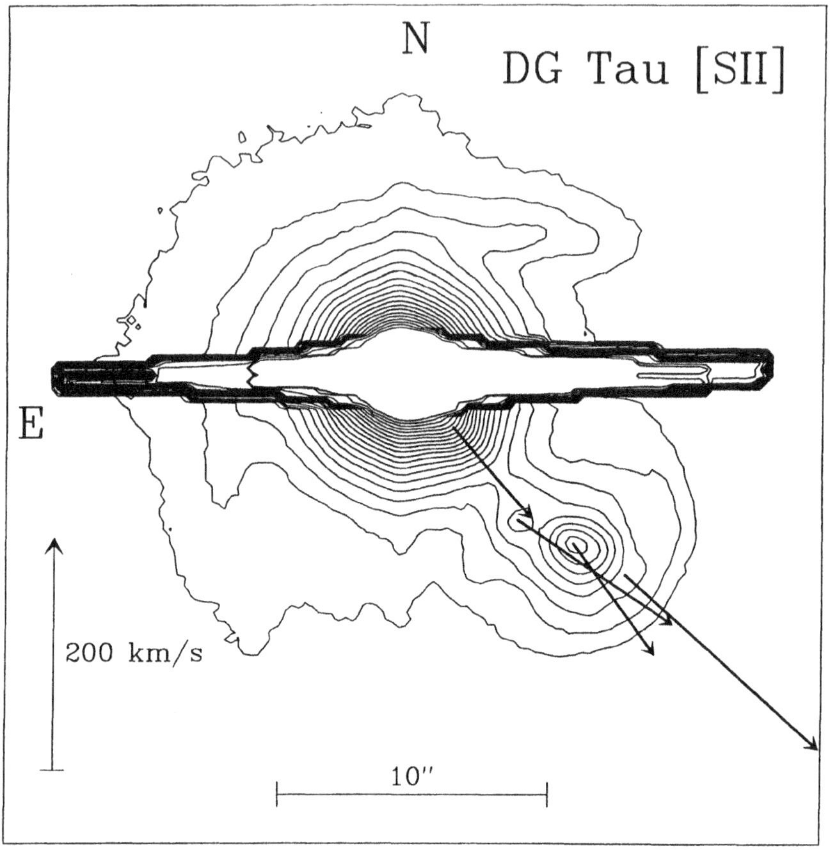

Fig. 1. Proper motions in the jet from DG Tau.

on Calar Alto, Spain. We used 3 to 10 images per object and a total epoch difference of 4 to 8 years to reach a measurement accuracy of about ±8% on average.

We find that all knots in the six investigated outflows move. Their tangential velocities range from close to the detection limit at about 15 km s^{-1} up to about 400 km s^{-1}. In many objects there are, however, marked changes in the tangential velocities of subsequent knots. As examples we show contour plots of DG Tau (Fig. 1) and DG Tau B (Fig. 2) with the proper motion vectors superimposed.

In the flow from DG Tau four knots were measured. They move at tangential velocities from about 100 km s^{-1} for the innermost knot to about 230 km s^{-1} for the outermost knot and show significant tangential velocity differences from knot to knot that amount to 60 – 100 km s^{-1}, much larger than the average measurement error of about 15 km s^{-1}. From these tangen-

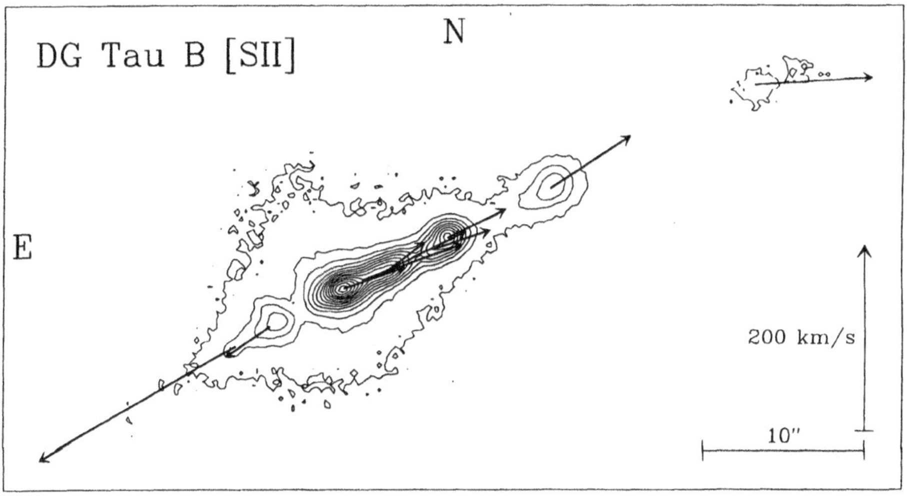

Fig. 2. Proper motions in the jet and counter-jet from DG Tau B.

tial velocities and radial velocity measurements (Mundt, Brugel & Bührke 1987) we find this jet pointing towards us at an angle of 51° with respect to the plane of the sky. An even larger angle has been found for T Tau – HH 155, while all other studied outflows lie close to the plane of the sky.

The knots in the north-western jet of DG Tau B have tangential velocities between 50 km s^{-1} and 130 km s^{-1}, again with significant differences between successive knots. In the south-eastern counterjet only two knots could be measured on our [SII] images. The tangential velocity of the inner knot is about 60 km s^{-1}, while the more distant knot is markedly faster at about 250 km s^{-1}.

The detailed tangential velocity fields of the knots in the flow from DG Tau B (and also in the flow from FS Tau B) show an interesting structure in the motion of the knots relative to the flow speed. We find evidence that only half of the knots move at close to the full flow speed, while the other half move at a (pattern) speed which is about 50% of the full flow speed (Fig. 3). From that Figure it is evident that strong variations in the pattern motion occur from knot to knot. Variations of the same magnitude are also seen in the pattern motion of the knots in the flow from FS Tau B. This knot pattern motion is a serious constraint on models of jets with internal working surfaces. In these models it is assumed that variability of the jet source causes variations in the jet velocity, which in turn leads to "parcels" of faster moving jet gas catching up with slower moving gas ahead of it and the creation of a shock wave in the form of a small bow shock in the flow. The bow shock is interpreted as a jet knot, which should then move along at close to the flow

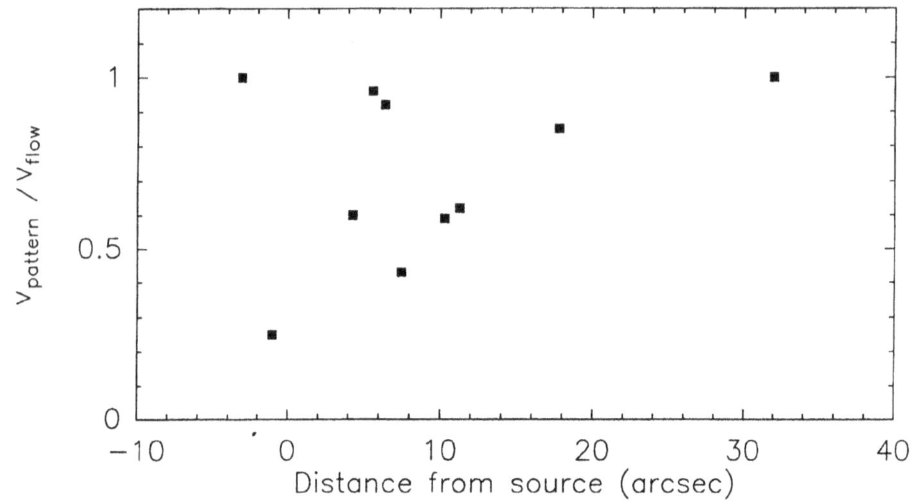

Fig. 3. Pattern motion of the knots in the outflow from DG Tau B.

speed (Rees 1978; Raga et al. 1990). However, this seems to be contrary to the observed pattern motions in the jet knots from DG Tau B and FS Tau B and some other jets (see Eislöffel & Mundt 1992, 1994).

Acknowledgements

JE acknowledges support from the Human Capital and Mobility programme of the European Union under grant ERBCHBGCT920205.

References

Eislöffel, J.: 1995, *Disks and Outflows Around Young Stars,* H.J. Staude and A.M. Quetz (eds.), Springer:Berlin, in press.
Eislöffel, J., Mundt, R.: 1992, *Astron. Astrophys.*, **263**, 292.
Eislöffel, J., Mundt, R.: 1994, *Astron. Astrophys.*, **284**, 530.
Eislöffel, J., Mundt, R., Böhm, K.H.: 1994, *Astron. J.*, **108**, 1042.
Heathcote, S., Reipurth, B.: 1992, *Astron. J.*, **104**, 2193.
Mundt, R., Brugel, E.W., & Bührke, T.: 1987, *Astrophys. J.*, **319**, 275.
Raga, A.C., Cantó, J., Binette, L., Calvet, N.: 1990, *Astrophys. J.*, **364**, 601.
Rees, M.J.: 1978, *Mon. Not. R. astr. Soc.*, **184**, 61P.

NEAR-INFRARED IMAGING IN H_2 OF MOLECULAR (CO) OUTFLOWS FROM YOUNG STARS

JOCHEN EISLÖFFEL
Laboratoire d'Astrophysique, Observatoire de Grenoble, B.P. 53X, F-38041 Grenoble Cedex, France

and

CHRISTOPHER J. DAVIS
Max-Planck-Institut für Astronomie, Königstuhl 17, D-69117 Heidelberg, Germany

Abstract. We have imaged several known molecular (CO) outflows in H_2 v=1-0 S(1) and wide-band K in order to identify the molecular shocks associated with the acceleration of ambient gas by outflows from young stars. We detected H_2 line emission in all the flows we observed: L 1157, VLA 1623, NGC 6334I, NGC 2264G, L 1641N and Haro 4-255. A comparison of the H_2 data with CO outflow maps strongly suggests that "prompt entrainment" near the head of a collimated jet probably is the dominant mechanism for producing the CO outflows in these sources.

Key words: Shock Waves – ISM: jets and outflows – ISM: kinematics and dynamics

1. Introduction

Extended H_2 emission has been observed in a number of Herbig-Haro outflows and also in a few CO outflows. In Herbig-Haro objects H_2 traces even weaker shocks than the low-excitation optical features (i.e. those strong in [SII]$\lambda\lambda$6716,6731 or [OI]$\lambda\lambda$6300,6363). Therefore, comparison of the various optical and H_2 line images allows us to examine the varying levels of excitation throughout each system (see HH 46/47 for an example; Eislöffel et al. 1994). However, in molecular CO outflows that are still deeply embedded in the parental cloud H_2 emission is often the only useful tracer of shocks in the flow that can readily be observed from the ground. In the few molecular outflows investigated so far the H_2 emission appears to be closely associated with the CO outflows. Therefore we attempted to investigate this relationship further.

2. Detection of H_2 Emission in CO Outflows

We imaged the outflows L 1157, VLA 1623, NGC 6334I, NGC 2264G, L 1641N and Haro 4-255 FIR in H_2 v=1-0 S(1) and wide-band K. We observed H_2 line emission in all of these flows and found that the emission features are closely related to the gross features in the CO outflows. Here we show H_2 images of the L 1157 and VLA 1623 molecular outflows with, superimposed, contour

Astrophysics and Space Science **233**: 59–62, 1995.

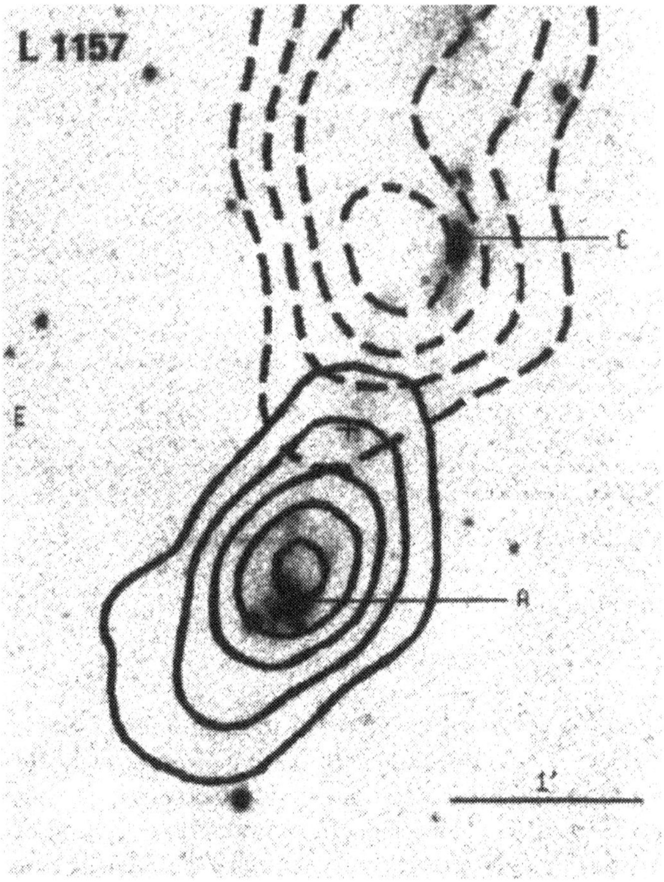

Fig. 1. A H_2 v=1-0 S(1) + continuum image with, superimposed, a CO J = 1–0 map (Umemoto et al. 1992) of the L 1157 outflow region. The position of IRAS 20386+6751 is marked with a cross. The blue-shifted CO contours are full, the red dashed. The main H_2 line emission features are labelled A and C.

plots of the associated CO outflows. For images of the other objects and for a more detailed discussion the reader is refereed to our paper (Davis & Eislöffel 1995).

In L 1157 two bright, bow-shaped H_2 features are observed about 1′ north and south of the IRAS 20386+6751 source (Fig. 1). The apices of these "bow shocks" point away from the source, as would be expected for shocks driven into the ambient cloud (or slower outflow gas situated further away from the source) by a collimated outflow. Moreover, these "bow shocks" are situated very close to peaks in the CO outflow maps, suggesting a causal relationship.

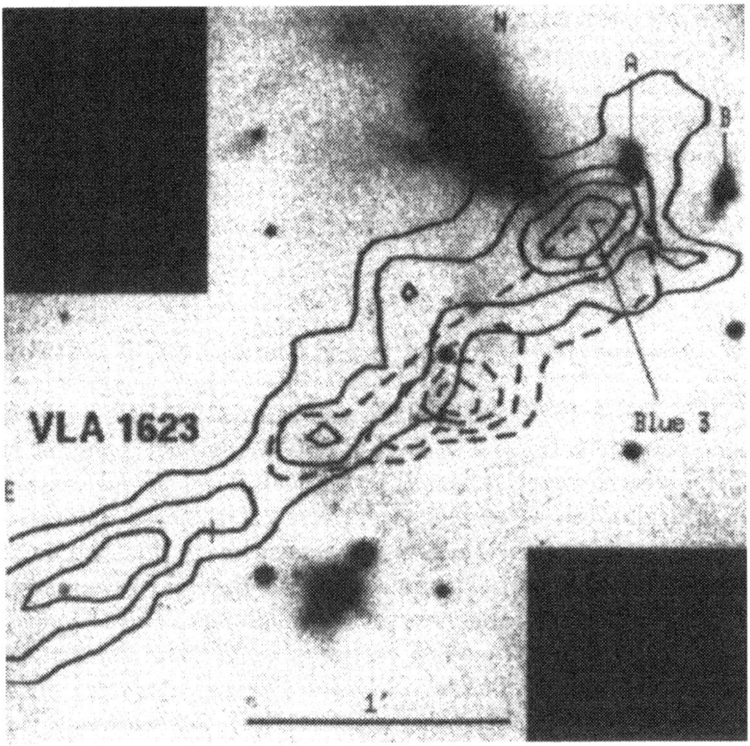

Fig. 2. H_2 v=1-0 S(1) + continuum image of the VLA 1623 region with contours of the CO J=2-1 outflow, from André et al. (1990). The compact clump in the CO outflow, BLUE 3, lies in the wake of the bright H_2 knot A. The dashed lines are red-shifted CO, the full blue-shifted CO.

In the VLA 1623 outflow (Fig. 2) the peak in the blue-shifted CO flow lies just behind of (i.e. closer to the source than) a bright, compact, and somewhat "bow-shaped" H_2 knot (labelled A). The head of the bow faces away from the VLA 1623 source, as is expected if the bow shock were associated with the outflow. A number of additional, fainter H_2 knots trace much of the north-western CO flow lobe and are similarly observed near peaks in the CO outflow.

Although much of the emission in the near-IR images of the NGC 6334I cluster is continuum, two compact line-emission knots, that are again somewhat bow-shaped in appearance, are observed just ahead of the peaks in the blue- and redshifted CO outflow lobes.

H_2 line-emission is observed in both the blue- and red-shifted CO lobes in the NGC 2264G outflow. A compact cluster of H_2 knots coincides with the peak in the blue-shifted CO flow, while a knotty filament of H_2 emission outlines the northern edge of the red-shifted CO flow.

In the L 1641N molecular outflow faint knots of H_2 emission fill much of the space occupied by the blue shifted CO lobe. The overall H_2 morphology suggests a cavity associated with the outflow.

In the Haro 4-255 FIR outflow, a faint bow-shaped H_2 knot is observed near to the peak in the blue-shifted CO flow. Additional H_2 knots are found closer to the FIR source of the flow; these features clearly delineate the flow direction.

3. Relationship between the H_2 and the CO Outflows

In the L 1157, VLA 1623, NGC 6334I and Haro 4-255 FIR outflows we see bow-shaped shocks in H_2 and in NGC2264G a compact group of knots just ahead of, i.e. downstream of, peaks in the CO flows. In all cases the "bow shocks" are orientated as would be expected if they were associated with working surface regions in the flows. It seems likely that the CO peaks in these flows represent clumps of molecular gas that have been swept up by the bow shocks, in a way analogous to a snow-plough. *The CO outflow hence would appear to consist of material entrained by the bow shock at the head of the collimated jet and not of a low collimation flow ejected from the source.*

In NGC 2264G, and perhaps also in L 1641N, H_2 emission outlines also the outer edges of the flows. We interpret these as evidence for a turbulent shock layer between the flows and the ambient gas in which additional ambient material is entrained into the flow.

Acknowledgements

We wish to thank Mike Smith and Reinhard Mundt for much time spent discussing jets and outflows. JE acknowledges support from the Human Capital and Mobility programme of the European Union under grant ERBCH-BGCT920205.

References

André, Ph., Martín-Pintado, J. Despois, D., Montmerle, T.: 1990, *Astron. Astrophys.*, **236**, 180.
Davis, C.J., Eislöffel, J.: 1995, *Astron. Astrophys.*, in press.
Eislöffel, J. Davis, C.J., Ray, T.P., Mundt, R.: 1994, *Astrophys. J.*, **422**, L91.
Umemoto, T. et al.: 1992, *Astrophys. J.*, **392**, L83.

NIR SPECTROSCOPY AND MOLECULAR SHOCKS IN HERBIG-HARO 1

A. NORIEGA-CRESPO
IPAC, California Institute of Technology, JPL, Pasadena, CA, 91125
Email: alberto@ipac.caltech.edu

and

P.M.GARNAVICH
Center for Astrophysics,
Email: peterg@cfanewton.harvard.edu

Abstract. Near Infrared spectroscopic observations (1.2-2.4 μm) of two condensations (HH 1A & HH 1G) of Herbig-Haro object 1 are used to explore the nature of its molecular shocks and extinction. The H_2 emission lines display a more complex behaviour than that accounted for C or J type shock models. The extinction, determined from the ratio of [Fe II] 1.257 & 1.644μm, is higher than what is expected from optical [S II] measurements, and it may indicate real discrepancies between the optical and IR regions of the ISM extinction curve.

1. INTRODUCTION

Shock excited Herbig-Haro (HH) objects and jets emit a considerable fraction of their energy at infrared wavelengths for shocks with velocities $\leq 50\ \mathrm{km\,s^{-1}}$. The emission arises mostly from molecular rotational-vibrational transitions of H_2, CO and H_2O, as well as from atomic and ionic fine structure (mostly forbidden) transitions, e.g. [Fe II] 1.64μm, [O I] 63μm, [C II] 157 μm and [Si II] 35μm (see e.g. Hollenbach & McKee 1979).

Low velocity molecular shocks can be classified into two kinds (see e.g. Draine & McKee 1993) J-shocks and C-shocks. In a J-type shock, like in an atomic shock, there is a discontinuous change in the density, temperature and velocity across the front. The magnetic fields and ionization fractions associated with them are relatively small ($\sim$ 100 μG), and large ($\geq 10^{-5}$), respectively. In a C-type shock stronger magnetic fields and lower ionization fractions are involved. Under these conditions the ion and the neutral matter in the molecular material react differently to a disturbance. Essentially, Alfvén waves prepare the neutral fluid ahead the shock through ambipolar diffusion, so changes in density, temperature and velocity are continuous. In the case of H_2, for instance, C-shocks allow for higher shock velocities ($\sim$ 50 km s^{-1}) without dissociating the molecules.

In these paper we present low resolution NIR (1.2 – 2.4 μm) observations of condensations HH 1 A and HH 1 G of the bowshock-like working surface

Astrophysics and Space Science **233**: 63–66, 1995.

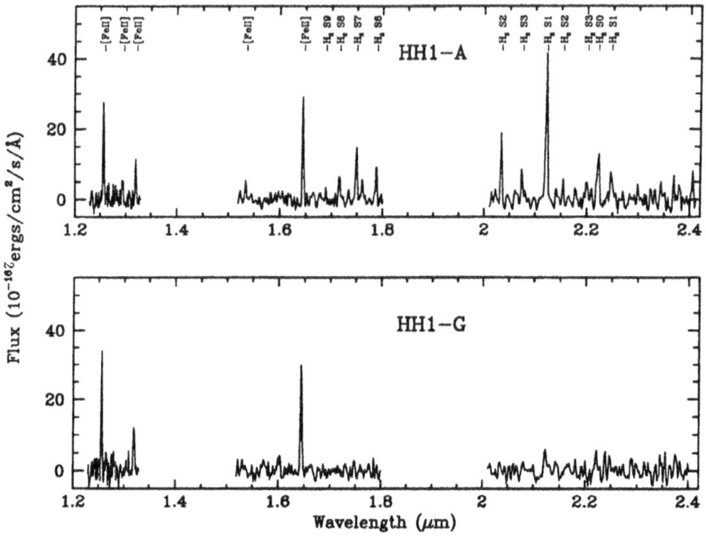

Fig. 1. NIR spectra of HH 1A and HH 1 G

of HH 1 (Noriega-Crespo & Garnavich 1994). The observations were carried out on 30/1/94 at the 1.5m telescope at CTIO, using OSIRIS. OSIRIS was used in its cross-dispersed mode which allowed spectra in the J, H and K bands to be recorded simultaneously with a resolving power of 400, using a 3″ wide slit. Figure 1 shows the spectra for both condensations; the errors in our fluxes are $\pm 0.4 \times 10^{-13}$ erg cm^{-2} s^{-1} for the bright lines and $\pm 0.6 \times 10^{-13}$ erg cm^{-2} s^{-1} for the weaker ones.

2. DISCUSSION

The H_2 lines provide a unique opportunity to search for the signature of magnetic molecular shocks in star forming regions. As mentioned in the introduction, depending of the pre-shock parameter space, both C-type and J-type shocks are expected to take place. A comparison of the observations with shock models should always be regarded just as a guide of the pre-shock conditions, however, since in general shocks cannot be described by a unique temperature, density and ionization fraction. The H_2 column densities can be used to compare with the theoretical predictions from shock models (Smith 1994a). Using Boltzmann factors for the population levels, it is possible to determine the excitation temperatures (or shock velocities) assuming local equilibrium between the levels (Smith 1994b). The models depend in detail on various parameters (Smith 1994a), for instance, the initial preshock density (3×10^5 cm^{-3}), ionization fraction per hydrogen atom (10^{-5} for J-shocks,

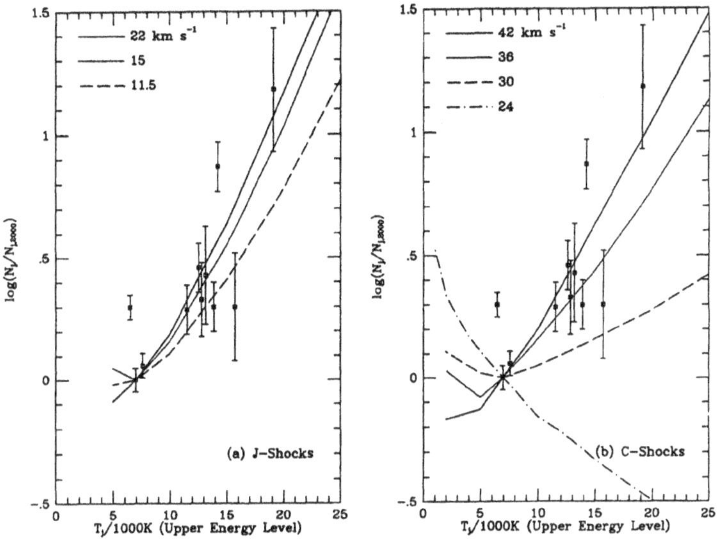

Fig. 2. The normalized H_2 column densities for different lines as a function of their upper level energy are compared with molecular J and C shock models (following Smith 1994b)

and 10^{-7} for C-shocks). And for the C-shocks, an Alfvén velocity ($V_A = 2$ km s^{-1}, which implies a magnetic field ~ 600 μG), with the magnetic field parallel to the shock (Smith 1994b).

In Fig 2, we show the comparison between our measurements and the shock models. For display, the column densities have been normalized (Smith 1994b) first by dividing them by a factor $exp(-T_j/2000K)$ where T_j is the upper temperature (energy) for level j, and then with respect to the 1−0 S(1) 2.121μm column density. At a glance, J and C type shock models seem to do an adequate job in explaining the observations (see Fig 2a,b). J-type shocks velocities of 11 to 22 km s^{-1} are "equivalent" to C-type shocks of 30 to 42 km s^{-1}.

Besides the H_2 lines, some strong [Fe II] are observed in HH 1A and HH 1G. Particularly useful are the bright [Fe II] lines at 1.26μm and 1.65μm, which have a common upper level, and can be used to estimate the absorption along the line of sight to the [Fe II] emission region. The [Fe II] lines are as bright in HH 1A as in HH 1G, which already indicates that the asymmetry in their H_2 emission may not be due to extinction.

The observed ratio for [Fe II] 1.26μm to [Fe II] 1.65μm is 0.7±0.1 while the predicted value is 1.36 (Oliva *et al.* 1990). Using the IR extinction law derived by Rieke & Lebofsky (1985), the observed ratio is equivalent to 6.7±1.4 magnitudes of visual extinction or an E($B-V$) of 2.2±0.4 magnitudes. This latter value is $\sim$ 5 times larger that which has been estimated optically using the [S II] *auroral* (4068, 4072 Å) to *transauroral* (10318, 10336, 10370 Å) line ratios, which give an average value $E(B-V) = 0.43 \pm 0.04$ (Solf, Böhm & Raga 1988). It is interesting to note (Table 1) that this effect manifests itself in other HH objects, suggesting discrepancies between the optical and IR absorption laws. Such differences may parallel the well known variations between optical and UV absorption (see e.g. Mathis 1990).

TABLE I

Optical & IR Extinctions

Object	E(B−V) from [Fe II]	E(B−V) optically	Refs
HH 1	< 2.0	0.43	1,2
HH 43	< 1.5	0.18	3,4
HH 111	< 2.0	0.6 - 1.0	5,6

(1) [Fe II] from this paper; (2) Optical, [S II] measurements (Solf et al. 1988); (3) [Fe II] from Gredel (1994); (4) Optical (Schwartz et al. 1985); (5) [Fe II] from Gredel & Reipurth (1993); (6) Optical (Morse et al. 1993; Noriega-Crespo et al. 1993).

We thank Salvador Curiel, Bruce Draine, Alejandro Raga, Roland Gredel and Michael Smith for their useful conversations and comments. ANC research is supported by NASA LTAP, through JPL-CalTech.

References

Draine, B.T., & McKee, C.F.: 1993, *Ann. Rev. Astron. Astrophys.* **31**, 373
Hollenbach, D.J., & McKee, C.F.: 1979, *Astrophys. J. Suppl.* **41**, 555
Gredel, R.: 1994, *Astron. Astrophys.* **292**, 580
Gredel, R., & Reipurth, B.: 1993, *Astrophys. J.* **407**, L29
Morse, J., Heathcote, S., Hartigan, P., & Cecil, G.: 1993, *Astron. J.* **106**, 1139
Mathis, J.S.: 1990, *Ann. Rev. Astron. Astrophys.* **28**, 37
Noriega-Crespo, A., Garnavich, P.M., & Raga, A.C.: 1993, *Astron. J.* **106**, 1133
Noriega-Crespo, A., & Garnavich, P.M.: 1994, *Astron. J.* **108**, 1432
Oliva, E., Moorwood, A.F.M., & Danziger, I.J.: 1990, *Astron. Astrophys.* **240**, 453
Rieke, G.H., & Lebofsky, M.J.: 1985, *Astrophys. J.* **288**, 618
Schwartz, R.D., Dopita, M.A., & Cohen, M.: 1985, *Astron. J.* **90**, 1820
Smith, M.D.: 1994a, *Mon. Not. R. astr. Soc.* **266**, 238
Smith, M.D.: 1994b, *Astron. Astrophys.* **289**, 256
Solf, J., Böhm, K.H., & Raga, A.C.: 1988, *Astrophys. J.* **334**, 229

H_2 IN MOLECULAR SHOCKS: OBSERVATIONS FROM 2 TO 12 MICRONS

M. J. RICHTER and J. R. GRAHAM
University of California, Berkeley

G. S. WRIGHT
Joint Astronomy Centre

and

D. M. KELLY and J. H. LACY
The University of Texas at Austin

Abstract. We present spectroscopy of the molecular shock in the supernova remnant IC 443. The H_2 level populations in IC 443 are nearly identical to Orion Peak 1 (Brand et al. 1988; Parmar et al. 1994) and are consistent with a partially dissociative J–shock. The similarity with Orion suggests the shock processes which dominate the emission in both regions are essentially identical.

Key words: Interstellar Molecules – Supernova Remnants – Infrared Spectroscopy

1. Introduction

H_2 observations of molecular shocks represent a potentially powerful diagnostic of conditions in the dense interstellar medium. Molecular shocks are common: they have been seen associated with star forming regions, supernova remnants, planetary nebulae, Herbig-Haro objects, and active galactic nuclei. Proper interpretation of these data will teach us about the shock mechanism and the pre-shock medium.

A supernova remnant, such as IC 443, is an ideal testing ground for shock models. The overall geometry is generally planar. No hot, central star produces photons capable of radiatively exciting H_2. Finally, the extent of the remnant guarantees interactions with a variety of pre-shock conditions.

2. Observations and Results

We obtained spectra of a collection of bright H_2 emission knots located towards the SE of IC 443, near the peak found by Burton (1987). Observations from 2 – 4.7 μm were made using CGS4 at UKIRT on 1992 January 21 and 22 (Richter et al. 1995a). Mid-infrared observations of the H_2 0-0 S(2) line ($\lambda = 12.28$ μm) were carried out at the IRTF on 1994 October 24 using Irshell, a mid-infrared cryogenic echelle spectrometer (lacy et al. 1989).

The observed line transitions are identified in Table 1 along with the air wavelength, upper state energy, flux per pixel, and observed column density

Astrophysics and Space Science **233**: 67–70, 1995.

TABLE I
Line Fluxes

Line Transition	Wavelength (air) [μm]	Energy of Upper State [K]	Observed Flux [10^{-19} W m^{-2} pix^{-1}]	Dereddened Column Density [10^{19} m^{-2}]
2-1 S(3)	2.0729	13914	106 (1)	5.3
1-0 S(1)	2.12125	6963	1351 (5)	107.3
2-1 S(2)	2.1536	13173	42 (1)	2.0
3-2 S(3)	2.2008	19119	20 (1)	0.9
1-0 S(0)	2.22268	6482	333 (1)	33.9
2-1 S(1)	2.2471	12571	116 (1)	59.0
3-2 S(2)	2.2864	18417	11 (1)	0.5
4-3 S(3)	2.3439	23996	11 (1)	0.6
2-1 S(0)	2.3550	12115	31 (1)	2.0
3-2 S(1)	2.3858	17849	28 (1)	1.3
1-0 Q(1)	2.40593	6159	1462 (5)	79.6
1-0 Q(2)	2.41278	6482	511 (3)	39.3
1-0 Q(3)	2.42307	6963	1317 (5)	109.8
1-0 Q(4)	2.43682	7597	401 (3)	34.9
1-0 Q(5)	2.4541	8379	442 (3)	39.7
0-0 S(17)	3.483	25531	170 (30)	1.0
1-0 O(6)	3.497	7597	290 (30)	37.6
0-0 S(15)	3.6236	21443	240 (30)	1.9
1-0 O(7)	3.8064	8379	490 (40)	88.1
0-0 S(13)	3.8451	17476	700 (40)	8.3
0-0 S(9)[a]	4.6933	10280	3600 (200)	107.9
0-0 S(2)[b]	12.275	1682	70 (20)	19000

[a] Observed with 2$''$.5 by 5$''$.0 pixels
[b] Observed with 0$''$.83 by 0$''$.83 pixels

for each line. Line fluxes were corrected for an extinction of $A_{2.12}$=1.7 magnitudes, as determined from line flux ratios involving levels with a common upper state. The lack of Brγ emission and the strength of high J pure rotation transitions demonstrate that radiative excitation of H_2 is unimportant.

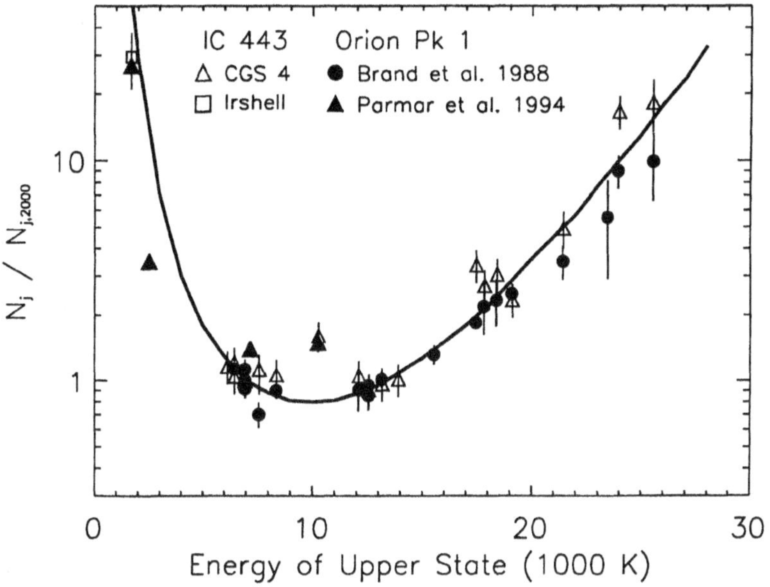

Fig. 1. A population level diagram for the H_2 emission in IC 443 normalized to an LTE level population at 2000 K. The upper state energies cover a very large range: $E/k = 1700$ K to 26,000 K. Error bars represent 1σ formal errors and do not account for atmospheric effects. The curve is taken from the partially dissociative J–shock model presented by Brand et al. (1988).

3. Population Level Diagrams and Shock Models

The H_2 level populations are consistent with a range of temperatures, such as expected behind a J-shock, but inconsistent with a uniform temperature gas, such as expected from a planar C–shock (Fig. 1). Brand et al. (1988) found nearly identical results for Orion Peak 1 and matched the data by modeling a J–shock followed by radiative cooling from H_2 and CO as well as H_2 dissociative cooling (Fig. 1). The nearly identical relative level populations found in IC 443 and Orion Peak 1 indicate the shock processes producing the bulk of the H_2 emission in these regions must be very similar.

The partially dissociative J–shock model has been questioned because it omits cooling from H_2O. Recently, it has been shown that H can be a very effective collision partner for H_2; the critical density for populating the lower vibrational levels of H_2 in mixed H–H_2 gas at T=2000 K is $\leq n_H = 10^4$ cm^{-3} (Mandy & Martin 1993), two orders of magnitude less than previously considered. At these densities, the omission of H_2O cooling from the partially dissociative J–shock model is a reasonable simplification.

Models of low Alfvén Mach number bow shocks are also consistent with the Orion Peak 1 population level diagram (Smith et al. 1991), but are inappropriate for IC 443. First, the strong magnetic fields necessary, 52 mG in the case of Orion, are unlikely to be present over the large spatial extent ($24'' = 0.2$ pc) sampled in IC 443. Second, the absence of Brγ precludes bow velocities much greater than 60 km s^{-1} while the Orion fit has bow velocity of 250 km s^{-1}. Finally, the driving pressure in IC 443 is on the order of $nT = 1 \times 10^8$ cm^{-3} K (Brown et al. 1988), while the pressure needed for the low Alfvén model is roughly $nT = 1 \times 10^{13}$ cm^{-3} K.

We wish to point out that partially dissociative J–shock models cannot account for all the properties of H_2 emission in Orion and IC 443. In particular, shock velocities more than roughly 30 km s^{-1} should totally dissociate H_2 (Chang & Martin 1991), yet line widths greater than that are seen in Orion and IC 443 (Geballe et al. 1986; Richter et al. 1995b)

Acknowledgements

We gratefully acknowledge CalSpace for its support through grant CS-70-93. MJR thanks the NASA Graduate Student Researchers Program for support through grant NGT-51082-GRAHA. DMK was supported by USAF contract F19628-93-K-0011.

References

Brand, P. W. J. L., Moorhouse, A., Burton, M. G., Geballe, T. R., Bird, M., & Wade, R.: 1988, *Astrophys. J.* **334**, L103

Brown, L. W., Woodgate, B. E. & Petre, R.: 1988, *Astrophys.J.* **334**, 852

Burton, M. G.: 1987, *Q. Jl. R. Astr. Soc.* **28**, 269

Chang, C. A., & Martin, P. G.: 1991, *Astrophys. J.* **378**, 202

Geballe, T. R., Persson, S. E., Simon, T., Lonsdale, C. J., & McGregor, P. J.: 1986, *Astrophys. J.* **302**, 693

Lacy, J. H., Achtermann, J. M., Bruce, D. E., Lester, D. F., Arens, J. F., Peck, M. C., & Gaalema, S. D.: 1989, *P.A.S.P.* **101**, 1166

Mandy, M. E., & Martin, P. G.: 1993, *Astrophys. J. Suppl. Ser.* **86**, 199

Parmar, P. S., Lacy, J. H., & Achtermann, J. M.: 1994, *Astrophys. J.* **430**, 786

Richter, M. J., Graham, J. R., & Wright, G. S.: 1995, *Astrophys. J.* submitted,

Richter, M. J., Graham, J. R., & Wright, G. S.: 1995, *in preparation* ,

Smith, M. D., Brand, P. W. J. L., & Moorhouse, A.: 1991, *Mon. Not. R. astr. Soc.* **248**, 730

SHOCKED MOLECULAR HYDROGEN IN THE LKHα234 REGION

A. S. B. SCHULTZ
MS 245-6, NASA-Ames Research Center, Moffett Field, CA 94035 USA

and

D. RANK, P. TEMI and D. HARKER*
University of California-Santa Cruz

Abstract. We present near-IR images of the shocked gas in in HH 167 (the LkHα234 jet) and HH 103. The H_2 1-0 S(1) (2.122 μm) and the 1.644 μm [FeII] lines were observed. The relative spatial distribution of these two lines provides some insight into the nature of the shocks producing the objects.

1. Introduction

A major puzzle in the study of Herbig-Haro objects is the seeming spatial coincidence between the atomic (generally optical) emission, and that of molecular hydrogen; the shocks which excite the optical emission are thought to be strong enough to dissociate the hydrogen.

In order to study this, we have begun a program of imaging HH objects in the near-infrared, specifically in the 1-0 S(1) (2.1218 μm) and 2-1 S(1) (2.2477 μm) molecular hydrogen lines. We also image the objects in the 1.644 μm line of [FeII]. This line is produced by shocks similar to those which produce the optical atomic lines (especially [SII]), yet it is not as susceptible to dust extinction. It therefore allows a more accurate comparison between the shocked molecular and atomic gas.

Here we present the first results of this program: observations of the LkHα234 jet and HH 103. LkHα234 is a Herbig Be star in the NGC 7129 region. Ray et al (1990) found a well-collimated optical jet (now designated HH 167) emanating from the star and extending out to 40" (= 0.19pc). The region also contains several HH objects, among them HH 103. These objects are colinear on a line that includes LkHα234, so they were once thought to be products of an outflow from that star. The proper motion studies of Ray et al show that this is very unlikely.

Wilking et al (1990), using a single-beam detector, found H_2 1-0 S(1) emission at several locations in this region, including LkHα234 and HH 103, but to our knowledge these are the first H_2 images of these objects.

These images were taken with the LIRC II, a 256$\times$256 NICMOS array, mounted on the Lick Observatory 3m telescope, on 25 May and 15 August,

* current address: University of Wyoming

Astrophysics and Space Science **233**: 71–74, 1995.

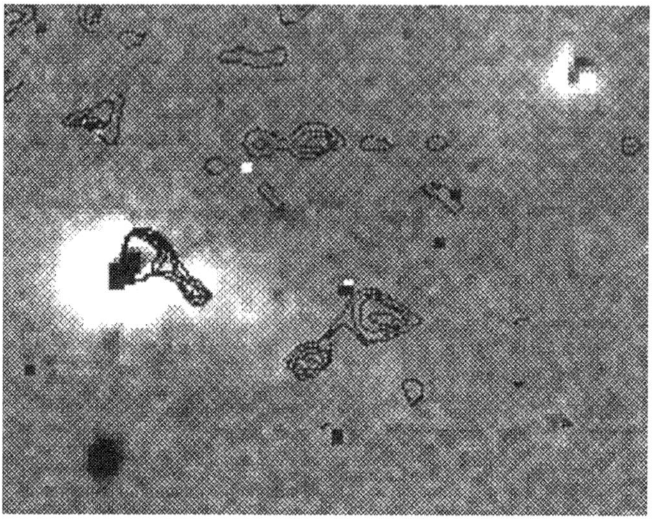

Fig. 1. LkHα234 and jet (grayscale: H_2 1-0 S(1), contours: [FeII])

1994. Filters used were 1-0 S(1) H_2 2.1218 μm, and [FeII] 1.644 μm. (The 2-1 S(1) 2.2477 μm line was also observed, but no emission was detected.) Continuum subtraction was effected using a CO filter (2.297 μm) for the 2 μm images and a broadband H filter for the [FeII] images. The images have a plate of scale 0.3"/pixel.

2. Results

2.1. LkHα234

Figure 1 shows LkHα234 and HH 167. The grayscale image is H_2 1-0 S(1), and the contours are [FeII]. We emphasize that the contrast levels have been chosen to show the *extent* of the emission, and should not be thought of as representing the relative strengths of the emission lines. We also emphasize the preliminary nature of this data. The images have not been flux-calibrated.

The H_2 1-0 S(1) emission extends to the southwest of the star; its distribution corresponds roughly to that of the [SII] emission (see figure 1*c* of Ray et al. 1990), although it is confined to a far smaller region near the star. They find that the [SII] jet is about 40" in length, while we find the H_2 jet is only about 9" long (although there are indications that that there is much more extensive faint emission).

The [FeII] emission is largely coincident with that of the H_2 1-0 S(1), and very faint emission can be traced to the end of the bright 1-0 S(1) emission. However, the [FeII] seems more confined to the region close to the star.

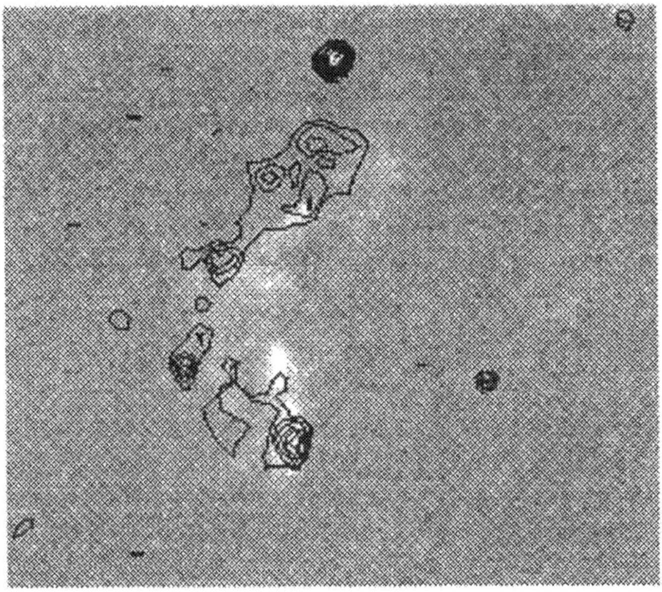

Fig. 2. HH103 (grayscale: H_2 1-0 S(1), contours: [FeII])

These results may be suspect since the continuum-subtraction was effected with a broadband filter, rather than a narrow-band as was done with the H_2 lines. The appearance of the emission is also strongly dependent upon the registration. Therefore the distribution of the [FeII] in this region should be regarded with caution.

The non-dectection of H_2 2-1 S(1) would seem to confirm that the emission is shock, rather than fluorescently, excited.

2.2. HH 103

We present images of HH 103 in Figure 2. Again, the grayscale image represents H_2 1-0 S(1) and the contours [FeII] 1.644 μm emission. No H_2 2-1 S(1) frames were taken of this object. As for LkHα234, the images are only meant to show the relative distributions of the line emission; they have not been flux-calibrated.

Except for a knot at the southern tip of the object, which is strong in both lines, the [FeII] and H_2 emission seem to occupy quite separate regions. The bulk of the [FeII] emission lies in a C-shaped region to the east of most of H_2 emission. It should be noted that the paucity of stars with which to register the two images may make this spatial distribution somewhat suspect.

This object was studied by Eiroa et al (1992), who did deep CCD images of it. Their figure 2 indicates that the distribution of the Hα and [SII] emission resembles that of the [FeII], although it is difficult to be certain without being able to compare the optical and IR images directly. The knot at the southern end of the object, where the [FeII] and H_2 are coincident, may also be Eiroa et al's knot C, where both $H\alpha$ and [SII] are strong.

If the $H\alpha$, [SII], and [FeII] data are spatially coincident, and displaced from the H_2 emission, this could indicate separate areas of a shock. Since the proper motion vectors point toward the southwest (Ray et al 1990), the object could be a shocked cloudlet, in which the optical emission represents shocked jet or wind material, and the H_2 shocked cloud material.

3. Conclusions

We have presented infrared images of HH 103 and the LkHα234 jet (HH 167) in the H_2 1-0 S(1) and [FeII] 1.64 μm lines. These images show that extended emission exists in both these objects. In LkHα234, the H_2 emission seems coincident with the optical jet found by Ray, et al 1990). The [FeII] emission in HH 103 may be coincident with the optical emission found by Eiroa et al (1992), whereas the H_2 emission comes from a different region of the object. If so, this may indicate a shocked cloudlet.

Further study of these objects is warranted. Higher S/N is needed to bring out the faint emission in LkHα234, where there is a great deal of scattered continuum. H_2 2-1 S(1) observations are also planned for HH 103.

References

Eiroa, C., Gómez de Castro, A.I., & Miranda, L.F.: 1992, *Astron. Astrophys. Suppl.* **92**, 721

Ray, T.P., Poetzel, R., Solf, J., & Mundt, R.: 1990, *Astrophys. J.* **357**, L45

Wilking, B.A., Schwartz, R.D., Mundy, L.G., & Schultz, A.S.B.: 1990, *Astron. J.* **99**, 344

SMOKE RINGS AROUND SUPERNOVA 1987A

J. MEABURN, M. BRYCE and A.J. HOLLOWAY
Department of Physics & Astronomy,
The University of Manchester
Oxford Road, Manchester M13 9PL

Abstract. Detailed observations of the [O III]5007 Å emission from the elliptical rings around SN 1987A suggest a model wherein the two faint, outer rings are due to emission from two circular toroids moving outwards (at ≤ 25 km s^{-1}) along a bipolar cone centred on the site of the supernova. The brighter, central ring is expanding radially outward at 8.3 km s^{-1}. The rings must have been created ~ 2–3×10^4 years before the supernova explosion and are thought to be a consequence of the interaction of stellar winds emanating from the progenitor system during the final stages of its evolution to a supernova.

Key words: supernovae: individual: SN 1987A – ISM: supernova remnants

1. Introduction

The explosion of Supernova 1987A in the halo of the 30 Dor nebula in the LMC has stimulated considerable interest in the circumstellar environment of the progenitior blue supergiant star. Ground based imagery by Wampler et al (1990) revealed a ring of ionized gas centred on the explosion site with two faint, outer arcs of emission. Recent publically released post-COSTAR HST imagery shows that these two faint arcs are also ring structures. The rings, which appear elliptical due to the inclined line of sight, are sketched in Fig. 1. Crotts & Heathcote (1991) have observed the kinematical structure of the central feature. Their results are compatible with a circular torus expanding radially outward from SN 1987A at 10.3 km s^{-1} viewed at an angle of 43° to the plane of the ring. Without the benefit of the HST images, they interpreted the emission observed from the north and south rings within a model involving a flash of Lyman photons from the SN explosion.

2. Observations and Results

We have now observed spatially resolved [O III]5007 Å emission line profiles over all three rings using the Manchester echelle spectrometer on the 3.9m Anglo-Australian telescope. Profiles were obtained from the four longslit positions marked in Fig. 1. The effective sampling area in the prevailing seeing conditions was 1.2 $''$ in the EW direction and 0.75 $''$ in the more critical NS direction. The data were wavelength calibrated and then the continuous spectra from field stars and the two extended features due to emission from two large sheets of gas, at $V_{\rm hel} = 258$ and 281 km s^{-1}, in the halo of 30

Astrophysics and Space Science **233**: 75–79, 1995.

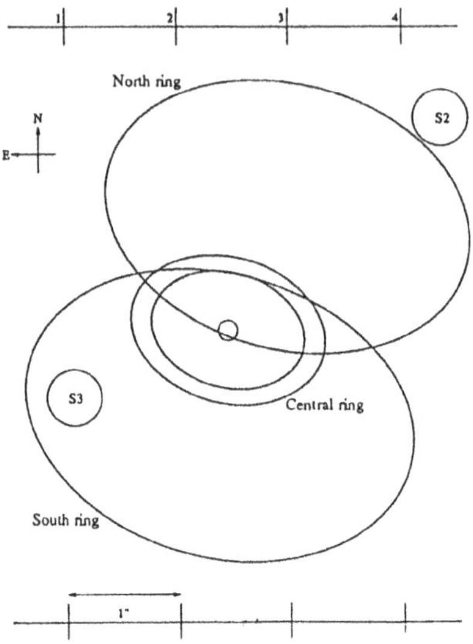

Fig. 1. A sketch showing the elliptical structure of the three ionized rings around SN 1987A. The stars S2 and S3 (West et al. 1987) and the centres of the four NS longslit positions are also marked.

Dor (Meaburn 1990) were removed. Contour plots of the resultant positon-velocity arrays from the four slits are shown in Fig. 2. The data were further analysed by fitting single Gaussians to the observed line profiles. The much fainter emission from the north and south rings (A and B in Fig. 2) is close to the mean radial velocity (286.5 ± 1 km s^{-1}) of the central ring.

The central ring can be interpreted as a torus of 0.38 pc diameter (assuming a distance of 50 kpc to the LMC), tilted at $43° \pm 3°$ to the plane of the sky (where 0° is face on) with a radial expansion velocity of 8.3 km s^{-1}. If the expansion velocity has always been constant, this implies that the central ring originated 2.2×10^4 years before the supernova explosion.

The inclination is such that the northern edge of the central ring is towards us. We propose that the fainter northern and southern rings have also been ejected from the central progenitor system and are travelling outward along a common polar axis whilst also expanding outwards radially in their own planes. This motion can be evisaged as smoke rings moving outward, away from the explosion site, along the surface of a bipolar cone, and is shown schematically in Fig. 3. The model accounts for the very low relative velocities observed from these rings. The orientation of the entire system to the line of sight means that the velocity components observed from A and B are almost tangential to the line of sight whereas the opposite edges of these two rings (C and D), which are largely obscured by the much brighter central ring, contain velocity components more closely aligned with the line of sight. An

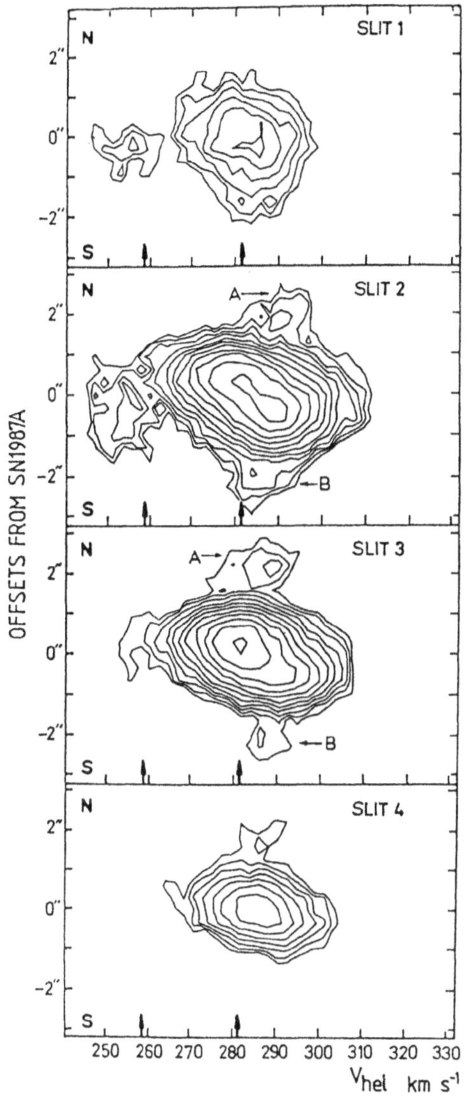

Fig. 2. Contour maps of the observed [O III]5007 Å emission from Slits 1–4. The two extended background velocity components (marked by arrows) have been subtracted from the data arrays as have stellar spectra. The logarithmic contours are equally spaced by 0.115 and the lowest contour levels are 0.5 for Slits 1 & 2 and 0.7 for Slits 3 & 4.

upper limit to the true velocities of these rings can be obtained by considering the widths of the observed profiles from the centre of the system, this limit is found to be 25 km s^{-1}, implying that the rings must have originated at least 3.0×10^4 years ago.

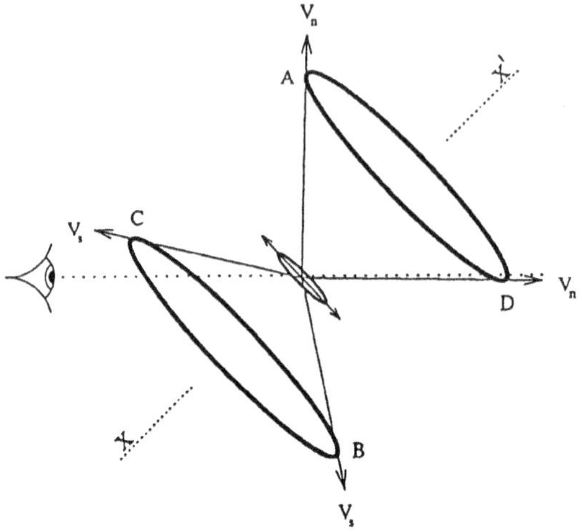

Fig. 3. A sketch of the model showing the three rings along a common axis XX'. The central ring is expanding radially outwards and the north and south rings are moving outward (with velocities V_n and V_s respectively) along the surfaces of the bipolar cone.

Circumstellar rings are a common feature in astrophysics, and are often associated with binary stars and bipolar structures. Indeed the central ring around SN 1987A is of a comparable size to the ring nebulae often observed around Wolf-Rayet stars in out own galaxy. A new hydrodynamical model by Lloyd, O'Brien and Kahn (1995 and in these proceedings) illustrates that when the fast wind from the final blue giant phase of the progenitor star interacts with bipolar lobes formed by earlier mass loss shaped by the presence of a binary compansion, ring-like structures are naturally formed, which are predicted to have the modest velocities reported in this work.

Acknowledgements

We would like to thank staff at the AAO for their assistance during the observing run. MB and AJH acknowledge support through a PPARC research associateship and a PPARC studentship respectively. The data were reduced and analysed using hardware and software provided by the STARLINK project.

References

Crotts A.P.S. & Heathcote S.R.: 1991, *Nature* **350**, 683
Lloyd H.M., O'Brien T.J. & Kahn F.D.: 1995, *Mon. Not. R. astr. Soc.* **273**, L19

Meaburn J.: 1990, *Mon. Not. R. astr. Soc.* **244**, 551
Wampler E.J., Wang L., Baade D., Banse K., D'Odorico S., Gouiffres, C. & Tareghi M.: 1990, *Astrophys. J.* **362**, L13

MODELS OF MOLECULAR OUTFLOWS

S. CABRIT*
DEMIRM, Observatoire de Paris
61 Avenue de l'Observatoire
75014 Paris, FRANCE
e-mail: Sylvie.Cabrit @ obspm.fr

Abstract. In this contribution, I present a broad "historical" review of the various hydrodynamical models that have been considered for explaining molecular outflows, and of their merits and failures when compared with observations. Wind-driven bubbles, viscous jet mixing-layers, and jet bowshocks, are discussed in turn. Most general properties of outflows can be understood in terms of a simple bowshock model. However, the detailed structure of outflows is more complex and not yet fully understood, given the presence of time variability in the jet velocity and/or direction. Finally, I discuss constraints on wind properties (momentum, mass-loss rate, radius) that can be derived from molecular outflows driven by jet bowshocks.

Key words: Outflows – Jets – Interstellar Shocks

1. Introduction

The large amount of mass in molecular flows from young stellar objects (YSOs) shows that they are mostly made of ambient material entrained by an underlying "primary wind" from the central source. Modelling outflows basically aims at understanding this dynamical entrainment process.

Molecular outflows share several observational characteristics that must be reproduced. Recent reviews are given, e.g., by Bachiller & Gómez-González (1992), and Masson & Chernin (1993). Before discussing specific hydrodynamical models, I will recall here two important "model-independent" constraints set by observations.

First, the small number of outflows with superimposed blueshifted and redshifted emission in their lobes, like e.g. B335 and RNO 43, has shown that, at least in the bulk of most flows, the transverse component of velocity must be small, ~ 5 km s^{-1}, compared with the axial component (Cabrit et al. 1988; Meyers-Rice & Lada 1991). Second, the broad emission line wings indicate that most of the mass is moving at low velocities, and that large intrinsic velocity gradients must be present in the swept-up gas (Masson & Chernin 1992). In particular a velocity that decreases with increasing polar angle θ could reproduce both the apparent linear acceleration ("Hubble-law") seen in edge-on flows and the apparent deceleration seen in pole-on flows (Moriarty-Schieven & Snell 1988; Cabrit 1989; Shu et al. 1991).

* On leave from Observatoire de Grenoble

Astrophysics and Space Science **233**: 81–96, 1995.

Two classes of dynamical models have been considered to explain these features in a coherent way. The first one (chronologically) involves the interaction of a loosely collimated wind with ambient material, giving rise to *wind-driven bubbles*, while the second, recently supported by observations, assumes the flow is driven by a highly-collimated jet. As a large body of work has been done on this subject over more than a decade, I will not give an exhaustive review, but rather try to discuss the basic flow properties for each of these two classes of models, and see how they compare with observations.

2. Models of Wind-Driven Bubbles

This first class of model was motivated by the fact that (1) outflows were initially thought to be only moderately collimated (e.g. Lada 1985), (2) the ionized component of YSO jets fails by 1 to 2 orders of magnitude to provide sufficient momentum for driving molecular flows (e.g. Mundt et al. 1987; Cabrit & Bertout 1992 and references therein). A natural possibility was that outflows were driven by a loosely collimated neutral wind, physically distinct from the jets. Detection of extremely high-velocity (EHV; $V \sim 150$ km s^{-1}) HI and CO emission in several flows further lent support to this atomic wind hypothesis (e.g. Lizano et al. 1988), although spatial resolution was insufficient at the time to assess its collimation.

The expansion of spherical wind-bubbles, investigated in the context of OB star winds by Weaver et al. (1977), was generalized to the case of molecular outflows by Dyson (1984) and Kwok & Volk (1985). A dense shell of swept-up gas is formed, with two possible expansion regimes: a momentum-conserving regime, where the wind cools rapidly after impacting the shell and imparts only its momentum, and an energy-conserving regime, where the shocked wind remains hot and further pushes the shell with its thermal pressure. The latter is more efficient at accelerating the bubble, by a typical ratio $V_{\rm wind}/V_{\rm shell} \sim 10$. As YSO winds could be in either regime (e.g. Dyson 1984), both cases have been investigated.

Wind-driven bubbles possess two interesting features. First, they are naturally collimated by the anisotropic pressure distribution around YSOs. As this distribution should have a minimum along the polar axis — e.g., in a rotationally flattened dense core, or in a structure supported by a dipolar magnetic field — prolate outflow lobes will result, as observed (Königl 1982). Second, the shell will expand faster toward the poles, in the direction of least resistance, and more slowly toward the equator, where it is confined. This type of latitude gradient is exactly what is needed to explain the differing velocity gradients in pole-on and edge-on flows (see Sect. 1). However, other important features of outflows are not well explained. Examples illustrating these problems follow.

2.1. Energy-Driven Bubbles

Stone & Norman (1992) have made MHD numerical simulations of the magnetic collimation of an adiabatic, spherical wind. Starting from a magnetically supported dense core, a shell with length-to-width ratio $q \sim 3$ is reached. However, the timescale is very long, typically 2×10^5 yrs. This conflicts with the fact that even outflows as young as $< 10^4$ yrs can be very collimated (e.g. Bachiller et al. 1990). In addition, the thermal pressure of the hot wind exerted on the shell walls produces a large transverse velocity component. Masson & Chernin (1993) pointed out that this conflicts with the small transverse velocities inferred in molecular outflows (cf. Sect. 1). An energy-conserving bubble thus seems ruled out.

2.2. Radially Expanding Momentum-Driven Bubbles

A first semi-analytical model was presented by Sakashita & Hanami (1986), who studied the radial expansion of a momentum-driven bubble into a hydrostatic disk. Collimation ratios comparable to the energy-driven case were reached.

Shu et al. (1991) and Masson & Chernin (1992) derived fully analytical predictions in the case of a radial density distribution $\propto r^{-2}$, assuming the wind thrust and the ambient density have a latitude dependence given by powers of $\cos\theta$. Masson & Chernin (1992) found that to reproduce both the collimation of molecular outflows, and the observed power-law decrease of swept-up mass with velocity ($m(v) \propto v^{-\gamma}$ with $\gamma \sim 1.8$) one would need a wind that is spherical or even stronger at the equator and a density distribution that has a marked polar cavity. While the latter condition could be fulfilled in the environment of YSOs over scales ≤ 0.1 pc, the former one conflicts with the observation of EHV molecular components along the axes of several molecular outflows (see Sect. 3), which suggests that the driving neutral wind is stronger toward the poles. Masson & Chernin (1992) argue that this problem is intrinsic to models of radially expanding momentum-driven bubbles and conclude they are not adequate to describe molecular outflows.

2.3. Steady-state Configuration

The approximation of radial expansion is not completely valid if the bubble is very collimated, as expected in a strongly anisotropic density distribution. In that case the wind refracts through an oblique shock and "slides" supersonically along the shell walls, exerting an additional centrifugal pressure.

Cantó (1980) showed that a steady-state configuration can be found. If the central star is located in the midplane of a stratified disk, a bipolar cavity with acute tips is formed (Barral & Cantó 1991). In this steady-state model, the molecular shell would be much older than its dynamical time, thus reducing momentum requirements on YSO winds (Dyson et al. 1988). However, wind material should eventually converge toward the tip of the cavity and shock against itself there, forming a dense, high-velocity jet (Cantó et al. 1988). This jet will then emerge from the confined shell, and presumably create a separate *jet-driven* outflow.

Detailed numerical computations of the confinement of isotropic winds by infalling envelopes, including radiative cooling, have been recently carried out by Mellema and Frank (this volume), and they confirm the formation of an axial jet analytically predicted by Cantó et al. Therefore such a model is probably more relevant to describe the initial collimation of jets close to the star, or the very late asymptotic stages of outflow evolution (e.g. after many jet direction changes), than to model the main phase of outflow formation.

3. Models of Jet-Driven Outflows

Besides the problems encountered by wind-driven bubbles, several recent observational findings have made this model the most promising; briefly, these include (1) the realization that optical jets have a very low ionization fraction and hence a much larger momentum rate than previously thought (e.g. Raga 1991; Hartigan et al. 1994); and (2) direct evidence from sensitive high-resolution maps of EHV CO and SiO that the flow is accelerated in situ by a collimated, neutral jet component (e.g. Bachiller et al. 1990; Masson et al. 1990; Guilloteau et al. 1992; Richer et al. 1992). The problem of entrainment of ambient gas by astrophysical jets has motivated numerous modelling efforts. Two basic processes can be distinguished (DeYoung 1986):

- "steady-state" entrainment that occurs via Kelvin-Helmholtz (KH) instabilities at the interface between the jet beam and the ambient medium.
- "prompt" entrainment that occurs at the jet head, in a curved *bowshock* that accelerates and sweeps aside ambient gas.

Both have been recently modelled in the context of molecular flows, and I summarize the main results below.

3.1. Viscous Mixing-Layers

In "steady-state" entrainment, KH instabilities ultimately lead to the formation of a turbulent viscous mixing layer. This layer grows both into the environment and into the jet, and eventually the whole flow becomes turbulent.

The possibility that molecular outflows could represent such turbulent viscous mixing-layers around YSO jets has been investigated by Cantó & Raga (1991) and Stahler (1994). Cantó & Raga (1991) assumed a linear Couette-flow velocity profile across the layer, and a "mixing-length" parameterization of the entrainment velocity, that allowed them to derive the growth of layer width with distance. Stahler (1994), on the other hand, focussed on the fully turbulent region downstream.

Chemical reactions and thermal balance in molecular mixing-layers were recently calculated by Taylor & Raga (1995). Despite the rather high temperatures inferred, exceeding 10^4 K, H_2 molecules are subthermally excited and survive over a significant length of the jet. The highly-collimated H_2 vibrational emission detected in some optical jets (e.g. Gredel & Reipurth 1993) could in fact be well explained with such a layer. However, this type of model encounters several important problems in explaining the bulk of emitting material in molecular outflows.

First, the predicted collimation ratio is much too large. As HH-objects and EHV jet tracers are often observed out to the end of outflow lobes, the outflow cannot typically extend much further than the point x_t where the whole jet becomes turbulent. Canto & Raga (1991) find that, at this distance $x_t \sim 0.55$ pc, the length-to-width ratio q of the mixing-layer is $q = 2.8(M_j)(c_j/c_l)^2$, where M_j is the jet Mach number and c_j and c_l are the sound speed in the jet and in the mixing-layer respectively. An M_j of 20 would then yield $q \sim 56$ for the mixing-layer, typical of the value in optical jets but 10 times larger than in molecular outflows.

Raga et al. (1993) showed that turbulent mixing-layers could be enhanced by the sideways ejection of overpressured jet material; this ejection would occur at velocity discontinuities along the jet, where internal "jet working surfaces" develop (see Section 3.3.1). However, q would still be very high, around 10, compatible only with the most collimated outflows known to date. Another drawback is the predicted velocity gradient: After providing strong initial acceleration near the origin, the ejected jet momentum is progressively shared among larger and larger amounts of ambient gas, and the mean velocity in the turbulent envelope gradually decreases with distance from the star. In contrast, most molecular outflows show ramps of velocity linearly increasing with distance from the star, followed by strong "deceleration" (e.g. L1448 and Orion B, Bachiller et al. 1990; Richer et al. 1992).

One concludes that turbulent "steady-state" entrainment as currently modelled cannot dominate the formation of molecular outflows at scales $\sim$ 1 pc. However H_2 images suggest this process is nevertheless active close to the jet injection point, at least in some jets, and perhaps also further along as the cocoon behind the bowshock comes into contact with the jet beam (see Section 3.2.2).

3.2. Bowshocks

Bowshocks in stellar jets have been extensively modelled with the goal of explaining optical observations of Herbig-Haro (HH) objects. Radiative models with a parameterized bowshock shape have been very successful in reproducing their velocity-position diagrams and line profiles (e.g. Hartigan et al. 1990). Hydrodynamical numerical simulations with various prescriptions for radiative cooling have also been carried out (e.g. Raga 1988; Blondin et al. 1990; Stone & Norman 1993a, 1994; Gouveia dal Pino & Benz 1993). However, most involve low ambient densities < 100 cm^{-3} not representative of the environment of YSOs driving outflows.

High ambient densities meet two computational limitations. First, time evolution is slower; momentum balance predicts that the bowshock propagates at velocity V_{BS} given by

$$V_{BS} \sim \beta\ V_j(1 + 1/\sqrt{\eta})^{-1}, \tag{1}$$

where $\eta = n_j/n_o$ is the jet to environment density ratio, and β is an efficiency factor ~ 0.7 (e.g. Stone & Norman 1994). If $n_j \ll n_o$ ("light jet") then $V_{BS} \ll V_j$. This generally restricts simulations of the whole jet to $\eta \geq 0.3$. Another limitation is set by the very small cooling distance $d_{\rm cool}$. The whole structure of the bowshock depends heavily on the value of the cooling parameter $\chi = d_{cool}/r_j$ (e.g. Blondin et al. 1990; Masson & Chernin 1993):

$$\chi = \frac{d_{\rm cool}}{r_j} \sim 0.36 \left(\frac{n_{pre}}{10^3{\rm cm}^{-3}}\right)^{-1} \left(\frac{r_j}{10^{16}{\rm cm}}\right)^{-1} \left(\frac{V_s}{300{\rm km\ s}^{-1}}\right)^4, \tag{2}$$

where n_{pre} is the preshock density, and V_s is the shock velocity (V_{BS} in the bowshock, where ambient material is accelerated, and $V_j - V_{BS}$ in the jet shock). Equs. 1-2 show that cooling will be stronger in the denser component (i.e. in the jet shock for a heavy jet, in the bowshock for a light jet). If a typical YSO jet has $n_j \sim 10^3$ cm^{-3} and $V_j \sim 300$ km s^{-1}, one of the two shocks will be highly radiative ($\chi < 10^{-3}$) unless η is close to 1.

As a result, one is currently limited either to studying the general flow dynamics, without resolving adequately the cooling regions (≤ 10 grid points), or to describing accurately the thermal and radiative properties of post-shock gas, but in a much more limited parameter space. The next two sections discuss recent simulations exploring these two extremes in the context of molecular outflow models.

3.2.1. *General Dynamical and Geometrical Properties of Bowshocks*

A very nice general feature of bowshocks is that most of the swept-up mass is at low velocity, and that the velocity gradient is in good qualitative agreement with that inferred to be present in molecular flows. In essence, while wind bubbles propagate nearly normal to their surface so that swept-up gas is at rest with respect to the expanding shell, bowshock wings propagate in the jet direction and drive an oblique shock into ambient gas. Swept-up gas then slides back with respect to the bowshock surface, and decelerates as more static ambient gas is intercepted. This produces an apparent acceleration toward the jet end, which agrees well with observations (Masson & Chernin 1993, Padman & Richer 1994).

To explore conditions more appropriate to the environment of outflow sources, Gouveia dal Pino & Benz (1993) and Chernin et al. (1994) used an SPH numerical code to simulate the propagation of a supersonic jet into an ambient density reaching $n_o \sim 10^3$ cm^{-3}. Chernin et al. found that terminal velocity vectors in the dense bowshock wings make an angle of less than 30° from the jet axis, as required by the mostly forward-directed motions inferred from observations of molecular outflows. This is caused by rapid cooling in the bowshock, which quickly inhibits transverse pressure gradients. These results apply mainly to the bulk of the flow, however, given the low resolution and simplified treatment of the cooling (see, e.g., Stone & Norman 1993a for a comparison of various cooling prescriptions).

The predicted overall shapes for bowshocks are also in relatively good agreement with outflow observations. While a bowshock propagating in a uniform medium is broadest at the injection point, it is not true in a more realistic situation where density increases sharply toward the source, because shell expansion will be very slow close to the star. A simplified analytical bowshock model with a density gradient $n_o \propto r^{-1}$ predicts a roughly cylindrical shell shape, in better agreement with observations (Masson & Chernin 1993).

On the other hand, transverse sizes of molecular flows are larger than expected from the radii of optical jets near YSOs. Namely, radiative bowshock models predict a maximum shell diameter $D \sim 7r_j$ (see e.g. Blondin et al. 1990; Masson & Chernin 1993; Fig. 1 in this paper). CO outflows showing a simple bowshock behavior have typical lobe widths $D \sim 0.06 - 0.2$ pc (for lobe lengths of $L \sim 0.2 - 0.7$ pc). One would then infer jet radii $r_j \sim 3 - 9 \times 10^{16}$cm, roughly 10 times larger than measured in the optical sample of Mundt et al. (1987). Several possible explanations may be found. First, we note that the inferred jet length-to-width ratio $q = L/2r_j$ would be ~ 30, hence such radii could be explained by free jet expansion or the growth a hot viscous mixing-layer (see Sect. 3.1). Next, the effective size of the bowshock may increase if the jet head wiggles, because of rapid precession, or fragmentation of dense post-shock gas at the working surface (Stone & Norman 1994). Finally, long-term variability in the jet velocity and/or direction, which are known to be present, could increase the overall outflow width (Masson & Chernin 1993; Raga & Cabrit 1993; see Section 3.3).

In conclusion, the overall velocity field, decreasing amount of mass with velocity, and moderate collimation in molecular outflows are all naturally understood in terms of a bowshock model, though transverse sizes seem to require spatial or temporal variations in the jet.

3.2.2. *Detailed Structure of a Molecular Bowshock*

The detailed density and thermal structure at the jet head can only be accurately studied in simulations that fully resolve cooling regions. This is necessary e.g. to calculate the predicted emission spectrum and to investigate the structure of the low density cocoon inside the bowshock.

Such a high resolution simulation was recently presented by Raga et al. (1995) using an adaptive grid technique. For the first time in hydrodynamical calculations, the ambient medium is taken to be molecular and the code explicitely follows the destruction of H_2 and CO, ionization and recombination of H and S II, and radiative cooling by H_2 and other atomic and ionic coolants. The jet has $n_j = 50$ cm^{-3}, $\eta = 10$, and $V_j = 130$ km s^{-1}, more typical of objects like HH 111.

It can be seen from the CO abundance map in Figure 1 that a substantial amount of molecules survive destruction in the bowshock wings. As the bowshock is only weakly radiative, some gas has re-expanded from the shock front, creating a low-density molecular cocoon ("wake") that comes into contact with the jet beam, where it reaches longitudinal velocities of up to 30 km s^{-1}. This may allow a molecular mixing-layer to develop in this region (cf. Sect. 3.1), even though most of the ambient gas has been swept aside in a dense shell.

The predicted velocity-position diagrams for Hα and H_2 1-0 S(1), shown in Figure 2, illustrate well the steep increase in velocity toward the jet head typical of a bowshock structure. Note that while Hα emission is strongly peaked at the jet head, H_2 emission dominates in the extended, low-velocity wings of the bowshock. Therefore, even a fast bowshock may have significant molecular signatures (cf. Smith et al. 1991; Wolfire & Königl 1991). Rotational CO emission, which traces cooler gas than H_2, would be predicted to peak slightly upstream from 2.12μm 1-0 S(1), as observed (e.g. Davis & Eislöeffel, 1995).

As this simulation does not include molecule reformation, there is less and less molecular emission at high velocity. However, plane-parallel shock models show that CO will reform in post-shock column densities of $N \sim 10^{20}$ cm^{-2}, and H_2 at $N \sim 4 \times 10^{21}$ cm^{-2} (e.g. Hollenbach & McKee 1989). The warm, narrow EHV CO features observed in outflows at velocities $\sim 50 - 150$ km s^{-1} could be nicely explained by molecules reformed in the dense stagnation region just behind the jet head (Chernin & Masson 1994), or inside an internal jet working surface (see below).

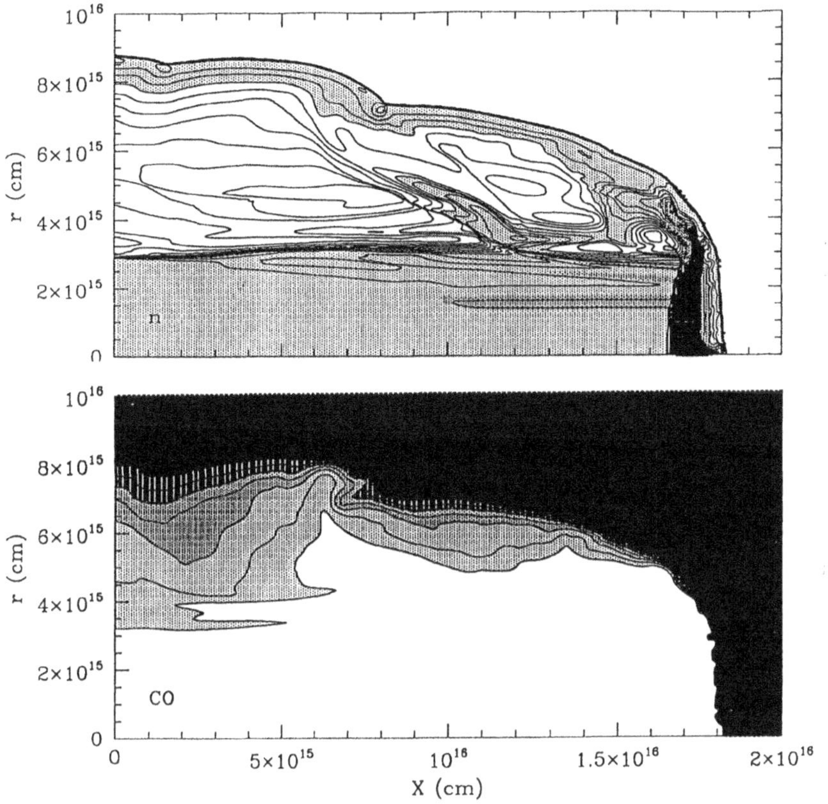

Fig. 1. Density (top) and fractional CO abundance (bottom) in the bowshock of a 130 km s^{-1} jet moving into molecular environment, from Raga et al. (1995). Density contours increase by factors of $2^{1/2}$ while contours of X[CO] are at constant intervals of 10^{-5}. As substantial molecular fraction is present in the bowshock wings and in the low-density cocoon.

3.3. Effect of Jet Time Variability

Intrinsic jet variability is indicated by the fact that optical jets often contain *several* successive "bowshocks". Similarly, molecular flows often present *several* EHV components roughly equally spaced along the flow axis, with a high degree of symmetry between the blueshifted and redshifted lobes (e.g. Bachiller et al. 1990). The outflow structure is then probably more complex than a single bowshock.

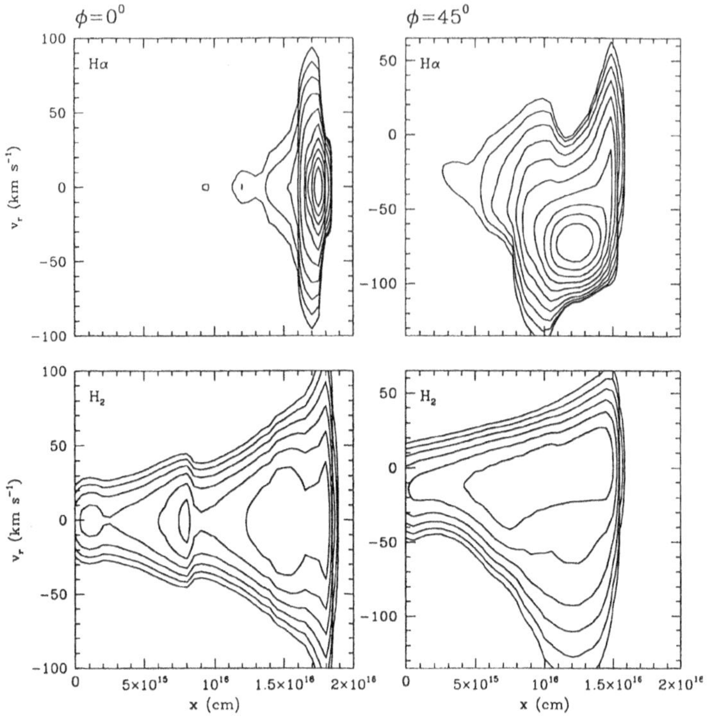

Fig. 2. Computed velocity-position diagrams in Hα (top) and the H_2 1-0 S(1) 2.12 μm line (bottom) for the simulation shown in Figure 1, both for an edge-on view ($\phi = 0°$) and for $\phi = 45°$. Intensity is contoured by factors of 2. The bowshock propagates at 100 km s^{-1}. H_2 emission is very extended and dominates at low velocity.

3.3.1. *Velocity Variability*

In several cases, the jet velocity increases between each bowshock, e.g., in the HH46/47 jet. Raga et al. (1990) showed that this can be explained by an intrinsic jet velocity variability of supersonic amplitude.

As faster jet material catches up with slower, previously ejected gas, a two-shock structure develops ("internal working surface"), which propagates along the jet (see Raga 1993 for a review). These internal jet working surfaces contain high-pressure gas that is "squeezed out" of the jet beam and drives a shock into the low-density cocoon. An example of a simulation showing the sideways expansion of such secondary "bowshocks" is shown in Figure 3, taken from Biro & Raga (1994). This high-pressure gas is expected to be mostly neutral, and possibly molecular if the postshock column density is large enough (cf. Sect. 3.2.2).

To investigate how entrainment of ambient gas could be enhanced by internal jet working surfaces, Raga & Cabrit (1993) calculated analytically the equilibrium shape and maximum size r_c reached by the shock driven into

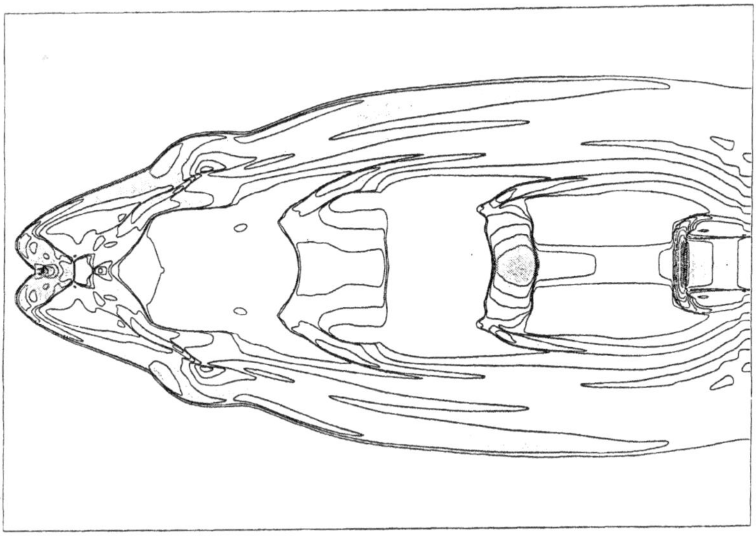

Fig. 3. Pressure stratification (contoured by factors of 2) in a jet with time-variable ejection velocity (V_j (km s^{-1}) $= 100 + 50\sin(2\pi t/380yrs)$), from Biro & Raga (1994). Each internal working surface ejects material sideways, which drives a shock into the ambient cocoon.

the ambient cocoon, and the structure of the associated wake (note that this "bowshock" structure is not exactly equivalent to a true bowshock at the jet head, and is more easily modelled). They find

$$r_c \sim \frac{\pi}{2^{3/4}} r_j \sqrt{\frac{\rho_{iws}}{\rho_{co}}} \left(\frac{V_o}{V_{iws}}\right) \tag{3}$$

where V_{iws} is the propagation velocity of the working surface (of order V_j), $\rho_{iws} \sim 10 - 100\rho_j$ is the density of shocked jet material ejected from it, V_o (typically 20 km s^{-1} for a radiative shock) is the sideways ejection velocity, and ρ_{co} is the density of the surrounding cocoon. This formula agrees well with results of numerical simulations (Biro & Raga 1994; Stone & Norman 1993b).

Clearly, if the cocoon is of sufficiently low density, the shock will reach a size of several times r_j and "scrape" the dense swept-up shell created by the leading jet head. Raga & Cabrit speculated that this could explain the enhanced velocities of swept-up gas at positions of EHV features. Unfortunately, it is not yet clear that enough shell material will be re-accelerated to give a detectable enhancement. This effect cannot be studied in the numerical simulations made so far, as the main swept-up shell is broader than would be the case in a realistic density distribution.

3.3.2. *Direction Variability*

Recent high-resolution observations show evidence for direction changes in several optical HH jets, and more recently in CO outflows as well. RNO 43 in particular shows individual molecular bowshocks associated with optical jet working surfaces propagating in different directions.

The propagation of a jet with continuously variable ejection direction was investigated analytically by Raga, Cantó, & Biro (1993), and in adiabatic and non-adiabatic numerical 2D simulations by Biro et al. (1995). The direction variability generates bends in the jet beam that propagate sideways into the environment, creating at each bend an oblique jet-shock/bowshock structure. As for internal working surfaces, these secondary bowshocks will clearly enhance entrainment of surrounding gas, but the effect on the main shell formed by the leading jet head has not yet been quantitatively studied. Obviously this will depend very much on the jet to environment density ratio, and on the timescale and amplitude of direction variations.

A different effect is obtained if both the jet velocity and the jet direction vary. As shown analytically and numerically by Raga and Biro (1993), direction changes then cause the jet beam to slip and "break" at the position of internal working surfaces, giving rise to independent "jet-bullets" travelling in different directions. Each "bullet" will then interact with the environment through its own individual bowshock. The result of a numerical simulation is shown in Figure 4. This scenario would also be appropriate if the jet is truly discontinuous, i.e. if it is ejected during short eruptions separated by quiescent stages, each ejection occuring at a different angle.

Detailed comparison of models with high-resolution observations of flows like RNO 43 should help us determine which of these two types of variability is at work in the driving source.

4. Constraints on Wind Parameters for Jet-Driven Flows

An important question is whether or not constraints on the driving wind must be significantly revised if molecular outflows are formed in jet bowshocks, instead of wind bubbles as assumed previously.

Since the interaction is probably momentum conserving, the time-averaged mass-loss rate $\dot{M}_{\rm w}$ is in both cases related to the momentum in the outflow P_{CO} by

$$P_{CO} = \varepsilon\ \dot{M}_{\rm w}\ V_j\ t_{\rm flow} \tag{4}$$

where V_j is the jet velocity, $t_{\rm flow}$ is the flow age, and ε is the efficiency of momentum transfer (the ratio of momentum in swept-up ambient gas to momentum provided by the wind over its lifetime).

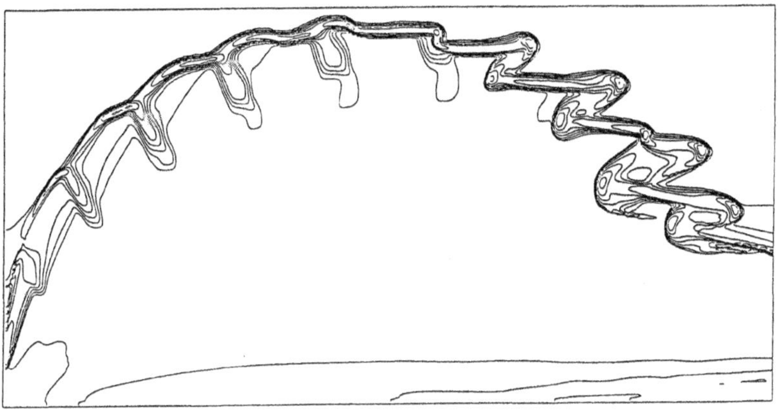

Fig. 4. Density structure (contoured every factor of 2) of a heavy jet ($\eta = 10$) with time variable velocity amplitude and direction, from Raga & Biro (1993). Note the formation of individual "bullets" with associated bowshocks.

While ε remains ~ 1 in a momentum-conserving wind-bubble (e.g. Dyson 1984), it is highly dependent on the density ratio η in the case of a jet-driven flow. Chernin et al. (1994) verified from their SPH simulations that, for jet Mach numbers $M_j > 30$, ε agrees with the theoretical value expected for pure momentum transfer at the jet head, namely

$$\varepsilon_{BS} \sim \frac{\rho_o V_{BS}^2}{\rho_j V_j^2} \sim \frac{1}{(1+\sqrt{\eta})^2}. \tag{5}$$

Since stellar jets are expected to be highly supersonic, ε should decrease from ~ 1 in very light jets ($\eta \ll 1$), which are almost entirely stopped by the environment, to $\varepsilon \sim \eta^{-1} \ll 1$ in heavy jets ($\eta \gg 1$), which travel almost unimpeded through ambient gas. The outflow momentum thus gives only a *lower limit* to the total momentum provided by the jet. From a statistical sample, Masson & Chernin (1994) estimated that *at least* 10% of the mass of the central object must be ejected during the outflow lifetime. Obviously a very efficient wind mechanism is implied.

Mass-loss rates are more uncertain as they also involve the flow age $t_{\rm flow}$, which also depends on the jet/cloud interaction. In a simplified model, we can write the outflow age as

$$t_{\rm flow} = f_t R_{CO}/V_{BS} \tag{6}$$

where R_{CO} is the full length of the outflow, V_{BS} is the bowshock velocity at its tip, and the unknown factor f_t takes into account possible variations of V_{BS} with distance from the source.

The uncertainty in $t_{\rm flow}$ may not be so severe as one would fear. Richer et al. (1992) developed a simple model of jet propagation for the Orion B outflow using direct estimates of the ambient density distribution and assuming a reasonable range of jet densities and opening angles. Their results suggest that f_t will typically introduce an uncertainty of at most a factor of 3 on $t_{\rm flow}$. If furthermore the bowshock velocity V_{BS} is close to the maximum observed CO velocity $V_{\rm max}$ at the outflow tip, as expected for a simple bowshock (see e.g. the velocity-position diagram in Figure 2), then the dynamical timescale $R_{CO}/V_{\rm max}$ will give a reasonable estimate of $t_{\rm flow}$ (e.g. 2 $\times 10^4$ yr in Orion B, instead of 4-6 $\times 10^4$ yr from the models of Richer et al. 1992).

Therefore, as long as the jet is light (ε close to 1) traditional estimates of mass-loss rates from P_{CO} using momentum conservation over the dynamical timescale should not be widely in error. In the case of Orion B, a value $\dot{M}_{\rm w}$ $\sim 10^{-6} M_\odot$ yr $^{-1}$ is inferred.

In the case of a heavy jet, however, the traditional approach is not valid as most of the jet momentum is still in the jet and not in the flow (e.g. Chernin & Masson 1995 for HH 34). Relatively accurate mass-loss rates could nevertheless be derived if η is known (e.g. from optical line ratios at the jet head), using the general formula (from Eqns. 4-6)

$$\dot{M}_{\rm w} = \sqrt{\eta}(1+\sqrt{\eta})\ P_{CO}/R_{CO} f_t^{-1}. \tag{7}$$

This equation remains valid if the CO outflow is detected only over a portion R_{CO} of the jet length, provided the flow momentum is not advected very far and is dominated by the last jet episode. In HH 34 ($\eta \sim 10$), redshifted CO parameters measured by Chernin & Masson (1995) yield $\dot{M}_{\rm w} \sim 10^{-7} M_\odot$ yr $^{-1}$ for $f_t \sim 3$, in agreement with mass-loss estimates from optical lines (Hartigan et al. 1994).

Finally, it is interesting to note that the jet radius r_j can be roughly estimated *independently* from η or other jet properties, provided V_{BS} and the ambient density ρ_o at the end of the flow lobe can be measured; replacing $\dot{M}_{\rm w}$ in Eqn. 4 by $\pi r_j^2 \rho_j V_j$, and $\rho_j V_j^2$ by $\rho_o V_{BS}^2/\varepsilon$ (definition of ε), one finds

$$\pi r_j^2 \sim \frac{P_{CO}}{R_{CO}} \frac{1}{\rho_o V_{BS}} f_t^{-1} \tag{8}$$

Applied to the Orion B outflow, this yields

$$r_j \sim 2 \times 10^{16} cm\ (30 kms^{-1}/V_{BS})^{0.5}, \tag{9}$$

similar to the maximum jet radius inferred from the outflow width $D \sim 0.1$ pc (see Sect. 3.2.1) and from the H_2 image (Richer et al. 1992).

5. Conclusions

Momentum-driven wind bubbles may be relevant to describe the collimation of YSO winds into jets, or the integral effect of many outflow episodes. However, several arguments show that young flows are most likely jet-driven.

Jet bowshock models provide a coherent framework that explains most general properties of molecular flows: apparent acceleration, the decreasing amount of mass with velocity, moderate collimation ($q \sim 2-10$), extended H_2 emission peaking downstream from CO, and discrete extremely-high velocity CO features. Viscous jet mixing-layers may well be present at the jet/cocoon interface, explaining the jet-like H_2 and SiO emission seen in some flows.

The bowshock hypothesis yields several constraints on jet properties. Both observed outflow widths and momentum balance would suggest jet radii that are large, typically $\sim 1/30$ of the flow length. This may imply free jet expansion, or the growth of a hot viscous mixing-layer, or rapid direction changes at the jet head. Mass-loss rates derived from CO may not need dramatic revision, unless the jet is much denser than the environment. A very efficient mechanism is still implied.

The structure of real flows is more complex than a single bowshock. Several of them contain multiple acceleration sites, indicating time variability in the jet velocity, ejection direction, or both (see, e.g., Bachiller et al. 1995). The advent of high-resolution maps, and comparison with more realistic numerical simulations (e.g. inclusion of molecular cooling, shock chemistry, ambient density gradients...) should soon give us a more detailed understanding of flow structure, and reveal unique clues to the mass-loss history of very young objects (including information on precession).

Finally, our improved understanding of outflow structure should allow related important questions to be addressed on a firmer basis. These include e.g. the interplay of infall and outflow motions in dense, magnetized circumstellar gas (e.g. Fiege & Henriksen 1995), and the impact of outflows on basic molecular cloud physics (e.g. time-dependent chemistry, grain destruction and/or re-formation, turbulence replenishment).

References

Bachiller, R., et al.: 1990, *Astron. Astrophys.* **231**, 174
Bachiller, R., Gómez-González, J.: 1992, *Astron. Astrophys. Rev.* **3**, 257
Bachiller, R., et al.: 1995, *Astron. Astrophys.* , in press
Barral, J.F., Cantó, J.: 1981, *Rev.Mex.Astr.Ap.* **5**, 101
Biro, S., Raga, A.C.: 1994, *Astrophys. J.* **434**, 221
Biro, S., Raga, A.C., Cantó, J.: 1995, *Mon. Not. R. astr. Soc.* , in press
Blondin, J.M., Fryxell, B.A., Königl, A.: 1990, *Astrophys. J.* **360**, 370
Cabrit, S.: 1989, in B. Reipurth, ed(s)., *Low Mass Star Formation and pre-Main Sequence Objects*, ESO: Garching, p.119
Cabrit, S., Bertout, C.: 1992, *Astron. Astrophys.* **261**, 274
Cabrit, S., Goldsmith, P.F., Snell, R.L.: 1988, *Astrophys. J.* **334**, 196
Cantó, J.: 1980, *Astron. Astrophys.* **86**, 327
Cantó, J., Raga, A.C.: 1991, *Astrophys. J.* **372**, 646
Cantó, J., Tenorio-Tagle, G., Rózyczka, M.: 1988, *Astron. Astrophys.* **192**, 287
Chernin, L.M., Masson, C.R.: 1995, *Astrophys. J.* , in press

Chernin, L.M., Masson, C.R., Gouveia dal Pino, E.M., Benz, W.: 1994, *Astrophys. J.* **426**, 204
Davis, C.J., Eislöffel, J.: 1995, *Astron. Astrophys.* , in press
DeYoung, D.S.: 1986, *Astrophys. J.* **307**, 62
Dyson, J.E.: 1984, *Astrophys. Sp. Sci.* **106**, 181
Dyson, J.E., Cantó, J., Rodriguez, L.F.: 1988, in L. Bianchi and R. Gilmozzi, ed(s)., *Mass Outflows from Stars and Galactic Nuclei*, Kluwer: Dordrecht, p. 299
Fiege, J.D., Henriksen, R.N.: 1995, *Astrophys. J.* , submitted
Gouveia dal Pino, E.M., Benz, W.: 1993, *Astrophys. J.* **410**, 686
Gredel, R., Reipurth, B.: 1993, *Astrophys. J.* **407**, L29
Guilloteau, S., Bachiller, R., Fuente, A., Lucas, R.: 1992, *Astron. Astrophys.* **265**, L49
Hartigan, P., Raymond, J., Meaburn, J.: 1990, *Astrophys. J.* **362**, 624
Hartigan, P., Morse, J., Raymond, J.: 1994, *Astrophys. J.* **436**, 125
Hollenbach, D., McKee, C.F.: 1989, *Astrophys. J.* **342**, 306
Königl, A.: 1982, *Astrophys. J.* **261**, 115
Kwok, S., Volk, K.: 1985, *Astrophys. J.* **299**, 191
Lada, C.J.: 1985, *Ann. Rev. Astron. Astrophys.* **23**, 267
Lizano, S., et al.: 1988, *Astrophys. J.* **328**, 763
Masson, C.R., Chernin, L.M.: 1992, *Astrophys. J.* **387**, L47
Masson, C.R., Chernin, L.M.: 1993, *Astrophys. J.* **414**, 230
Masson, C.R., Chernin, L.M.: 1994, D. Clemens and R. Barvainis, ed(s)., *Clouds, Cores, and Low Mass Stars*, PASP Conference Series: San Francisco
Masson, C.R., Mundy, L.G., Keene, J.: 1990, *Astrophys. J.* **357**, L25
Meyers-Rice, B.A., Lada, C.J.: 1991, *Astrophys. J.* **368**, 445
Moriarty-Schieven, G.H., Snell, R.L.: 1988, *Astrophys. J.* **332**, 364
Mundt, R., Brugel, E.W., Bührke, E.: 1987, *Astrophys. J.* **319**, 275
Padman, R., Richer, J.S.: 1994, *Astrophys. Sp. Sci.* **216**, 129
Raga, A.C.: 1988, *Astrophys. J.* **335**, 820
Raga, A.C.: 1991, *Astron. J.* **101**, 1472
Raga, A.C.: 1993, *Astrophys. Sp. Sci.* **208**, 163
Raga, A.C., Biro, S.: 1993, *Mon. Not. R. astr. Soc.* **264**, 758
Raga, A.C., Cabrit, S.: 1993, *Astron. Astrophys.* **278**, 267
Raga, A.C., Cantó, J., Binette, L., Calvet, N.: 1990, *Astrophys. J.* **364**, 601
Raga, A.C., Cantó, J., Biro, S.: 1993, *Mon. Not. R. astr. Soc.* **260**, 163
Raga, A.C. et al.: 1993, *Astron. Astrophys.* **276**, 539
Raga, A.C., Taylor, S.D., Cabrit, S., Biro, S.: 1995, *Astron. Astrophys.* **296**, 833
Richer, J.E., Hills, R.E., Padman, A.P.G.: 1992, *Mon. Not. R. astr. Soc.* **254**, 525
Sakashita, S., Hanami, H.: 1986, *Publ. Astron. Soc. Pac.* **38**, 879
Shu, F.H., Ruden, S.P., Lada, C.J., Lizano, S.: 1991, *Astrophys. J.* **370**, L31
Smith, M.D., Brand, P., Moorhouse, A.: 1991, *Mon. Not. R. astr. Soc.* **248**, 451
Stahler, S.: 1994, *Astrophys. J.* **422**, 616
Stone, J.M., Norman, M.L.: 1992, *Astrophys. J.* **389**, 297
Stone, J.M., Norman, M.L.: 1993a, *Astrophys. J.* **413**, 198
Stone, J.M., Norman, M.L.: 1993b, *Astrophys. J.* **413**, 210
Stone, J.M., Norman, M.L.: 1994, *Astrophys. J.* **420**, 237
Taylor, S., Raga, A.C.: 1995, *Astron. Astrophys.* **296**, 823
Weaver, R. et al.: 1977, *Astrophys. J.* **218**, 377
Wolfire, M.G., Königl, A.: 1991, *Astrophys. J.* **383**, 205

MULTIFLUID MODELS OF SHOCKS IN MAGNETIZED INTERSTELLAR MOLECULAR CLOUDS

T.W. HARTQUIST
Max-Planck-Institut für extraterrestrische Physik
D-85740 Garching bei München, Germany

Abstract. In a weakly ionized magnetized medium the dissipation in a shock moving at a speed sufficiently below the speed of high frequency hydromagnetic waves occurs on a lengthscale that is very large compared to any of the collision lengths. Chemistry affects the structures of these continuous or 'C-type' shocks which are sometimes unstable. In dark cloud shocks grain-neutral friction is an important dissipation process, and the adoption of a fluid description of grain dynamics elucidates various effects which include the onset of an $E \times B$ drift-driven runaway in some perpendicular shocks. The imperfect nature of the conductivity in the clouds may cause significant rotation of the magnetic field about the shock propagation direction in a dense cloud shock. Models of shocks in which continuous dissipation is important have been developed in studies of the origins of CH^+ in diffuse clouds, H_2 emission features with widths of around 100 km s^{-1} in Orion and OH masers around young stars, but CH^+ and such broad H_2 features may arise in boundary layers between molecular regions and fast flows.

1. Introduction

The responses of diffuse plasma in a galaxy to large scale gravitational perturbations (including density waves) and to momentum and energy sources (such as stellar winds and supernovae) determine the locations and nature of star formation. The magnetization but weak ionization of most of the diffuse plasma in the Milky Way lead single fluid hydrodynamic models of the responses of the plasma to be inadequate for many interesting circumstances. Thus, Mullan (1971), Draine (1980) and Draine, Roberge and Dalgarno (1983) presented a multifluid hydromagnetic formulism for the treatment of shocks in interstellar clouds.

In the next section, I review some of the different types of shocks that this formulism describes. Section 3 is an introduction to the effects of chemistry in interstellar molecular clouds on the structures of shocks in them. A magnetic Rayleigh-Taylor type of instability of importance in some of the shocks in interstellar clouds is described in section 4. Section 5 concerns the so-called 'switch-on' and 'switch-off' shocks. Sections 6 and 7 are about effects arising in some shocks in dark interstellar regions because grain-neutral drag contributes significantly to the dissipation and grains are important carriers of charge and current. Shock models developed to address the origin of interstellar CH^+ (considered by some to be the outstanding problem in interstellar chemistry), the source of broad H_2 emission features in star forming regions

Astrophysics and Space Science **233**: 97–109, 1995.

(including the Orion Kleinmann-Low Becklin-Neugebauer region in particular), and the production of large abundances of OH and CH_3OH in some maser regions are reviewed in section 8. Section 9 presents alternative explanations for the high CH^+ abundances and broad H_2 emission features.

2. Basic Types of Multifluid Hydromagnetic Shocks

For convenience, in this section I focus only on the so-called perpendicular shocks, those that propagate in a direction that is perpendicular to the upstream magnetic field. I assume that far upstream the plasma and magnetic fields are uniform. In this section the charged particles are taken to be electrons and ions of one species only. The electrons and ions are supposed to be well-coupled to the magnetic field.

In an interstellar cloud the thermal pressures of the ions and the electrons are negligible compared to the magnetic pressure. This implies that small amplitude MHD waves propagating perpendicular to the magnetic field with frequencies above roughly the frequency associated with the transfer of momentum of an ion to the neutrals (this frequency is given by the ion-neutral reduced mass divided into the product of the ion-neutral momentum transfer rate coefficient and the neutral density) have speeds very close to V_{Ai}, the Alfvén speed of a decoupled wave calculated on the assumption that only the ions and electrons provide inertia (i.e. that to lowest order the neutrals are not present).

In a diffuse cold interstellar cloud the total thermal pressure of ions, electrons and neutrals is not negligible compared to the magnetic pressure but also does not exceed it by more than a factor of a couple. The speeds of the small amplitude MHD waves propagating perpendicular to the magnetic field with frequencies much lower than the frequency associated with the transfer of the momentum of a neutral to the ions (This frequency is given by the ion-neutral reduced mass divided into the momentum transfer collision rate coefficient times the ion mass density and will be written as ν_{n}.) are of the order of V_{An}, the Alfvén speed of well-coupled waves calculated on the assumption that all ions, electrons and neutrals provide the inertia and have well-coupled motions. In a dark cold interstellar cloud the magnetic pressure is much greater than the total thermal pressure, and the speeds of the low amplitude coupled waves propagating perpendicular to the magnetic field are nearly V_{An}.

Perpendicular shocks propagating at speeds greater than V_{Ai} may be treated with single fluid MHD theory. The flows of the different species in such a shock possess discontinuities that are nearly cospatial.

Perpendicular shocks propagating at speeds less than V_{Ai} have extended dissipation regions. At each point in the dissipation region of a shock with a speed less than V_{Ai} the charged fluids move at velocities that are closer to the shock velocity than the velocity of the neutral fluid is. Ion-neutral friction arising from the difference in the velocities of the ions and the neutrals is a major source of dissipation. If the neutrals are heated sufficiently that (in a frame moving with the shock) they undergo a transition from supersonic to subsonic motion, a viscous subshock in which neutral-neutral collisions provide most of the viscosity often (but not always) develops in the neutral fluid. When the shock speed exceeds V_{Ai} or if it is less than V_{Ai} but a viscous subshock exists, the shock is said to be of the jump – or J-type (Draine 1980). For a sufficiently low shock speed radiative cooling prevents the dissipation from heating the neutral gas to a high enough temperature for it to undergo a supersonic to subsonic transition. Shocks in which such a transition does not occur contain no jumps in flow parameters and are said to be of the continuous – or C-type (Draine 1980). The maximum speed of a C-type shock exceeds V_{An} and is not much below V_{Ai} if the neutral gas radiates efficiently enough. The length of the dissipation zone in a C-type shock is determined by the rough equality of the gradient in the magnetic pressure to the frictional force per unit volume due to ion-neutral collisions and is given approximately by

$$L_D \simeq \frac{V_{An}^2}{2V_s\nu_n}, \tag{1}$$

as noted by Draine (1980) and where V_s is the shock speed.

There are shocks in which supersonic to subsonic transitions of the neutral flows take place via smooth passages through the respective sonic points (Chernoff 1987; Roberge and Draine 1990). These shocks are said to be of the C*-type and bear similarity to multicomponent stellar winds that pass smoothly through their sonic points. Models of perpendicular steady shocks in diffuse interstellar clouds show C*-type behaviour for $18\,\mathrm{km\,s^{-1}} < V_s < 30\,\mathrm{km\,s^{-1}}$ (Roberge and Draine 1990); hence, a shock of the C*-type is, perhaps, not just an oddity. Roberge and Draine (1990) have argued that the characteristic behaviour of the radiative cooling (which is negligible at low temperature but increases rapidly and then levels off somewhat as the temperature reachs and then exceeds thresholds associated with the excitation of key cooling features) is responsible for C*-type solutions existing for a nonnegligible range of steady shock model parameters.

In this review I will focus on C-type shocks in part because, as shown in section 8, C-type shock models have been studied in connection with a variety of observed sources. C*-type shock models have not been used in such applications while J-type shock models can often be based on single fluid treatments on which the scientific literature is extensive.

3. The Effects of Chemistry

ν_n depends on the number density of ions and L_D, the dissipation lengthscale in a C-type shock, is inversely proportional to ν_n (cf. eqn 1). Thus, the shock structure depends on the chemistry that determines the ionization structure.

In diffuse molecular clouds the most abundant ions are C^+, formed by photoionization, and H^+, formed as a consequence of cosmic ray induced ionization. At typical diffuse cloud temperatures of less than about 100 K, C^+ and H^+ do not react with H_2. However, in diffuse molecular cloud shocks with speeds of about $6 - 7\,\mathrm{km\,s^{-1}}$ and more the neutral temperature and ion-neutral relative speed are high enough for the reaction of

$$C^+ + H_2 \rightarrow CH^+ + H \quad (2)$$

to proceed. CH^+ reacts with H_2 which initiates a chemical sequence that produces CH_2^+, CH_3^+ and other species. Most molecular ions recombine with electrons at rates that are four to six orders of magnitude more rapid than the recombination rate of C^+ with electrons. Thus, diffuse cloud shocks that are fast enough for reaction (2) to be an important removal mechanism of C^+ (but are not fast enough for collisional dissociation of H_2 to be important) actually have higher net recombination rates per ion than do cold diffuse cloud regions. Consequently, in many diffuse cloud shocks the fractional ionization is significantly lower in the dissipation region than in the upstream gas, and the dissipation region is considerably longer than one would calculate by assuming that the fractional ionization remains constant (Flower, Pineau des Forêts and Hartquist 1985). The reaction of H^+ with H_2 to form H_2^+ does not proceed rapidly in shocks that may exist most commonly in diffuse clouds, and in the dissipation zones of some diffuse cloud shocks H^+ is probably the most abundant ion (e.g. Hartquist, Flower and Pineau des Forêts 1990).

In cold molecular clouds the most abundant ions are metallic ions including Mg^+ and Na^+ (Oppenheimer and Dalgarno 1974). They recombine primarily on grain surfaces, and I postpone further mention of the ionization structures in dense dark molecular cloud shocks until sections 6 and 7 in which the importance of grains for such shocks is treated more fully.

In diffuse molecular and dark dense molecular clouds the chemistry produces molecules that contribute significantly in many cases to the cooling by radiative losses (Draine, Roberge and Dalgarno 1983). The cooling of shocked neutral molecular cloud gas with temperatures above about 400 K is dominated by emission from H_2, OH and H_2O. Because OH and H_2O are formed rapidly in such regions by reactions that are slow in cold molecular cloud gas, the shocks in diffuse and dark molecular clouds have higher combined OH and H_2O fractional abundances than obtain in cold cloud gas.

4. An Instability

For computational simplicity most models of C-type shocks have been based on the assumption that the flow is steady. Wardle (1990; 1991a) considered the stability of perpendicular steady C-type shocks to small amplitude periodic perturbations with associated magnetic field disturbances parallel to the shock propagation direction. The ion-neutral friction retards the ion motion along the perturbation magnetic field component so that the ion density increases in the regions where the field bows in the upstream direction but decreases in the regions where the field bows in the downstream direction. The ions in the regions of increased density experience a higher ion-neutral drag per unit volume towards the upstream direction than the ions in the regions of decreased density; this tends to increase the amplitude of the bowing.The tendency to increase the amplitude of the bowing must overcome the tendency of ion-neutral friction to damp oscillatory perturbations if the amplitude increase actually is to occur. Wardle (1990) found that steady C-type shocks with $V_s/V_{An} > 5$ are unstable with respect to the type of perturbations that I have been describing and calculated linear growth rates of the perturbations.

The instability was also found to take place in oblique C-type shocks (Wardle 1991b), those C-type shocks that propagate at angles other than 0 or $\pi/2$ relative to the upstream magnetic field. The nonlinear development of the instability has been followed with a multidimensional, time-dependent code, and growth ceases when the density contrast relative to the immediately surrounding shocked gas is about unity (Tóth 1995).

5. Switch-On and Switch-Off Shocks

The single fluid MHD Rankine-Hugoniot relations admit several types of shocks (e.g. Kantrowitz and Petschek 1966; Kennel, Blandford and Coppi 1989). I consider an arbitrary propagation direction with respect to a uniform upstream magnetic field in a uniform upstream medium and denote the propagation speeds of the fast-mode, intermediate-mode (also referred to as the Alfvén-mode) and slow-mode low frequency, well-coupled, small amplitude waves as c_f, c_A and c_s respectively. The velocity regimes $u > c_f$, $c_f > u > c_A$, $c_A > u > c_s$ and $c_s > u$ will be referred to respectively as regimes 1, 2, 3 and 4 with u being the flow velocity measured in the frame comoving with the shock. Shocks with the upstream value of u in regime 1 and the downstream value of u in regime 2 are called fast-mode shocks; in one the component of the magnetic field perpendicular to the propagation direction increases through the shock and the magnetic field can be considered to be 'refracted' away from the shock propagation direction. 3→4 type shocks are slow-mode shocks; in one the component of the magnetic field perpendicular to the shock direction decreases through the shock and the magnetic field can be considered to be 'refracted' towards the shock propagation direction. In a shock the far downstream magnetic field lies in the plane of the propagation velocity and the far upstream magnetic field.

The single fluid MHD Rankine-Hugoniot relations show that in some fast-mode shocks propagating parallel to the upstream magnetic field, the field is refracted so that (though the upstream magnetic field has a vanishing component perpendicular to the propagation direction) the downstream field has a nonzero component perpendicular to the propagation direction. Such fast-mode shocks are called switch-on shocks and exist only when the twice the upstream magnetic pressure is greater than the upstream thermal pressure times the ratio of specific heats and only for a limited range of the ratio V_s/c_f between 1 and a value (which depends on the ratio of upstream pressures) of up to a few (e.g. Kantrowitz and Petschek 1966).

The single fluid MHD Rankine-Hugoniot relations show that a slow mode shock is a switch-off shock (one in which the component of the upstream magnetic field perpendicular to the shock propagation direction is finite but the component of the downstream magnetic field perpendicular to the shock propagation direction is zero) when the upstream thermal pressure times the ratio of the specific heats exceeds two times the upstream magnetic pressure and $V_s = c_A$.

As emphasized by Kennel et al. (1989) and by Kennel, Blandford and Wu (1990) in their studies of intermediate-mode single fluid MHD shocks (i.e. those in which the velocity changes are of the 1→3, 1→4, 2→3 and 2→4 types) the kinds of shocks that actually exist depend on the structures of the dissipation regions. Smith (1992) has argued that the ion-neutral frictional dissipation that occurs in C-type shocks is a dissipation mechanism that allows the existence of switch-on and switch-off shocks.

6. Grain-Neutral Friction in Perpendicular Shocks in Dark Clouds

Baker (1979) and Elmegreen (1979) argued that in dark interstellar clouds grain-neutral friction is of comparable or of more importance than ion-neutral friction in determining the coupling between the magnetic field and the neutral gas. Draine (1980) introduced an approximate treatment of grain-neutral friction in dark cloud shocks; this approximate treatment was used by him and his collaborators in later work (e.g. Draine, Roberge and Dalgarno 1983) and is adequate in many circumstances. However, in some cases the grain dynamics must be followed more thoroughly, and, as a consequence, Pilipp, Hartquist and Havnes (1990) included grains as a separate fluid in which each grain possesses properties identical to those of each other grain and carries the average charge per grain. Pilipp et al. (1990) focussed on perpendicular shocks in dark molecular regions.

Grains play an important role in establishing the ionization structure in such regions. They generally carry a negative average charge (because gas phase electrons have higher thermal speeds than do gas phase ions and, hence, a neutral grain collides with gas phase electrons more often than it does with gas phase ions) which (in regions where the gas phase electron number density is much higher than the product of the grain number density and the average number of negative charges carried by a grain) is proportional to the electron temperature (if the electron and ion temperatures are equal and the grain motion relative to the ion fluid is very subsonic). Ion-grain and electron-grain collisions are important means of removing ions and electrons from the gas phase as well as the processes in dark clouds that establish the average charge carried by a grain. Pilipp et al. (1990) included chemical equations to evaluate the gas phase ion number density and electron number density and calculated the average charge per grain.

The dynamics of the grains in a shock depend on the gas phase ionization structure and the number density, charges and masses of grains as well as on the magnetic field and the grain-neutral friction. In turn, the contribution of grain-neutral friction to dissipation in the shock and, hence, the shock structure depend on the grain dynamics. The work of Pilipp et al. (1990) produced more refined steady models of perpendicular dark cloud shocks than had existed before.

It also revealed a curious $E \times B$ drift-driven runaway effect that operates in perpendicular shocks in dark cloud regions where the number density is above about $10^7\,\mathrm{cm}^{-3}$. In such shocks the neutral drag on grains along the shock propagation direction is in some places balanced primarily by a charge separation generated electric field component along the shock propagation direction. The electric field component is inversely proportional to the average charge on a grain. Clearly, the $E \times B$ drift of ions perpendicular to the plane of the shock propagation direction and the upstream magnetic field is also inversely proportional to the average charge on a grain. In some shocks the electron and ion temperatures and ion-grain relative motions become sufficiently high that recombination onto grain surfaces becomes so rapid that the ion and electron number densities and the magnitude of the average grain charge drop by large factors in the shock. The drop in the magnitude of the average grain charge results in a large $E \times B$ ion drift speed which increases the recombination rate of ions in collisions with grains which decreases the magnitude of the average charge of the grains. That decrease in that magnitude increases the $E \times B$ drift speed even more and in some shocks drives a runaway which must saturate when the magnitude of the average grain charge approachs order unity. The effects of the runaway on the ionization balance and average grain charge affect the dissipation in the shock.

7. Grain-Neutral Friction in Oblique Shocks in Dark Clouds

If all of the ion, electron and grain fluids were perfectly conducting and quasineutrality obtained, the $E \times B$ drift in a medium would have no net current associated with it. However, the grain fluid and the ion fluid (to a lesser extent) in a dark cloud shock cannot be considered to be perfectly conducting; as a consequence there is an $E \times B$ drift current in such a shock. In an oblique dark cloud compressive disturbance, even if it is of small amplitude, the presence of the $E \times B$ drift current results in the coupling of a mode in which the perturbation in the magnetic field lies in the plane of the propagation direction and upstream magnetic field to a mode in which the perturbation in the magnetic field is perpendicular to that plane. This coupling results in the rotation of the magnetic field about the shock propagation direction. To understand the coupling better, consider a planar steady shock; because the time derivative is zero the equation of charge conservation requires that the component of the total current along the shock propagation direction is zero. However, the $E \times B$ drift current has a component along the propagation direction of an oblique shock; this component of the $E \times B$ drift current must be balanced by a current moving along the magnetic field, which has a component along the shock propagation direction. The current moving along the magnetic field has a component parallel to the magnetic field component perpendicular to the shock propagation direction; using Ampére's Law one finds that magnetic field rotation around the shock propagation direction follows.

Pilipp and Hartquist (1994) have shown that in each of many cases significant magnetic field rotation occurs in a steady oblique dark cloud shock having a magnetic perturbation that far upstream is in the plane of the unperturbed magnetic field and shock propagation direction. The single fluid MHD Rankine-Hugoniot relations require that the distant downstream magnetic field be in the plane of the shock propagation direction and distant upsteam magnetic field, but they allow (and indeed require) the component of the field perpendicular to the propagation direction to undergo a sign change when the shock is of an intermediate-mode type. For a limited range of shock speeds Pilipp and Hartquist (1994) were able to construct steady models of shocks in each of which the magnetic field undergoes a rotation through π radians about the shock propagation direction. They failed to find (but did not rigorously demonstrate the nonexistence of) steady solutions for any shocks having speeds much above the Alfvén speed and in which the magnetic field undergoes no net rotation around the shock propagation direction.

Rather they found that rotation began to occur and that it led in each model to the component of the velocity parallel to the shock propagation direction of at least one of the charged fluids going to zero well before the neutral gas was decelerated (in the shock frame). Much further work is required, but the question of whether any steady dark cloud oblique fast-mode shock solutions exist does arise.

8. Applications

Steady C-type shock models have been developed for several applications.

Models of cold diffuse cloud chemistry fail by large factors to reproduce the CH^+ abundances measured in absorption against bright background stars. Partly because its features were amongst the earliest ones formed by interstellar molecules to be detected and partly because of its extreme simplicity, CH^+ is considered by many molecular astrophysicists to present the most pressing problem in molecular astrophysics. The failure to explain the existence of high abundances of such a simple molecule has long been taken to imply that some key dynamical property of diffuse interstellar clouds has been neglected or improperly understood. Elitzur and Watson (1978) first suggested that shocks, in which the endothermic reaction (2) proceeds far more rapidly than in cold clouds, are the sites of diffuse cloud CH^+. Their models were of nonmagnetic single fluid shocks and the chemistry within them. The production of CH^+ in shocks is constrained by the measured OH and excited H_2 rotational level abundances, and it was clear that single fluid models could not meet the constraints while giving rise to sufficient CH^+ production (Hartquist and Dalgarno 1982). Because OH production is low and the excitation of the $J = 5$ and 6 rotational levels of H_2 is slow unless the neutral temperature is high, the possibility that the $C^+ + H_2$ reaction could be driven by ion-neutral streaming in C-type shocks while the neutral gas remained at lower temperatures than in J-type shocks stimulated interest in the exploration of C-type shocks as the sources of CH^+ (Draine 1986; Draine and Katz 1986a,b; Pineau des Forêts et al. 1986). However, even with favourable assumptions if the correct photodissociation products of CH^+ are used the shock models fail to reproduce the observed CH^+ abundances while meeting the constraints mentioned above and the additional constraints provided by the velocity data (Hartquist et al. 1990).

C-type shock models of the H_2 infrared emission region in the Orion Kleinmann-Low, Becklin-Neugebauer region have been developed. The early such models (Chernoff, Hollenbach and McKee 1982; Draine and Roberge 1982) were of a shock driven by the expansion of a stellar wind-blown bubble into an ambient medium with a number density of the order of $10^6\,cm^{-3}$ and a magnetic field strength of the order of 10^{-3} Gauss. A shock speed of almost 40 km s^{-1} was selected by each group as that giving the best model fit to some of the data then in existence. However, the wings of the H_2 lines are far too wide to be compatible with a C-type shock propagating into an ambient dark molecular cloud medium. Chernoff et al. (1982) suggested that the broad wings might form in a dissociating J-type shock through which the stellar wind passes as it approachs the edge of the bubble; in this picture molecular formation occurs after postshock cooling occurs. Smith, Brand and Moorehouse (1991) have noted that the magnetosonic speed in a stellar wind might be sufficiently high that the bow shocks formed in the wind as it passes around clumps embedded in the wind-blown bubble may be C-type shocks with high enough speeds to account for the observed broad H_2 wings. Of course, in the Smith et al. (1991) scenario the wind must be molecular. Brand (1995) has reviewed work in this area.

C-type shocks probably play a role in the formation of OH and CH_3OH masers around young, bright stars. Around W3(OH) the OH and CH_3OH maser spots are spatially coincident (Menten et al. 1992), a fact that provides insight into the chemical-dynamical means of their formation. Hartquist et al. (1995) have suggested that a very slow (few km s^{-1}) C-type shock driven into dense ambient molecular material by the HII region around the star associated with W3(OH) causes the injection into the gas phase of H_2O and CH_3OH ices previously on the grain surfaces. (A calculation of the velocity of a fluid of small grains relative to the velocity of a fluid of larger grains in a low velocity C-type shock remains to be performed to demonstrate whether such relative motion drives grain-grain collisions that efficiently sputter ices. Alternatively, the liberation of H_2O and CH_3OH from grain surfaces may be due to heating by the absorption of infrared radiation emitted by dust nearer to the star.) As the gas approaches the HII region H_2O is photodissociated to form OH (as in the model proposed by Elitzur and de Jong 1978). The abundance of CH_3OH remains high where the OH abundance is high.

9. Boundary Layer Alternatives

That dynamic, heated regions are the sites of CH^+ formation receives support from the observational result that the detected CH features with which no detected CH^+ features are associated are significantly narrower than the detected CH features that have associated detected CH^+ features (Crane, Lambert and Sheffer 1995). Because shock models for CH^+ formation face difficulties, Duley et al. (1992) and Hartquist, Dyson and Williams (1992) have considered the possibility that CH^+ is formed in turbulent boundary layers between molecular clouds and hot interstellar gas. (Alex Dalgarno suggested to me in the middle 1970s that I consider cloud-hot gas interfaces as possible sites of CH^+ formation, but at that time reasonably quiescent conduction dominated boundary layers in which the gas is primarily atomic were the subject of our focus. The rates for the removal of CH^+ in reactions with H and with electrons and by photodissociation were very uncertain at that time, and the hope was that in somewhat heated gas all three removal mechanisms would be sufficiently slow for CH^+ to be abundant.) H_2 will reach parts of the outer boundary of pressure-bound molecular cloud because pressure differences existing along the boundary due to the flow of gas past it will result in cloud evaporation towards the lowest pressured parts of the boundary. There the dissipation of shear-driven turbulence and waves will heat molecular gas to several thousand degrees, resulting in reaction (2) being an efficient source of CH^+. In part because boundary layers are at about the same pressure as diffuse molecular clouds rather than being markedly overpressured with respect to them (as gas in a shock is) observational constraints that are problematic for shock models of CH^+ formation can apparently be met in the boundary layer models of CH^+ formation (Duley et al. 1992). Hartquist et al. (1992) have pointed out that the narrow component-broad component structure of an observed CH^+ feature (Crane, Hegyi and Lambert 1991) is consistent with the boundary layer model.

Malone, Hartquist and Dyson (1994) and Dyson et al. (1995) have considered the possibility that boundary layers are the sites of formation of the broad H_2 features originating in Orion and other sources. Their proposal is that the viscous couple (resulting from the dissipation of waves and turbulence driven by shear) between the wind of a young stellar object and a molecular clump will accelerate H_2 gradually enough that it does not pass through a dissociating shock. They have developed a two dimensional steady flow model of the dynamics of molecular gas being entrained by the wind. The rates at which mass, momentum and energy are transferred from the fast stellar wind to the molecular gas are taken as unknowns but satisfy basic inequalities which ensure that mass is not added without a minimal corresponding addition of momentum and that momentum is not transferred without a minimal transfer of energy. While in the most windward region of the clump the pressure must be roughly equal to the wind ram pressure, the pressure distribution elsewhere as well as the ratio of magnetic pressure to thermal pressure are assumed to be unknowns. Various combinations of values for the free parameters can be assumed in order to obtain model H_2 emission profiles similar to those observed. The aim is to obtain from the spectral data a knowledge of properties of a magnetized boundary layer, a task which is intractable unless pursued in a semiempirical fashion. As discussed by Brand (1995) and by Tedds (1995) recent high spatial and spectral resolution H_2 data for an individual clump in the Orion KL-BN H_2 emission region have become available; their existence will facilitate the attainment of this aim.

The proposal and exploration of boundary layer models of CH^+ absorption regions and of H_2 infrared emission regions should encourage all of us to hesitate briefly before attributing every unexplained observation in molecular astrophysics to shocks. The challenge of proving the presence of a shock should make each of many problems considerably more interesting.

References

Baker, P.L.: 1979, *Astron. Astrophys.*, **75**, 54.
Brand, P.W.J.L.: 1995, this volume.
Chernoff, D.F.: 1987, *Astrophys. J.*, **312**, 143.
Chernoff, D.F., Hollenbach, D.J. and McKee, C.F.: 1982, *Astrophys. J.*, **334**, L103.
Crane, P., Hegyi, D.J. and Lambert, D.L.: 1991, *Astrophys. J.* ,**378**, 181.
Crane, P., Lambert, D.L. and Sheffer, Y.: 1995, *Astrophys. J.*, in press.
Draine, B.T.: 1980, *Astrophys. J.*, **241**, 1021.
Draine, B.T.: 1986, *Astrophys. J.*, **310**, 408.
Draine, B.T. and Katz, N.S.: 1986a, *Astrophys. J.*, **306**, 655.
Draine, B.T. and Katz, N.S.: 1986b, *Astrophys. J.*, **310**, 392.
Draine, B.T. and Roberge, W.G.: 1982, *Astrophys. J.*, **259**, L91.
Draine, B.T., Roberge, W.G. and Dalgarno, A.: 1983, *Astrophys. J.*, **264**, 485.
Duley, W.W., Hartquist, T.W., Sternberg, A., Wagenblast, R. and Williams, D.A.: 1992, *Mon. Not. R. astr. Soc.*, **255**, 463.
Dyson, J.E., Hartquist, T.W., Malone, M.T. and Taylor, S.D.: 1995, *Rev. Mex. Astron. Astrofis.*, in press.
Elitzur, M. and de Jong, T.: 1978, *Astron. Astrophys.*, **67**, 323.
Elitzur, M. and Watson, W.D.: 1978, *Astrophys. J.*, **222**, L141.
Elmegreen, B.G.: 1979, *Astrophys. J.*, **232**, 729.
Flower, D.R., Pineau des Forêts, G. and Hartquist, T.W.: 1985, *Mon. Not. R. astr. Soc.*, **216**, 775.
Hartquist, T.W. and Dalgarno, A.: 1982, 'Galactic and Extragalactic Infrared Spectroscopy, SP-192' in M.J. Kessler, J.P. Phillips and T.D. Guyenne, ed(s)., *European Space Agency*, Paris, 29
Hartquist, T.W., Dyson, J.E. and Williams, D.A.: 1992, *Mon. Not. R. astr. Soc.*, **257**, 419.
Hartquist, T.W., Flower, D.R. and Pineau des Forêts, G.: 1990, 'Molecular Astrophysics - A Volume Honouring Alexander Dalgarno' in T.W. Hartquist, ed(s)., *Cambridge University Press*, Cambridge, 99
Hartquist, T.W., Menten, K.M., Lepp, S. and Dalgarno, A.: 1995, *Mon. Not. R. astr. Soc.*, **272**, 184.
Kantrowitz, A. and Petschek, H.E.: 1966, 'Plasma Physics in Theory and Application' in W.B. Kunkel, ed(s)., *McGraw-Hill Book Company*, New York, 147
Kennel, C.F., Blandford, R.D. and Coppi, P.: 1989, *J. Plasma Phys.*, **42**, 299.
Kennel, C.F., Blandford, R.D. and Wu, C.C.: 1990, *Phys. Fluids B*, **2**, 253.
Malone, M.T., Dyson J.E. and Hartquist, T.W.: 1994, *Astrophys. Sp. Sci.*, **216**, 143.
Menten, K.M., Reid, M.J., Pratap, P., Moran, J.M. and Wilson, T.L.: 1992, *Astrophys. J.*, **401**, L39.
Mullan, D.J.: 1971, *Mon. Not. R. astr. Soc.*, **153**, 145.
Oppenheimer, M. and Dalgarno, A.: 1974, *Astrophys. J.*, **192**, 29.
Pilipp, W. and Hartquist, T.W.: 1994, *Mon. Not. R. astr. Soc.*, **267**, 801.
Pilipp, W., Hartquist, T.W. and Havnes, O.: 1990, *Mon. Not. R. astr. Soc.*, **243**, 685.
Pineau des Forêts, G., Flower, D.R., Hartquist, T.W. and Dalgarno, A.: 1986, *Mon. Not. R. astr. Soc.*, **220**, 801.
Roberge, W.G. and Draine, B.T.: 1990, *Astrophys. J.*, **350**, 700.

Smith, M.D.: 1992, *Astrophys. J.*, **390**, 447.
Smith, M.D., Brand, P.W.J.L. and Moorehouse, A.: 1991, *Mon. Not. R. astr. Soc.*, **248**, 730.
Tedds, J.A. et al.: 1995, this volume.
Tóth, G.: 1995, *Mon. Not. R. astr. Soc.*, **274**, 1002.
Wardle, M.: 1990, *Mon. Not. R. astr. Soc.*, **246**, 98.
Wardle, M.: 1991a, *Mon. Not. R. astr. Soc.*, **250**, 523.
Wardle, M.: 1991b, *Mon. Not. R. astr. Soc.*, **251**, 119.

GRAIN DESTRUCTION IN INTERSTELLAR SHOCK WAVES

B.T. DRAINE
Princeton University Observatory, Peyton Hall, Princeton, NJ 08544, USA
draine@astro.princeton.edu

Abstract. Interstellar shock waves can erode and destroy grains present in the shocked gas, primarily as the result of sputtering and grain-grain collisions. Uncertainties in current estimates of sputtering yields are reviewed. Results are presented for the simple case of sputtering of fast grains being stopped in cold gas. An upper limit is derived for sputtering of refractory grains in C-type MHD shocks: shock speeds $v_s \gtrsim 50\,\mathrm{km\,s^{-1}}$ are required for return of more than 30% of the silicate to the gas phase. Sputtering can also be important for removing molecular ice mantles from grains in two-fluid MHD shock waves in molecular gas. Recent estimates of refractory grain lifetimes against destruction in shock waves are summarized, and the implications of these short lifetimes are discussed.

Key words: Dust Destruction – Shock Waves – Sputtering

1. Introduction

A shock wave passing through interstellar gas may create a sufficiently hostile environment that grains present in the preshock gas may be appreciably altered or destroyed. This is important for several reasons:

1. Many elements (e.g., Ca, Fe, Si) are normally found to be substantially depleted from the interstellar gas – locked up in interstellar grains. Return of this material to the gas can alter the gas phase composition, possibly resulting in significant changes in the rate of radiative cooling, or in the strengths of diagnostic emission or absorption lines.
2. In very fast ($v_s \gtrsim 500\,\mathrm{km\,s^{-1}}$) shocks, collisions of electrons with the grains may dominate the cooling of the gas until the grains are substantially destroyed (Ostriker & Silk 1973; Draine 1981). Detailed understanding of the cooling of the shocked gas then hinges upon understanding what happens to the grains.
3. The cool dense gas behind a radiative shock wave may provide conditions suitable for molecule formation, but this will depend on whether grains survive to catalyze the formation of H_2. Grain catalysis of H_2 can be important as a source of heat (the energy released in H_2 formation), or as a source of molecules which may be significant coolants or useful diagnostics.
4. The grains may be dynamically important: charged grains may dominate the coupling of magnetic fields to the gas when the fractional ionization is low, and fast grains may even account for a significant fraction of the energy and momentum flux under some conditions (McKee et al. 1987).

Astrophysics and Space Science **233**: 111–123, 1995.

5. Finally, shock-processing plays a major role in the overall evolution of interstellar dust! Our best estimates of the effects of interstellar shocks point to refractory grain lifetimes considerably shorter than the time scale for recycling the interstellar medium through stars; these estimates are difficult to reconcile with the observed extreme depletions of elements such as Fe and Ca.

Some contemporary views on the nature of the interstellar grain population are described in §2. Sputtering is thought to be the dominant grain destruction mechanism; current estimates for sputtering yields are reviewed in §3. The sputtering of grains being decelerated in cold gas is considered in §4, and in §5 the fate of grains in realistic shock models is discussed, including the sputtering of grains in two-fluid C-type shocks. The implications for our understanding of interstellar grain evolution are discussed in §6; a brief summary is provided in §7.

2. What are Interstellar Grains?

Any discussion of the effects of interstellar shocks on grains must begin with some assumptions about the properties of these grains – their size distribution, structure, composition, and overall numbers. Unfortunately, we remain quite uncertain on many of these points. Current grain models are discussed in several recent reviews (Mathis 1993; Draine 1995).

Models for interstellar grains are constrained to reproduce the observed extinction, scattering, polarization of starlight, and infrared emission, while not consuming elements in excess of what is allowed by cosmic abundances (minus the abundances observed to be present in the gas). These constraints are not sufficient to uniquely determine the model, and a number of different grain models are currently under consideration.

All models agree that:

- Approximately $^2/_3$ of the grain mass is in sizes $a > 0.05\mu$m – these are required to explain the observed extinction, polarization, and scattering at visible wavelengths.
- $\sim 20\%$ of the grain mass is in very small grains with $a < 0.025\mu$m, these being required to explain the steep rise in ultraviolet extinction.
- Most of the interstellar Si is locked up in silicate grains – these are required to explain the strong 10 and 18μm absorption/emission features.
- Most of the available Mg, Si, Fe is in grains, as well as a substantial ($^1/_3$–$^2/_3$?) fraction of the C, consistent with observed interstellar gas-phase abundances – see Figure 1.

There are substantial differences among the different grain models, some of which are as follows:

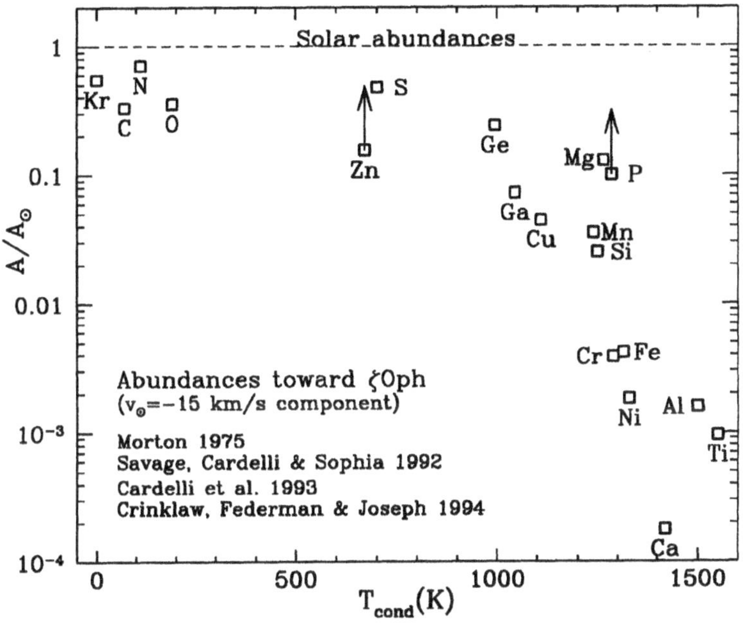

Fig. 1. Gas-phase abundances, relative to solar, toward ζ Oph, plotted against "condensation temperature" (see Field 1974).

1. A mixture of bare graphite and and bare silicate grains, with approximately power-law size distributions extending from $0.005 < a < 0.25\mu m$ (Mathis, Rumpl, & Nordsieck 1977; Draine & Lee 1984), supplemented by a population of ultrasmall grains (Draine & Anderson 1985) including polycyclic aromatic hydrocarbon (PAH) molecules (Léger and Puget 1984).
2. Silicate cores with a carbonaceous coating, plus a poulation of small bare graphite grains, plus PAH molecules (Désert, Boulanger, & Puget 1990).
3. Silicate cores with hydrogenated amorphous carbon coatings (Duley, Jones & Williams 1989).
4. Small pieces of graphite, silicate, and amorphous carbon – some as individual small grains, but mostly coagulated into mixed fluffy structures (Mathis & Whiffen 1989).

3. Grain Destruction Mechanisms

3.1. Sputtering

Sputtering appears to be the dominant grain destruction mechanism in the galaxy. Sputtering rates depend on the sputtering yield $Y(E, \theta)$, the average number of sputtered atoms (or molecules) per incident atom or ion. Refractory materials are believed to sputter one atom at a time; molecular ices (e.g., H_2O) will sputter molecule by molecule. Sputtering yields depend on the projectile type, energy, and incidence angle, and on the target composition. Experimental sputtering data have been reviewed by Andersen & Bay (1981) and Betz & Wehner (1983). Yields for H and He incident on candidate grain materials have been estimated by Barlow (1978a), Draine & Salpeter (1979a, hereafter DS79), and Tielens et al. (1994, hereafter TMSH94).

The threshold energy E_T is a critical parameter. Barlow found that $E_T \approx 4U_0$ for a wide range of projectile-target combinations; DS79 extended this to

$$E_T = \max\left(4U_0, \frac{(m_P + m_T)^2}{4m_P m_T} U_0\right) \tag{1}$$

to allow for inefficient energy transfer for large or small values of the projectile/target mass ratio m_P/m_T.

For $E > E_T$, Barlow took the sputtering yield to be a linear function of energy; DS79 and TMSH94 adopted more complicated fits, allowing for the decrease in sputtering yields at high energies, based upon the theory of Sigmund (1969; 1981). We recommend using either the DS79 or TMSH94 yields.

Estimated sputtering yields for H and He incident on solid carbon are shown in Figure 2. The estimate of DS79 uses the binding energy $U_0 = 7.37$eV appropriate for graphite; TMSH94 have noted that the surface of a graphite grain undergoing sputtering will be damaged, and adopt an effective binding energy $U_0 = 4$eV. The differences at low energy between the DS79 and TMSH94 estimates for carbon arise from this difference in adopted binding energy. No reliable data appears to be available for silicates; as a similar material, we show in Figure 3 theoretical and experimental results for SiO_2. In this case the DS79 and TMSH94 estimates are quite similar. Figure 4 shows predicted yields for $MgFeSiO_4$, with an average binding energy per atom $U_0 = 5.65$eV. Once again, the DS79 and TMSH94 estimates are very similar.

The above yields are for normal-incidence monoenergetic particles, and it is necessary to average over incidence angle and energy. For a neutral spherical target moving with drift velocity v_d, the rate of sputtering due to a projectile species with density n and mass m is (DS79)

$$\frac{dN}{dt} = \pi a^2 n \left(\frac{8kT}{\pi m}\right)^{1/2} \int_0^\infty dx\, x^2 e^{-x^2} \sinh(2sx) \langle Y(E = x^2 kT)\rangle_\theta \tag{2}$$

$$s^2 \equiv m v_d^2 / 2kT \tag{3}$$

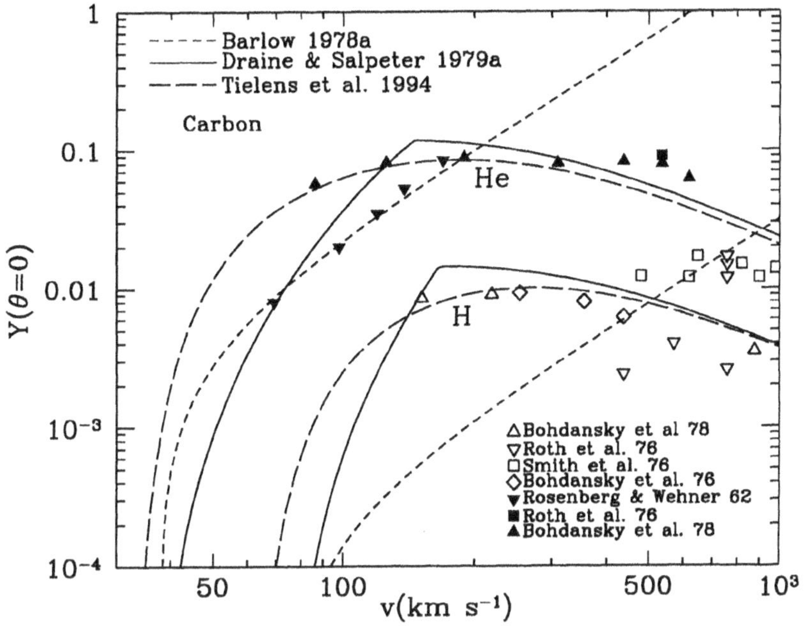

Fig. 2. Normal-incidence sputtering yield for H and He on solid carbon, vs. projectile velocity. Curves are theoretical estimates; symbols are experimental measurements.

where the angle-averaged sputtering yield

$$\langle Y(E)\rangle_\theta \equiv 2\int_0^{\pi/2} Y(E,\theta)\sin\theta\cos\theta d\theta \approx 2Y(E,\theta=0) \quad . \tag{4}$$

Sputtering rates should be summed over all important projectile species (H and He generally dominate for cosmic abundances). The effects of grain charging are readily included (DS79).

3.2. Grain-Grain Collisions

Grain-grain collisions at relative speeds exceeding a few $\mathrm{km\,s^{-1}}$ can destroy grains. The physics of grain-grain collisions has recently been discussed by TMSH94. For equal-sized refractory grains colliding at $v_s = 20\,\mathrm{km\,s^{-1}}$, they estimate $\sim 50\%$ vaporization.

3.3. Chemisputtering

Carbon grains may be subject to chemical attack by impinging H and O atoms. Barlow & Silk (1977) and Barlow (1978b) concluded that such chemisputtering would rapidly destroy carbon grains in HI or HII regions. However,

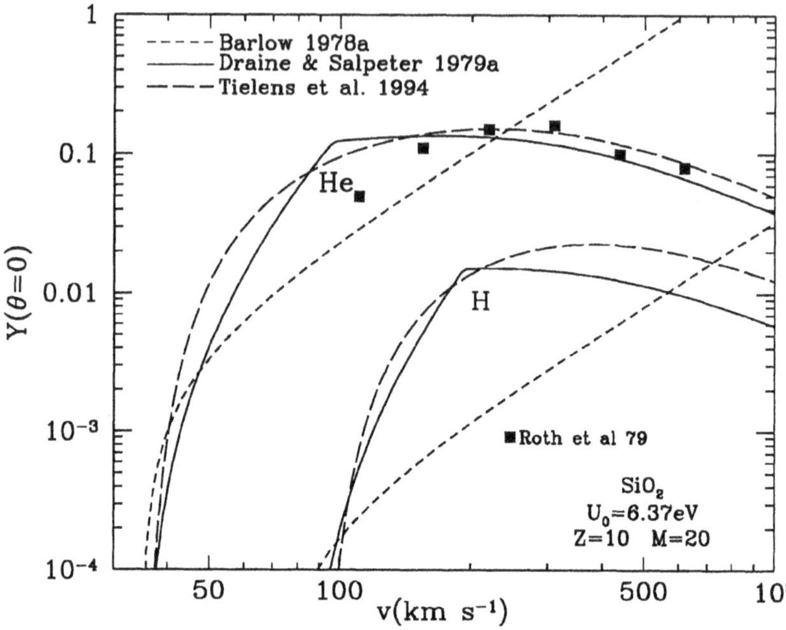

Fig. 3. Normal-incidence sputtering yields for H and He on SiO_2. Experimental data of Roth, Bohdansky, & Ottenberger (1979) is taken from Betz & Wehner (1983).

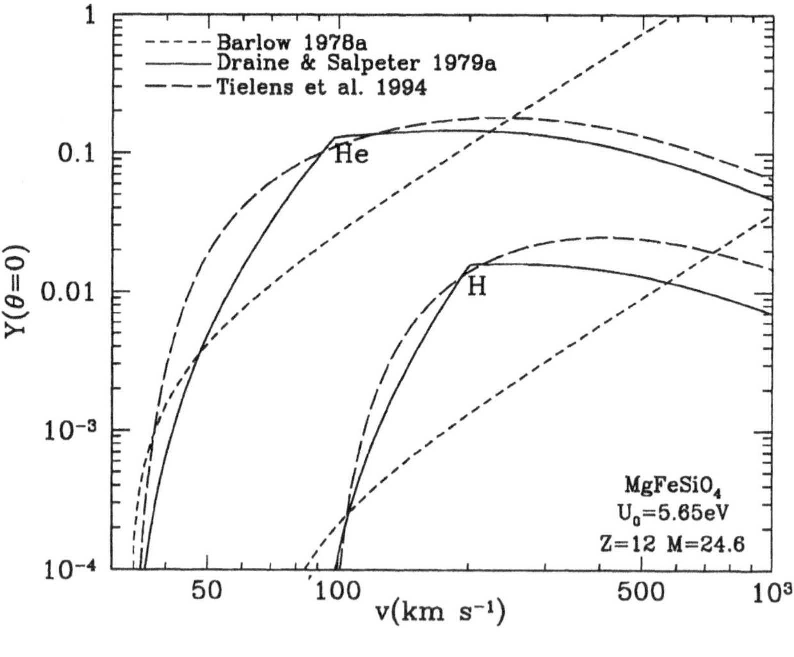

Fig. 4. Normal-incidence sputtering yields for H and He on $MgFeSiO_4$.

it appears that *removal* of carbon atoms by impinging H and O is thermally-activated, and is very inefficient at the low temperatures of interstellar grains (Draine 1979), so this does not appear to be an important destruction mechanism. Impinging hydrogen may, however, hydrogenate the carbon (cf. Furton & Witt 1993).

4. Sputtering of Hypersonic Grains

Sputtering of grains in single-fluid shocks has been discussed by many authors, including Shull (1977,1978), Cowie (1978), Barlow (1978a), Draine & Salpeter (1979b), Seab & Shull (1983), McKee et al. (1987), and Jones et al. (1994). Following Cowie (1978), it is instructive to consider a grain with an initial velocity v_0 in cold gas. It will decelerate by gas drag, and, if moving rapidly enough, will be subject to sputtering. If the motion is hypersonic, and He/H=0.1, then it is straightforward to show that the grain radius a and velocity v will vary as

$$4\pi\rho a^2 \frac{da}{dt} = -\pi a^2 v n_{\rm H} Y_{\rm eff}(v) m_T \tag{5}$$

$$\frac{4\pi}{3}\rho a^3 \frac{dv}{dt} = -\pi a^2 n_{\rm H} 1.4 m_{\rm H} v^2 \tag{6}$$

$$Y_{\rm eff}(v) \equiv \langle Y_{\rm H}(v)\rangle_\theta + 0.1\langle Y_{\rm He}(v)\rangle_\theta \tag{7}$$

where $\langle Y_{\rm H}(v)\rangle_\theta$ and $\langle Y_{\rm He}(v)\rangle_\theta$ are the angle-averaged yields for H and He with velocity v. Dividing (5) by (6) we obtain

$$d\ln a^3 = \left(\frac{m_T}{1.4 m_{\rm H}}\right) Y_{\rm eff}(v)\, d\ln v \tag{8}$$

This equation has been integrated numerically to compute the fractional volume sputtered in slowing a grain down from some initial velocity v_0; the results are shown in Fig. 5. It should be recognized that this is neither an upper nor a lower bound to the sputtering in an actual shock. Charged grains may be decelerated by plasma drag in addition to direct collisions, reducing the sputtering. On the other hand, thermal motions of the atoms and ions will tend to increase sputtering rates, and betatron acceleration can accelerate the grains (see below).

5. Shock Structure and Grain Dynamics

The structure of interstellar shocks is complicated by many issues, including collisionless processes, magnetic stresses, the possible acceleration of cosmic

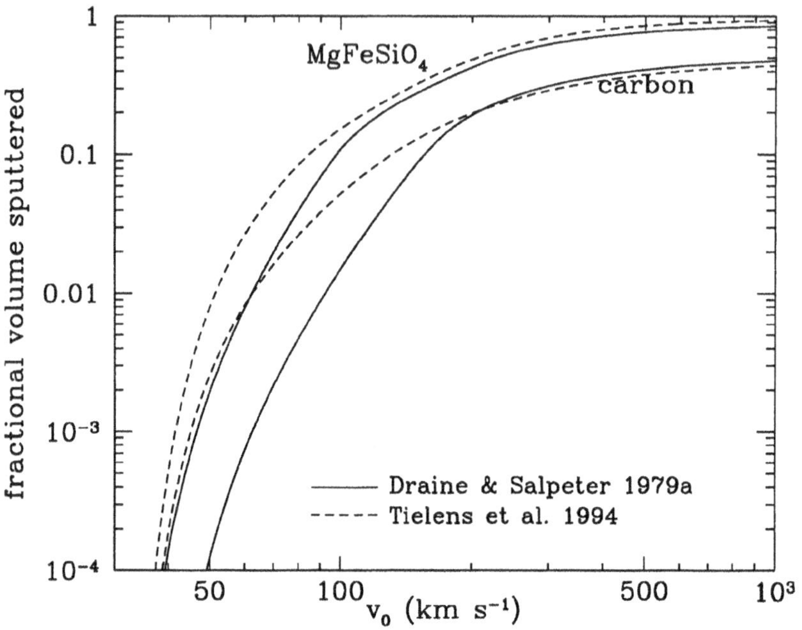

Fig. 5. Fractional volume sputtered from a grain, with initial velocity v_0, when stopped by cold gas.

rays to dynamically significant pressures, multifluid behaviour, and shock instabilities; these have been reviewed recently elsewhere (Draine & McKee 1993). Here we confine discussion to two simple cases.

5.1. Fast Single-Fluid Shocks

When the fractional ionization is sufficiently high, ions, neutrals, and magnetic field are well-coupled and flow together as a single fluid – this is the case in shocks in ionized gas and in HI clouds. In a strong shock, the gas is compressed by a factor of 4 at the jump front. In the "shock frame", the gas is decelerated to $v_s/4$, but the grains cross the jump front with their preshock velocity v_s, and hence have a velocity $0.75v_s$ relative to the gas. If a transverse magnetic field is present, the (charged) grains will begin to orbit around field lines with an orbital velocity $0.75v_s$. Calculations of grain erosion by sputtering must take into account these substantial gas-grain velocities, particularly with regard to sputtering by helium. Gas drag will tend to reduce the orbital velocity, but if the gas cools and the magnetic field lines are compressed on a time short compared to the time scale for gas drag to slow the grain (but on a time scale long compared to the grain gyroperiod), then Spitzer (1977) pointed out that the grains can actually be *accelerated* by the "betatron" mechanism.

Detailed calculations of grain destruction in shocks, including betatron acceleration, have been carried out (Draine & Salpeter 1979b; Seab & Shull 1983; McKee et al. 1987; Jones et al. 1994); these studies generally find substantial ($> 30\%$) sputtering of refractory grains in radiative shocks with $v_s \gtrsim 200\,\mathrm{km\,s^{-1}}$. The most recent work by Jones et al. finds sputtering to be more important than grain-grain collisions for $v_s \gtrsim 60\,\mathrm{km\,s^{-1}}$, with grain-grain collisions destroying a few percent of the grain volume at $v_s = 50\,\mathrm{km\,s^{-1}}$.

5.2. C-Type Shocks in Molecular Gas

When the fractional ionization is sufficiently low, shocks in magnetized gas develop a two-fluid structure, since the charged particles are coupled to the magnetic field, but the neutral particles "know" about the magnetic field only via infrequent collisions with the ions. The resulting shock is often "C-type" (Draine 1980), with large streaming velocities between the ions and the neutrals. Because the dust grains are charged, they are strongly affected by the magnetic field, which causes the dust grains to move through the neutral gas with a drift velocity which can approach the ion-neutral streaming velocity. This drift through the neutrals can lead to sputtering of molecular ices (Draine, Roberge & Dalgarno 1983; Flower & Pineau des Forêts 1994) and possibly even refractory grains (Flower & Pineau des Forêts 1995). Flower, Pineau des Forêts, & Walmsley (1995) have proposed that hot NH_3 observed toward Sgr B2 was injected into the gas by sputtering of ice mantles in a $v_s \approx 25\,\mathrm{km\,s^{-1}}$ shock.

In these C-type shocks the gas remains relatively cold, and sputtering will be determined largely by the motion of the grain through the gas. We have seen above the sputtering which occurs when a grain is stopped in cold gas (see Fig. 5); we see that even for an initial gas-grain velocity of $50\,\mathrm{km\,s^{-1}}$ only $\sim 0.2\%$ of a silicate grain will be eroded (using the DS79 sputtering yields). However, in an MHD shock the grains will be driven through the gas by the magnetic field. What happens here is that in the "shock frame" the grains are rapidly decelerated by magnetic forces; the neutral gas continues to stream until it is brought to rest by collisions with ions or charged grains. If the ionization is low, then collisions with grains may dominate, and each grain has to collide with many times its own mass of gas, leading to greater sputtering than would occur in the absence of magnetic field (in which case the results shown in Fig. 5 would apply).

For molecular ices, models of MHD shocks in dense molecular gas lead to erosion of $\sim 10\%$ of water ice mantles at shock speeds $v_s \approx 20\,\mathrm{km\,s^{-1}}$ (Draine, Roberge, & Dalgarno 1983). Flower & Pineau des Forêts (1994) found that shock speeds as small as $10\,\mathrm{km\,s^{-1}}$ were sufficient to fully erode ice mantles, but they assumed a sputtering threshold $E_T = 0.52$eV; for H_2O ice, with $U_0 = 0.53$eV, the sputtering threshold should be about 4 times larger [see eq. (1)] than the value adopted by Flower & Pineau des Forêts.

We may obtain an upper limit to magnetic-field-driven sputtering in C-type shocks by a fairly simple argument. Suppose we have grains of a single radius a. The grain erosion rate is given by

$$4\pi a^2 v_g \frac{da}{dx} = -\pi a^2 v_{\rm rel} n_{\rm H} Y_{\rm eff}(v_{\rm rel}) m_T \tag{9}$$

where v_g is the grain velocity and $v_{\rm rel} = v_n - v_g$ is the streaming velocity between the neutrals and the grains. In the shock frame, the momentum conservation equation for the cold neutral gas is

$$\frac{d}{dx}(\rho_n v_n^2) = F_{ni} + F_{ng} \tag{10}$$

where ρ_n and v_n are the density and velocity of the neutral gas, and F_{ni} and F_{ng} are the force per volume on the neutral gas due to collisions with streaming ions and dust grains, respectively. We have

$$F_{ng} = n_g \pi a^2 \rho_n v_{\rm rel}(v_g - v_n) \tag{11}$$

If we define

$$\alpha \equiv \frac{F_{ng}}{F_{ng} + F_{ni}} \tag{12}$$

we may divide (9) by (10) to obtain

$$d(a/a_i)^3 = R \frac{m_T}{1.4 m_{\rm H}} Y_{\rm eff}(v_{\rm rel}) d\ln v_n \tag{13}$$

$$R \equiv \alpha \frac{v_n}{v_n - v_g} \frac{\rho_n v_n}{(4\pi/3)\rho a_i^3 n_g v_g} \tag{14}$$

where a_i is the initial grain radius. If we now assume $v_g \ll v_n$ in the region of the flow where sputtering is important, then $v_{\rm rel} \approx v_n$. Since $\alpha < 1$, we have

$$R < \frac{\rho_n v_n}{(4\pi/3)\rho a_i^3 n_g v_g} = \frac{\text{initial gas mass}}{\text{initial grain mass}} \approx 140 \tag{15}$$

If we integrate eq.(13) with $R = 140$ and $v_{\rm rel} = v_n$ we obtain an upper limit on the the fractional volume sputtered in C-type shocks. Figure 6 shows the results of this calculation for carbon and silicate grains. It should be noted that this upper limit will be approached only in shocks where the fractional ionization remains very low, $n_e/n_{\rm H} < 10^{-6}$, since otherwise ion-neutral collisions are important, and $\alpha \ll 1$. Sputtering of refractory grains is negligible for $v_s < 35\,{\rm km\,s^{-1}}$.

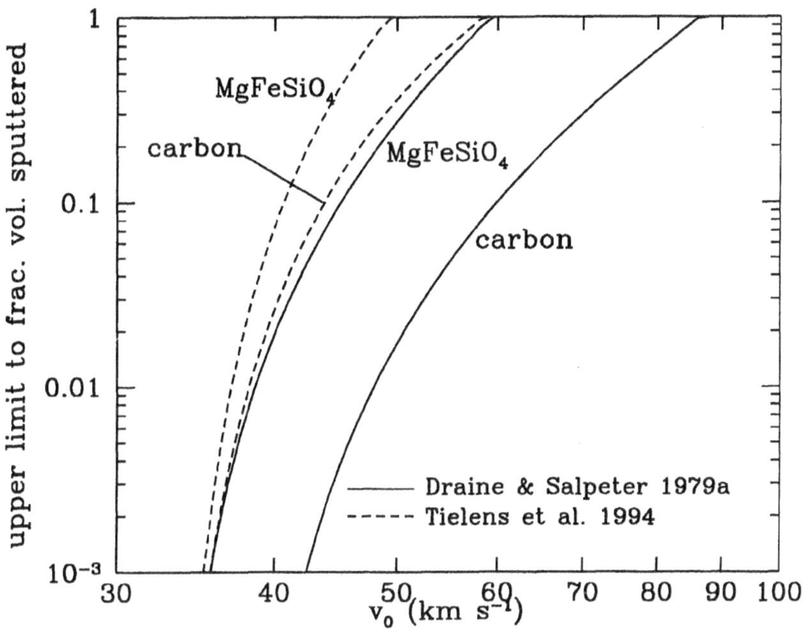

Fig. 6. Upper limit to volume sputtered in C-type MHD shock, using eq. (13) with $R < 140$ and $v_{rel} < v_n$. Here $v_0 = v_s(1 - C^{-1})$ is the total velocity change in the shock, where C is the overall compression ratio in the shock.

6. Implications for Grain Evolution

The interstellar medium is a violent place, and shock waves are ubiquitous. Current estimates for frequencies of supernova-driven shocks, together with calculations of grain destruction in shocks, lead to lifetimes of $\sim 3 \times 10^8$yrs for refractory grains (Draine & Salpeter 1979b; Jones et al. 1994). As discussed elsewhere (e.g., Draine 1990), this is a somewhat troubling result, since the timescale for replenishing the interstellar medium with "stardust" is considerably longer, of order $\sim 2 \times 10^9$yr. As a result we would expect that only $\sim$ 10% of the refractory elements (e.g., Si or Fe) should be locked up in "stardust". Since we observe 90% or more of some of these elements to be missing from the gas, we must conclude that they are locked up in solid material which was condensed *in the interstellar medium* rather than in stellar outflows. Since atoms and ions in the interstellar gas will collide with and may stick to preexisting grains, it is clear that a mechanism for such interstellar condensation exists – the problem is that the predicted *rates* for such condensation in diffuse clouds are simply too slow to explain the observed depletions. As a result, it seems to be necessary to postulate that most of this condensation takes place in dense molecular clouds; this, in turn, can only explain the depletions observed in diffuse clouds if there is quite rapid exchange of material between the diffuse and dense phases (Draine 1990).

In any case, the following conclusion appears inescapable: *interstellar grains are not (for the most part) stardust!*

7. Concluding Remarks

Shock waves can have substantial effects on interstellar grains, and in some cases the structure of the shock depends upon the grains mixed with the shocked gas.

Much of the physics involved in the modification of grains in shocked gas remains poorly understood. For example, sputtering yields, expecially at low energies, are quite uncertain – laboratory experiments with appropriate targets (e.g., silicates, amorphous carbon, graphite) and projectiles (H, H_2, He) are required for progress on this front.

Our modelling of shock waves in dusty gas leaves much room for progress. The grain trajectories are complicated by curving magnetic fields in oblique shocks, quantized changes in the grain charge, and grain-grain collisions.

Finally, the nature of the grains themselves is uncertain – in some models they are compact objects, in other models they are fluffy aggregates – and this lack of knowledge propagates into uncertainty on the fate of the grains in shock waves. Progress on this front will come slowly, as we gain a better understanding of interstellar grains from careful modelling of observed extinction, scattering, polarization, and emission.

This work was supported in part by NSF grant AST-9319283.

References

Andersen, H. H. & Bay, H. L.: 1981, in *Sputtering by Particle Bombardment I*, ed. R. Behrisch, Springer-Verlag: New York, p. 145

Barlow, M.J.: 1978a, *Monthly Not. Roy. Astr. Soc.* **183**, 367

Barlow, M.J.: 1978b, *Monthly Not. Roy. Astr. Soc.* **183**, 397

Barlow, M.J. & Silk, J.: 1977, *Astrophys. J.* **215**, 800

Betz, G. & Wehner, G.K.: 1983, in *Sputtering by Particle Bombardment II*, ed. R. Behrisch, Springer-Verlag, New York, p. 11

Bohdansky, J., Bay, H.L & Ottenberger, W.: 1978, *J. Nucl. Mater.* **76**, 163

Bohdansky, J., Roth, J. & Sinha, M.K.: 1976, in *Proc. 9th Symp. on Fusion Technology*, Pergamon: London, p. 541

Cardelli, J.A., Mathis, J.S., Ebbets, D.C. & Savage, B.D.: 1993, *Astrophys. J.* **402**, L17

Cowie, L.L.: 1978, *Astrophys. J.* **225**, 887

Crinklaw, G., Federman, S.R. & Joseph, C.L.: 1994, *Astrophys. J.* **424**, 748

Désert, F.X., Boulanger, F. & Puget, J.L.: 1990, *Astron. Astrophys.* **237**, 215

Draine, B.T.: 1979, *Astrophys. J.* **230**, 106

Draine, B.T.: 1981, *Astrophys. J.* **245**, 1021

Draine, B.T.: 1990, in *The Evolution of the Interstellar Medium*, ed. L. Blitz, *ASP Conf. Series*, **12**, 193

Draine, B.T.: 1995, in *Physics of the Interstellar Medium and Intergalactic Medium*, ed. A. Ferrara, C. Heiles, C.F. McKee & P.R. Shapiro, ASP Conf. Series

Draine, B.T. & Anderson, N.: 1985, *Astrophys. J.* **292**, 494

Draine, B.T. & Lee, H.-M.: 1984, *Astrophys. J.* **285**, 89

Draine, B.T. & McKee, C.F.: 1993, *Ann. Rev. Astron. Astrophys.* **31**, 373

Draine, B.T., Roberge, W.G & Dalgarno, A.: 1983, *Astrophys. J.* **264**, 485

Draine, B.T. & Salpeter, E.E.: 1979a, *Astrophys. J.* **231**, 77 (DS79)

Draine, B.T. & Salpeter, E.E.: 1979b, *Astrophys. J.* **231**, 438

Duley, W.W., Jones, A.P. & Williams, D.A.: 1989, *Monthly Not. Roy. Astr. Soc.* **236**, 709

Field, G.B.: 1974, *Astrophys. J.* **187**, 453

Flower, D.R. & Pineau des Forêts, G.: 1994, *Monthly Not. Roy. Astr. Soc.* **268**, 724

Flower, D.R. & Pineau des Forêts, G.: 1995, *Monthly Not. Roy. Astr. Soc.* , in press

Flower, D.R., Pineau des Forêts, G. & Walmsley, C.M.: 1995, *Astron. Astrophys.* **294**, 815

Furton, D.G. & Witt, A.N.: 1993, *Astrophys. J.* **415**, L51

Jones, A.P., Tielens, A.G.G.M., Hollenbach, D.J. & McKee, C.F.: 1994, *Astrophys. J.* **433**, 797

Léger, A. & Puget, J.L.: 1984, *Astron. Astrophys.* **137**, L5

Mathis, J.S.: 1993, *Rep. Prog. Phys.* **56**, 605

Mathis, J.S., Rumpl, W. & Nordsieck, K.H.: 1977, *Astrophys. J.* **217**, 425

Mathis, J.S. & Whiffen, G.: 1989, *Astrophys. J.* **341**, 808

McKee, C.F., Hollenbach, D.J., Seab, C.G. & Tielens, A.G.G.M.: 1987, *Astrophys. J.* **318**, 674

Morton, D.C.: 1975, *Astrophys. J.* **197**, 85

Ostriker, J.P. & Silk, J.: 1973, *Astrophys. J.* **184**, L113

Rosenberg, D. & Wehner, G.K.: 1962, *J. Appl. Phys.* **33**, 1842

Roth, J., Bohdansky, J. & Ottenberger, W.: 1979, Rpt. IPP 9/26, Max-Planck-Institut für Plasmaphysik, Garching

Roth, J., Bohdansky, J., Poschenrieder, W. & Sinha, M.K.: 1976, *J. Nucl. Mater.* **63**, 222

Savage, B.D., Cardelli, J.A. & Sophia, U.J.: 1992, *Astrophys. J.* **401**, 706

Seab, C.G. & Shull, J.M.: 1983, *Astrophys. J.* **115**, 227

Shull, J.M.: 1977, *Astrophys. J.* **215**, 805

Shull, J.M.: 1978, *Astrophys. J.* **226**, 858

Sigmund, P.: 1969, *Phys. Rev.* **184**, 383

Sigmund, P.: 1981, in *Sputtering by Particle Bombardment I*, ed. R. Behrisch, Springer-Verlag: Berlin, 9

Smith, J.N., Meyer, C.H. & Layton, J.K.: 1976, *Nucl. Technol.* **29**, 318

Spitzer, L.: 1976, *Comments Astrophysics* **6**, 177

Tielens, A.G.G.M., McKee, C.F., Seab, C.G. & Hollenbach, D.J.: 1994, *Astrophys. J.* **431**, 321 (TMSH94)

THE IMPORTANCE OF CHEMICAL DATA FOR THE STUDY OF INTERSTELLAR SHOCKS

EVELYNE ROUEFF and GUILLAUME PINEAU DES FORÊTS
DAEC, Observatoire de Paris
Place Janssen, 92190 Meudon, France

and

DAVID R. FLOWER
Dept. of Physics, Durham University
Durham DH1 3LE, U. K.

Abstract. Shock waves perturb the chemical state of the interstellar gas. We consider the effects of C-shocks on the composition of molecular clouds, for a range of values of the pre-shock gas density and magnetic induction. The time required to re-establish equilibrium in the post-shock gas depends on the initial conditions and can become very large. The significance of the two known chemical phases of dark clouds and of bistability is considered in this context.

Key words: - Interstellar Medium – Dark Clouds – Chemistry – Bistability – C-shock Waves

1. Introduction

Chemical reactions affect the fluid dynamical behaviour of a shock by modifying the fractional ionization, and hence the coupling between ionized and neutral fluids, and by altering the abundances of important coolants. In this review, we investigate the influence of the chemical composition of the pre-shock gas on the structure of a C-shock and on the characteristics of the post-shock gas.

As a result of the non-linearity of the chemical and thermal balance equations, models of quiescent dark clouds may present more than one steady-state solution. When these solutions overlap, the phenomenon is termed 'bistability'. In Section 2, we recall the chemical features associated with bistability, in the context of a model of dark clouds. This model serves to determine the conditions in the pre-shock gas when studying, in Section 3, the structure of a C-shock propagating in the medium. The properties of the post-shock gas are analyzed in Section 4, for various initial conditions. We emphasize the importance of the rate of dissociative recombination of H_3^+ in Section 5. Brief conclusions are given in Section 6.

Astrophysics and Space Science **233**: 125–137, 1995.

2. Dark Cloud Chemistry and Bistability

The occurrence of multiple solutions to the chemical rate equations is well known in non-linear chemical kinetics (Scott 1991). Such behaviour may also occur in models of dark interstellar clouds, where the chemical rate equations are non-linear (Le Bourlot et al. 1993, 1995).

Figs. 1a & b display the time evolution of the fractional ionization and the C:CO abundance ratio for an isochoric dark cloud of density $n_H = 5000\,\mathrm{cm}^{-3}$ and two different sets of initial conditions corresponding to slightly different temperatures (T = 24.5 K and T = 24.8 K); the cosmic ray ionization rate was taken to be $\zeta = 2\,10^{-17}\mathrm{s}^{-1}$. The model includes 95 species containing H, He, C, N, O and S, as well as a representative metal; there are approximately 700 chemical reactions. Cosmic rays directly ionize the gas in the cloud and also give rise to ultraviolet photons which dissociate and further ionize the gas. The thermal balance equations were included in the model.

It may be seen from Fig. 1 that two markedly different steady states are attained as t tends to infinity. Le Bourlot et al. (1993) labelled the two solutions according to their ionization properties. In the 'HIP' (High Ionization Phase), the fractional ionization $n_e/n_H \approx 10^{-6}$, whereas, in the more familiar 'LIP' (Low Ionization Phase), $n_e/n_H \approx 10^{-7}$. The C:CO abundance ratio follows the ionization fraction, as shown in Fig. 1b, and all the variables (including temperature) exhibit two solutions. The time at which steady state is attained depends on the initial conditions.

In interstellar clouds, many parameters can affect the chemical and physical equilibrium, including the density, the cosmic ray ionization rate, and the elemental depletions; these parameters control the evolution of the gas. The range in which bistability occurs can be established by solving the steady state equations, varying the value of one 'control' parameter. Figs. 2a & b display the steady-state values of the same quantities as in Fig. 1, i.e. the ionization fraction and the C:CO abundance ratio. The density n_H was varied between 500 and $10^5\mathrm{cm}^{-3}$ for constant $\zeta = 2\,10^{-17}\mathrm{s}^{-1}$, using the same chemical network and thermal balance as previously.

Each point in this Figure represents one model. A typical hysteresis curve is obtained, with two stable branches corresponding to the HIP and LIP solutions. Two critical points determine the range of density in which bistability occurs and where two stable solutions can be found for the same value of n_H; they are connected by an unstable branch. Another way of describing this phenomenon is in terms of the values of the variables at any given time in phase space. The stable solutions are associated with 'basins of attraction' in which lie the fixed points towards which the variables converge at infinite time. Initial conditions belonging to one basin of attraction will evolve to the corresponding stable fixed point. The evolutionary time scale will depend on the initial conditions. A frontier, called the 'separatrix', separates the basins

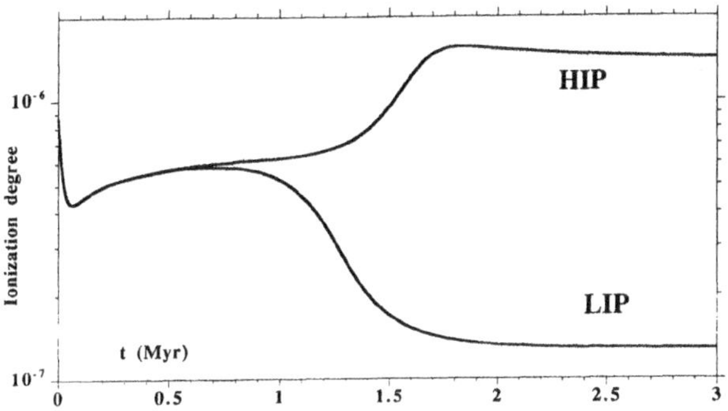

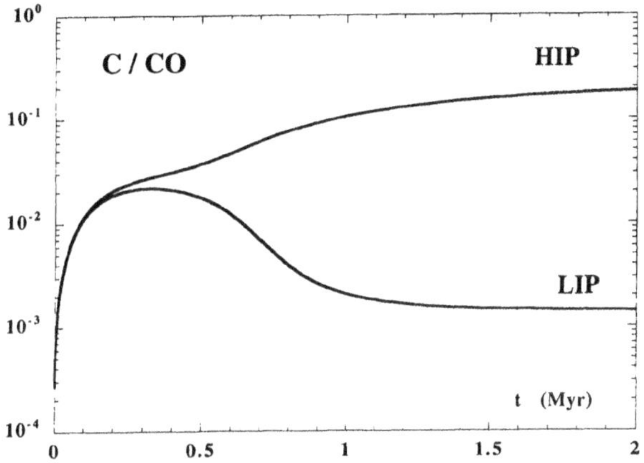

Fig. 1. Evolution to steady state of a cloud at constant density $n_H = n(H) + 2n(H_2) = 5000\ cm^{-3}$ and cosmic ray ionization rate $\zeta = 2\,10^{-17} s^{-1}$, for two differing sets of initial conditions: (a) fractional ionization; (b) C:CO abundance ratio.

of attraction (cf. Le Bourlot et al. 1993) and cannot be crossed unless the gas undergoes a major perturbation, such as a shock compression (see below).

It is worthwhile recalling the chemical properties of the solutions (Le Bourlot et al. 1993, 1995). In the HIP, the chemistry is dominated by charge transfer reactions involving the key atomic ions, H^+ and C^+. The molecular ion H_3^+ is then destroyed mainly by the electrons. Unsaturated species are favoured and consequently the C:CO abundance ratio is high. On the other hand, in the LIP, the chemistry is dominated by proton transfer reactions involving the H_3^+ ion. Then, saturated molecules are efficiently produced and C and C^+ are less abundant, as they are destroyed by O_2 and H_2O.

Le Bourlot et al. (1993, 1995) have shown that bistability can occur in the range of values of the control parameters (e.g. density, cosmic ray ion-

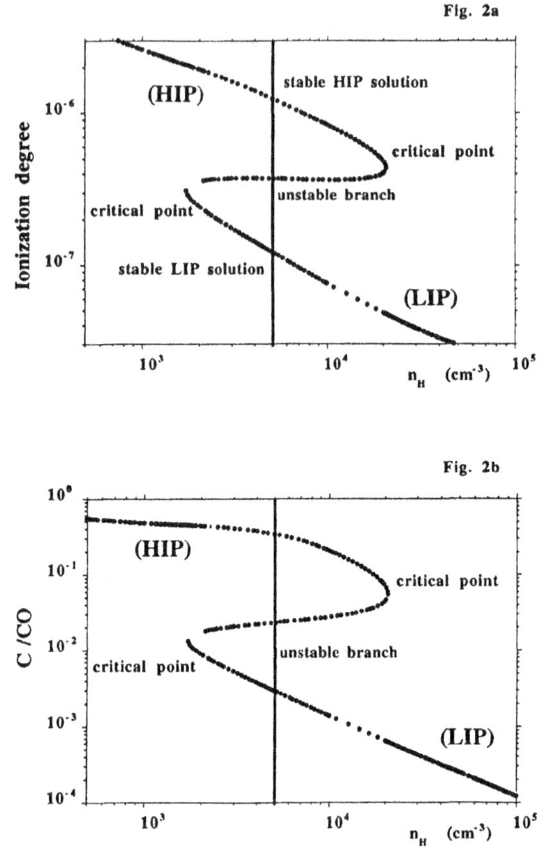

Fig. 2. Steady-state solutions for isochoric clouds of density $500 \leq n_H \leq 10^5\ \mathrm{cm}^{-3}$. Each point represents the steady-state solution at the corresponding density: (a) fractional ionization; (b) C:CO abundance ratio.

ization rate, elemental depletions) which are astrophysically relevant. The rate coefficients may also be taken as control parameters, in so far as many of them are poorly known. A particularly critical reaction in the chemical network is the dissociative recombination of the H_3^+ molecular ion (cf. Roueff & Pineau des Forêts 1993). This reaction governs the transition between HIP and LIP solutions, which depends on the dominant mechanism of destruction of H_3^+. Dissociative recombination is the main destruction channel in the HIP, whereas reactions with neutral species, such as O and CO, predominate in the LIP. We have adopted the value of the rate cofficient for dissociative recombination of H_3^+ which derives from the recent measurements of Sundström et al. (1994). The branching ratio towards dissociative ionization of H_2

by cosmic rays is also important, because it determines the rate of production of H^+.

3. Structure of The C-Shock

We now consider the effect of a MHD shock, propagating in moderately dense gas which is initially in a steady state. The previous figures have shown that there can exist quite distinct sets of pre-shock conditions, depending on the particular steady-state solution which is attained. We take a shock velocity of 10 km s^{-1}, a pre-shock density of 2000 cm^{-3} (to which there corresponds a unique solution for the chemical state of the gas), and an initial magnetic induction of 30 μG.

Figs. 3a and b show the velocities of the neutral and ionized fluids, the density of the neutrals, and the fractional ionization.

For the chosen initial conditions, the width of the shock attains about 0.2 pc, with a compression factor of about 10 for the neutral gas. The temperature of the neutral fluid reaches only about 300 K, whereas the ion temperature attains values of several thousand Kelvin within the shock. The fractional ionization undergoes large variations, which have already been discussed, for example, by Hartquist, Flower and Pineau des Forêts (1990), and by Hartquist (1995) at this conference. The ion density first increases, owing to the shock compression, and then decreases when the atomic ions C^+ and S^+ begin to react with H_2. These endothermic reactions are driven by the ion-neutral drift velocity, which attains approximately 6 km s^{-1}. Towards the rear of the shock, the fractional ionization increases because the velocity difference becomes insufficient to maintain the endothermic reactions. The small decrease in n_e/n_H at the trailing edge of the shock is associated with the compression of the neutrals, which occurs later than the compression of the ions. At a given density, the compression factor in the shock depends on the value of the magnetic field. It is often assumed (Myers 1990) that the magnetic induction B (μG) in gas of density n_H (cm^{-3}) is about ${n_H}^{0.5}$. On the other hand, observations of the Zeeman effect in 21cm HI spectra in dark clouds (Goodman and Heiles 1994) and in diffuse clouds (Myers et al. 1995) indicate an approximate equality between the kinetic and magnetic energy densities, which implies B (μG) $= 0.22\ {n_H}^{0.5}$. Accordingly, we shall consider a range of magnetic field strengths, given by $0.22{n_H}^{0.5} \leq B \leq {n_H}^{0.5}$.

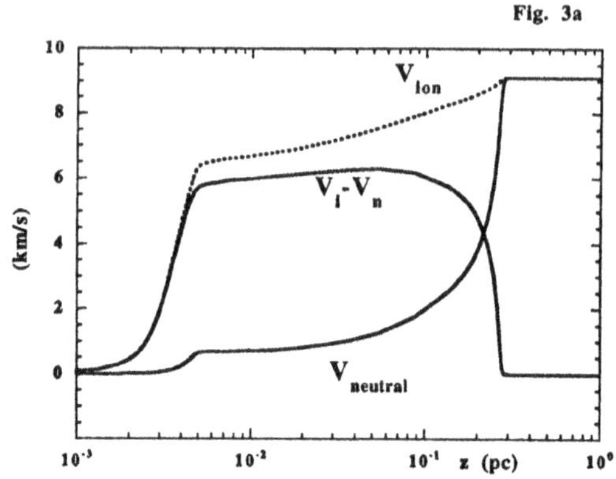

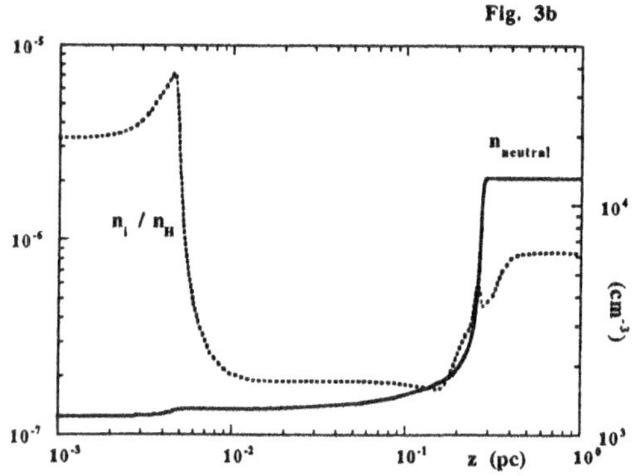

Fig. 3. The profile of a C-shock for an initial density $n_H = 2000\,\mathrm{cm}^{-3}$ and magnetic induction B = 30 μG. The shock velocity is $v_s = 10$ km s^{-1} : (a) flow velocities of the ions and the neutrals, and their relative velocity; (b) the fractional ionization n_e/n_H and the density of the neutrals.

4. Properties of The Post-Shock Gas

4.1. Case of a Single Pre-Shock Chemical State

We consider $n_H = 2000\,\mathrm{cm}^{-3}$ in the pre-shock gas, and a range of magnetic inductions, $10 \leq B \leq 45\mu$G. The compression factor increases with decreasing magnetic field strength, and the post-shock densities lie between $7\,10^4\,\mathrm{cm}^{-3}$ and about $10^4\,\mathrm{cm}^{-3}$.

Figs. 4a and b show the evolution of the C:CO abundance ratio and the fractional abundance of C_2H for three values of the initial magnetic induction, 10, 25 and 30 μG.

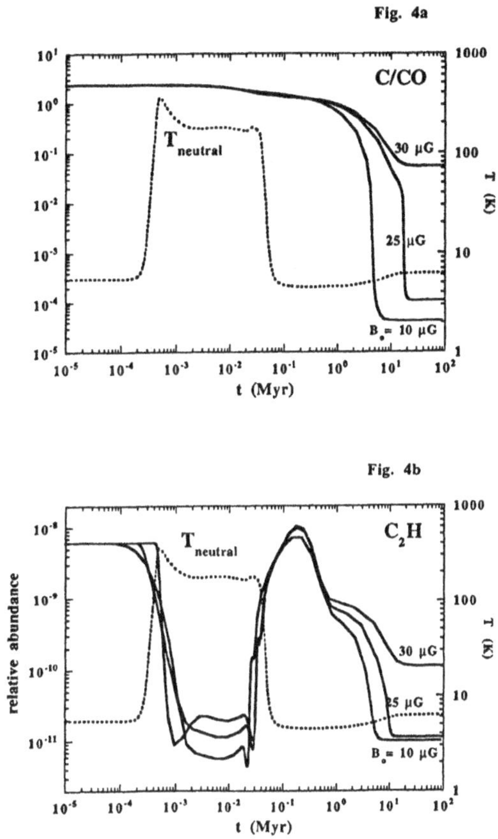

Fig. 4. The same C-shock model as in Fig. 3, but varying the initial magnetic induction: (a) the C:CO abundance ratio; (b) $n(C_2H)/n_H$.

The temperature profile delineates the shock. The most striking feature of these Figures is the presence, for several million years, of a large C:CO ratio in the post- shock gas. The fractional abundance of the carbon-chain molecule C_2H also remains high until the new steady state is finally reached. Steady state is attained at different times, ranging from about 5 to 15 Myr. Furthermore, two distinct steady states occur, characterized by high or low values of the C:CO ratio and the fractional abundance of C_2H; they are the HIP or LIP chemical states.

If we now concentrate on the post-shock gas, Figs. 5a and b show the evolution (on a linear time scale) towards steady state of the degree of ionization and of C:CO, for five values of the initial magnetic induction from 10 to 30 μG.

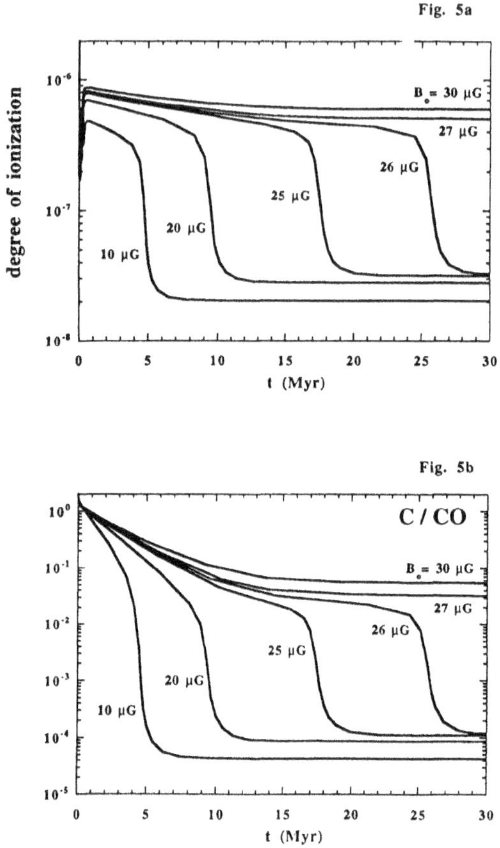

Fig. 5. Showing the evolution to steady state, as in Fig. 4 but with a linear time scale to emphasize the post-shock region. Note the two distinct steady-state solutions, attained for large t: (a) fractional ionization; (b) C:CO abundance ratio.

We see that the two distinct steady states are reached on time scales which vary considerably and can attain 25 Myr. A small change in the value of the initial magnetic induction may result in a quite different steady state, as the curves corresponding to B = 26 and 27 μG clearly show. This behaviour may be explained by means of the hysteresis curve which is displayed in Fig. 6.

There, we see that steady state at $n_H = 2000\,\text{cm}^{-3}$ in the pre-shock gas is characterized by a unique (HIP) solution. The arrows indicate the post-shock densities, which vary from $7\,10^4$ to $1.8\,10^4\,\text{cm}^{-3}$ with the values of the initial magnetic induction. For magnetic field strengths up to 26 μG, the compression of the shock is such that the final state of the gas corresponds to the LIP. For larger magnetic field strengths, two final states are, in principle, possible. In practice, the calculations show that the gas remains in the initial

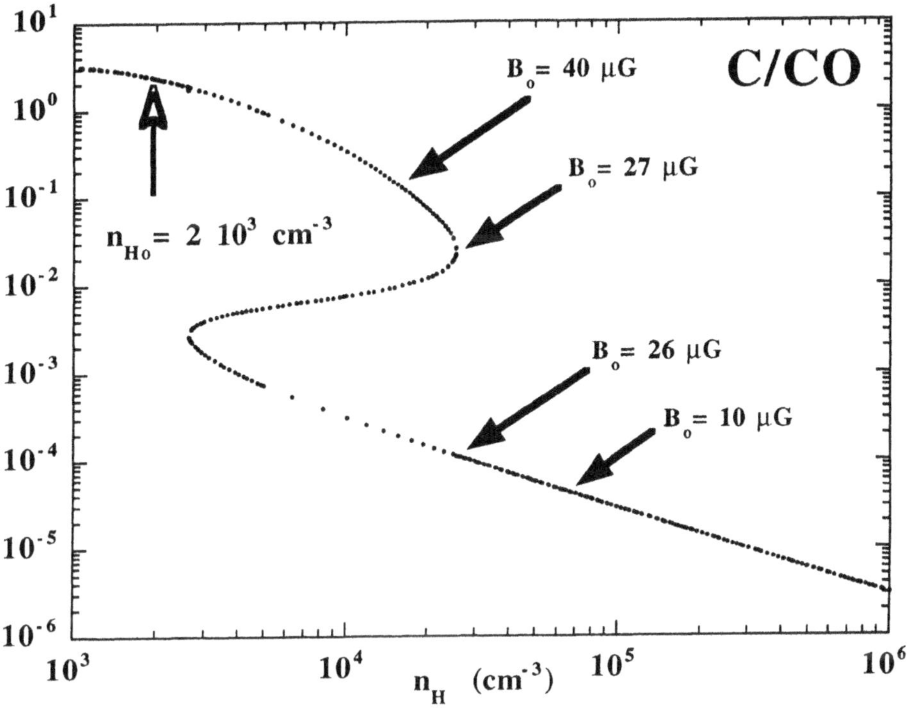

Fig. 6. Steady-state values of the C:CO abundance ratio. The arrows indicate the post-shock densities attained for the initial magnetic field strengths indicated. The pre-shock density is $n_H = 2000\ cm^{-3}$, for which a unique (HIP) solution obtains.

(HIP) phase (see Figs. 4 and 5). The long time scales may be understood in terms of the crossing of a critical point when moving from one basin of attraction to the other, as a consequence of the shock compression (cf. the curves corresponding to $B = 26\ \mu G$). This property of complex dynamical systems is significant in the context of MHD shocks because the post-shock gas can remain for a long time in a state which retains some of the chemical characteristics of the pre-shock gas, before ultimately reaching its steady state.

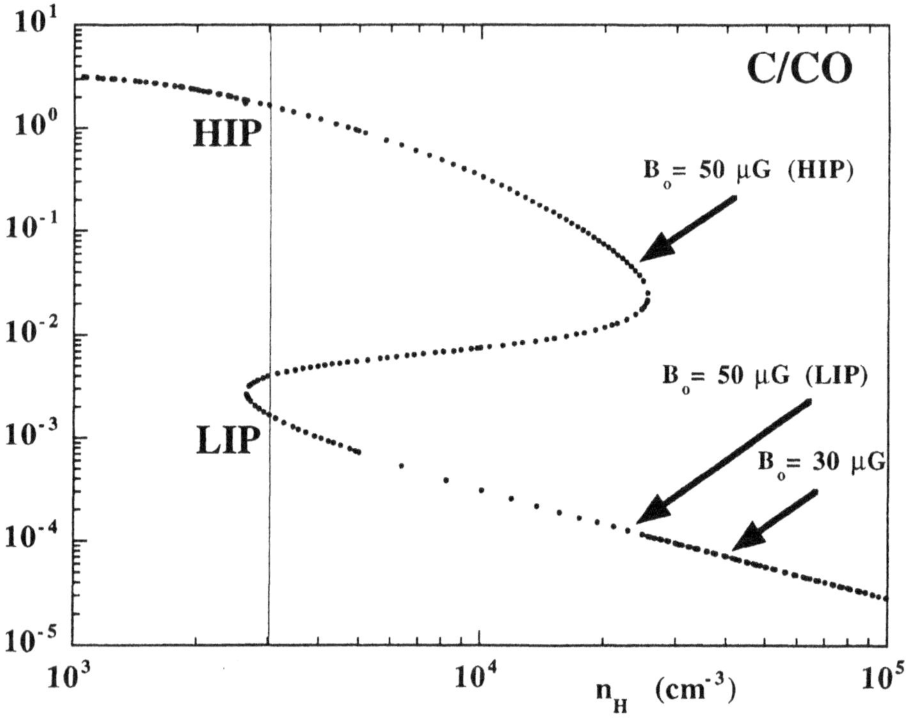

Fig. 7. As Fig. 6, but with reference to models in which the pre-shock density is $n_H = 3000\,cm^{-3}$ and two solutions (HIP and LIP) are initially possible.

4.2. Case of 2 Possible Pre-Shock Chemical States

We consider now a pre-shock density of 3000 cm^{-3}. At this density, we see (from Fig. 7) that 2 stable initial states may be found, with HIP and LIP characteristics.

The relevant range of magnetic field strengths is now between 12 and 55 μG. The arrows on the Figure indicate the final state of the gas, after the passage of the shock wave. For B = 50 μG, we find that the solution remains in the initial basin of attraction. If the pre-shock gas is in the HIP, the post-shock gas is also in the HIP, and similarly for the LIP. This situation is illustrated in Fig. 8a, where the C:CO ratio is plotted; the shock is delineated by the temperature profile. A single temperature profile is shown in the Figure, as the thermal structure is similar for both solutions. The evolution

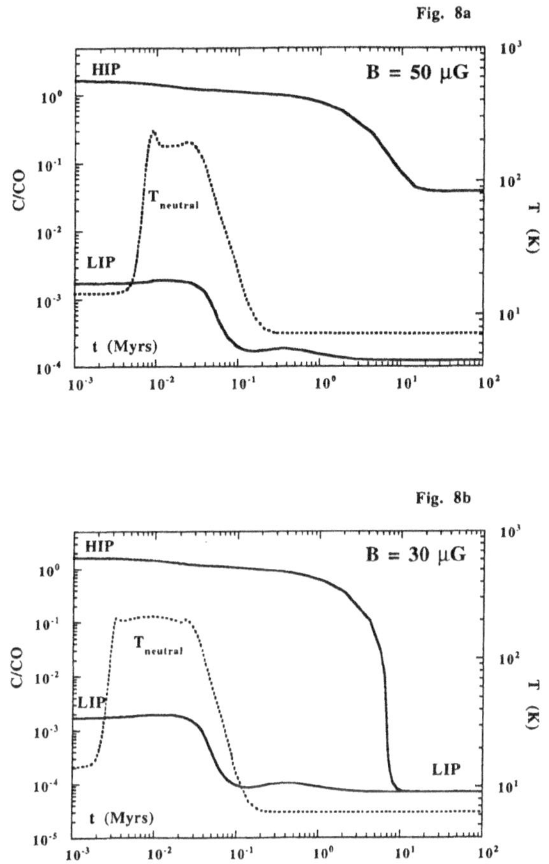

Fig. 8. The profile of a C-shock for an initial density $n_H = 3000\,\mathrm{cm}^{-3}$ and magnetic induction (a) B = 50 μG and (b) B = 30 μG. The initial conditions are those appropriate to the HIP or the LIP, as indicated.

time is much greater when the initial state is the HIP; this is again related to the fact that the evolution is taking place along the separatrix and close to the critical point. Fig. 8b shows the evolution of the C:CO ratio when only one (LIP) final steady state is attainable from both the pre-shock conditions.

A much longer time is required to reach steady state when a transition from the HIP to the LIP occurs. These Figures also show that the shock structure is not very dependent on the phase to which the initial conditions belong.

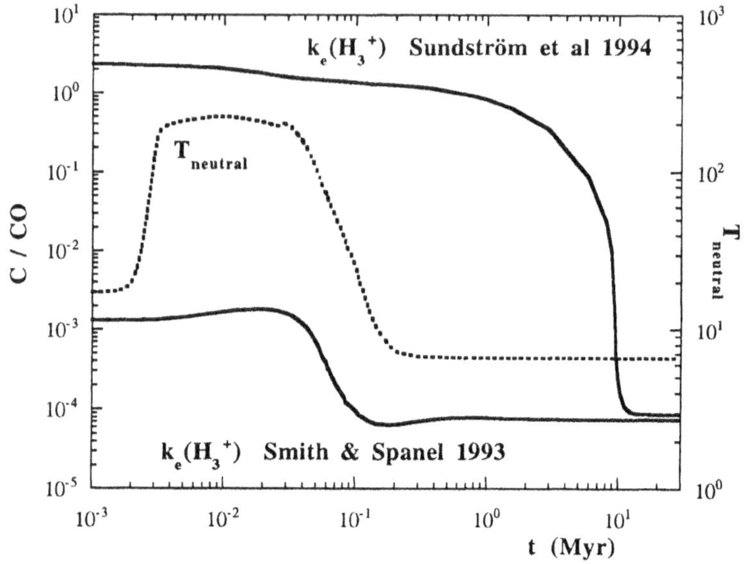

Fig. 9. Showing the sensitivity of the C-shock profile to the adopted value of the rate coefficient for dissociative recombination of H_3^+.

5. The Critical H_3^+ Dissociative Recombination Reaction

The 'accepted' value of the rate coefficient for dissociative recombination of H_3^+ has varied by orders of magnitude in the past decade (cf. the review by Dalgarno 1994). The controversy is not yet over: results from the two most recent experiments (Smith and Spanel 1993; Sundström et al. 1994) differ by about a factor of 10 at 300 K. It should be noted that the temperature dependence of the rate coefficient is also important, since the temperatures in dark clouds are of order 10K. Smith and Spanel used the afterglow technique to determine the rate coefficient at 300K; they obtained $k_e = 2\,10^{-8}\,cm^{-3}\,s^{-1}$. On the other hand, Sundström et al. measured the cross-section as a function of energy in a storage ring, and derived the rate coefficient by integrating over a Maxwellian distribution, obtaining $k_e = 1.15\,10^{-7}(T/300)^{-0.65}\,cm^{-3}\,s^{-1}$. Fig. 9 displays the C:CO evolution obtained with these two rate coefficients, for pre-shock gas of 2000 cm^{-3} in which B = 20 μG and v_s= 10 km s^{-1}.

The lower rate coefficient, which has the usual $T^{-0.5}$ temperature dependence, favours the LIP, and the solution remains in the corresponding basin of attraction. All reactions affect the bistability phenomenon to some degree, and this should be borne in mind when interpreting the results of the calculations.

6. Concluding Remarks

This paper has been concerned with the effects of the phenomenon of bistability on the structure of a C-shock and the post-shock gas. We have shown that, if, during its evolution, the state of the gas passes close to a critical point, the chemical time scale required to reach equilibrium in the post-shock gas becomes very large. Chemical properties of interstellar clouds (such as a high C:CO abundance ratio, large abundances of radicals or of long chain molecules) might be explained in this way. The observed spectra will not necessarily retain evidence of the passage of the shock wave, in view of the long time scales involved. The numerical results are sensitive to some of the reaction rate coefficients (e.g. for dissociative recombination of H_3^+ and dissociation by cosmic ray-induced photons) incorporated in the model. The relevance of bistability to the evolution of the 'real' interstellar gas remains to be unequivocally demonstrated, but it is already clear that the phenomenon cannot be simply ignored.

References

Dalgarno, A.: 1994, *Adv. Atom. Molec. Opt. Phys.* **32**, 57

Goodman, A.A. & Heiles, C.: 1994, *Astrophys. J.* **424**, 208

Hartquist, T.W.: 1995, this volume

Hartquist, T.W., Flower, D.R. & Pineau des Forêts G.: 1990, in T. W. Hartquist, ed(s)., *Molecular Astrophysics,* Cambridge University Press: Cambridge, 99

Le Bourlot, J., Pineau des Forêts, G., Roueff, E., & Schilke, P.: 1993, *Astrophys. J.* **416**, L87

Le Bourlot, J., Pineau des Forêts, G. & Roueff, E.: 1995, *Astronomy. Astrophys.*, in press

Myers, P.C.: 1990, in T.W. Hartquist, ed(s)., *Molecular Astrophysics,* Cambridge University Press: Cambridge, 328

Myers, P.C.,Goodman, A.A., Güsten, R. & Heiles, C.: 1995, *Astrophys. J.* **442**, 177

Roueff, E. & Pineau des Forêts, G.: 1993, in B.R. Rowe, J.B.A. Mitchell and A. Canosa, ed(s)., *Dissociative Recombination: Theory, Experiment and Applications,* Plenum Press: New York, 249

Scott, S.K.: 1991, *Chemical Chaos*, Clarendon Press

Smith, D. & Spanel, P.: 1993, *Chem. Phys. Lett.* **211**, 454

Sundström, G. et al.: 1994, *Science* **263**, 785

MOLECULAR MIXING LAYERS IN STELLAR OUTFLOWS

S.D.TAYLOR
University College, Gower St., London WC1E 6BT
and
A.C.RAGA
Instituto de Astronomía UNAM, México

Abstract. High velocity jets from young stars interact with the surrounding molecular environment and molecular outflows quite possibly are the result. This interaction can take place through the formation of a turbulent mixing layer. Models have been constructed (following Cant/'o and Raga) of a plane mixing layer in the boundary between a high velocity, atomic wind (*i.e.*, the stellar jet) and a stationary, molecular environment, computed considering a detailed chemical network.

The chemical composition of the mixing layer initially corresponds to the direct mixture of the (atomic) jet and (molecular) environmental material. However, we find that the mixing layer is hot (with temperatures exceeding 10^4 K), and the surprising only partial dissociation of H_2 means that a number of molecules are either created or survive in the high velocity gas. This contrasts with the slower, cooler flows that have tended to be termed a 'molecular outflow'.

The emission from such atomic jet/molecular environment mixing layers is dominated by emission in the rotational and vibrational lines of H_2. As a result of the high temperatures and velocities (ranging from zero to the jet velocity) of these mixing layers, the predicted H_2 emission line spectrum has interesting characteristics.

Key words: Herbig-Haro Objects – Stellar Outflows – Shock Waves

1. Introduction

Young stellar objects many times are the sources of both Herbig-Haro (HH) jets and molecular outflows. This coincidence was initially interpreted in terms of a "unified model" in which a partially ionized wind (associated with the optically detected HH emission) entrains molecular gas from the surrounding environment (this entrained gas being then associated with the observed molecular outflow). Although this work was called into question by claims that the jets did not have enough momentum to drive the associated outflow, leading to a "two-wind" model, new analyses that strongly increase the momentum estimate appear to once again validate the "unified" HH jet/molecular outflow model.

This interaction could take the form of entrainment of gas at the head of the jet or "lateral entrainment" through the jet beam/environment boundary. This paper is concerned with the second mechanism. It is clear from laboratory experiments that once the head of the jet has gone by, in the region close to the source the jet develops a turbulent mixing layer in which the environmental gas is entrained into the beam of the jet. The formation

Astrophysics and Space Science **233**: 139–144, 1995.

of this turbulent boundary has been studied in detail by Cantó and Raga (1991, hereafter CR), who developed a simple formalism (based on flow variables averaged across the cross section of the mixing layer) that successfully reproduces the results of hypersonic mixing layers, and then applied this to the non-adiabatic case appropriate for stellar outflows.

However, in order to obtain detailed predictions of the observational properties of such turbulent mixing layers it is necessary to incorporate a chemical network in the model calculation. This is done in the present paper, in which we have extended the work of Cantó and Raga (1991) to incorporate a detailed treatment of the chemical history of the entrained molecular gas.

2. The Turbulent Mixing Layer

The equations obtained by CR for this interaction were derived under three main assumptions. Firstly the mixing layer is assumed to be thin, so that there is an approximate pressure balance between the outer edge of the jet, the mixing layer and the (homogeneous) environment; secondly that the flow parameters are constant across the layer so that the average velocity is equal to one half the jet velocity; and finally that the entrainment velocity (with which the environment is slowly dragged into the mixing layer) is given by $v_e = \epsilon c_e$, where c_e is the environmental isothermal sound speed, and the "efficiency" ϵ is given by :

$$\epsilon \approx 0.03 \times c_e/c_l\,, \tag{1}$$

where c_e and c_l are the sounds speeds of the environment and the mixing layer.

This last assumption reflects the results obtained by CR from a fit of a "mixing length" parameterization of the entrainment velocity to hypersonic mixing layer experiments. Astrophysical contexts mostly imply a "mixing layer-limited" regime, which has been assumed here. The physical situation is shown in a schematic way in Figure 1. At the injection point, the mixing layer has a vanishing thickness, and it grows both into the jet and into the environment for increasing values of x (*i.e.*, away from the source). The width of the mixing layer measured into the environment is $h_1(x)$, and the width measured into the jet is $h_2(x)$, so that the total width of the mixing layer is $h = h_1 + h_2$.

In addition to the these equations we need to consider the rate of change of the abundances of the relevant species. If we consider the fractional abundance $y^a = n_a/n$ of a (molecular, atomic or ionic) species a (where n_a is the number density of species a and n is the number density of particles), then we have

$$v_l \frac{dy^a}{dx} = \frac{0.03\, c_l}{h} \left[(y^a{}_e - y^a) + 2(y^a{}_j - y^a) \right] + S_a\,, \tag{2}$$

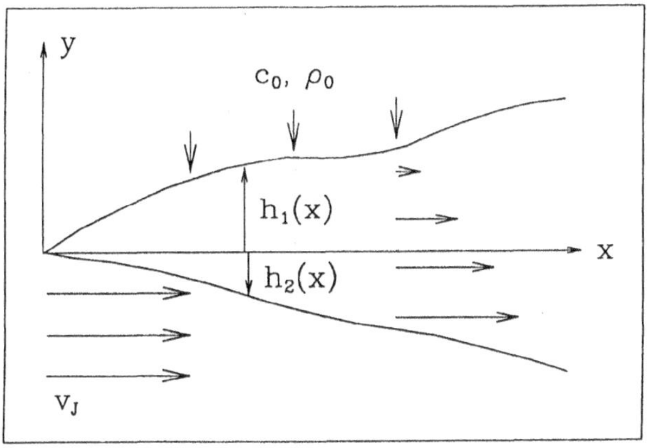

Fig. 1. Schematic diagram of turbulent mixing layer. The x-axis is the flow direction, and $y = 0$ corresponds to the outer jet boundary.

where $v_l = v_j/2$ is the average velocity along the mixing layer, ${y^a}_e$ and ${y^a}_j$ are the abundances of species a in the environment and in the jet (respectively), and S_a is the source/sink term associated with chemical or atomic reactions with other species.

3. Chemical Processes

In these calculations we have included a full description of the non-equilibrium chemistry and ionization level of the gas. Since the layer is found to be hot and relatively diffuse it is necessary to ensure adequate destruction of molecules by H atoms and most importantly to account for the supression of the collisional dissociation rate of H_2 at low densities (Roberge and Dalgarno 1982). The implementation of these processes can be found in Taylor and Raga (1995).

In order to compute mixing layer models, we have to specify the chemical compositions of the jet and the surrounding environment. We assume that the jet is purely atomic, with cosmic elemental abundances. For the environment, we use the chemical abundances at 10^6 yr of a one-point, initially atomic, dark cloud model. The environment computed in this way is overwhelmingly molecular, a fact that has fundamental consequences for the properties of the mixing layer.

4. Mixing Layer Structure And Chemistry

The initial chemical and atomic fractional abundances of the mixing layer are given by the $h \rightarrow 0$ limit of equation 2 ($y = (y_e + 2y_j)/3$). With these initial conditions, the equations describing the dynamical structure and the chemistry of the mixing layer are integrated downstream (away from the source). The chemistry is coupled to the dynamical structure of the mixing layer through the internal energy and the energy loss terms. The parameterization used by CR for hydrogen recombination, collisional ionization and line excitation is included here, but additionally we now have the collisional excitation of rovibrational transitions (especially important in the lower temperature regions of the mixing layer) and the dissociation of H_2 molecules. No other molecular coolants are important in these models. The H_2 cooling is calculated explicitly since the departures from LTE are likely to be large at the low densities found in mixing layers.

Figure 2 shows the flow variables along the mixing layer as a function of distance x from the source for a "standard model" of a mixing layer formed at the edge of an atomic jet of velocity $v_j = 100$ km s^{-1}and temperature $T_j = 10^4$K moving in a stationary, molecular environment of hydrogen nuclei number density $n_e = 2 \times 10^6$ cm^{-3} and temperature $T_e = 10$ K. The density of the jet ($n_j = 1000$ cm^{-3}) is chosen so as to have pressure balance with the environment. Possibly the most striking feature is the collapse of the mixing layer, as evidenced by the sudden rapid decrease of h_1 . This collapse occurs at the point where H_2 molecules are no longer dissociated. This is clear from Figure 3 which shows the abundances (with respect to hydrogen nuclei) of the most abundant molecular species together with some reference atomic species. As the width of the mixing layer grows for larger values of x, the hot layer becomes more chemically active, resulting in the dissociation of the H_2 molecules. At some distance along the flow the temperature of the mixing layer becomes too low for H_2 dissociation to occur, so that the H_2 abundance starts to grow back to the value corresponding to the direct mixture of the jet/environment gases. The increase in abundance increases the cooling which lowers the temperature, in turn lowering the dissociation rate leading to a runaway effect.

The combination of a high H_2 abundance and a high ($T_l > 10^4$ K) temperature leads to the production of large amounts of simple unsaturated hydrated molecules like CH, CH_2, CH_3, OH, NH, NH_2 and HS via hydrogen abstraction reactions. As the ionization fraction increases (to a peak of $\approx 1\%$ in this model) there is a subsequent partial conversion to ionized species.

These results may be modified if the reactivity is reduced as a result of subthermal excitation. An analysis of the high temperature oxygen chemistry by Wagner and Graff (1987) suggested that all rate coefficients were affected by a factor 2-4 in the $300 < T < 5000$ K temperature range if the molecules

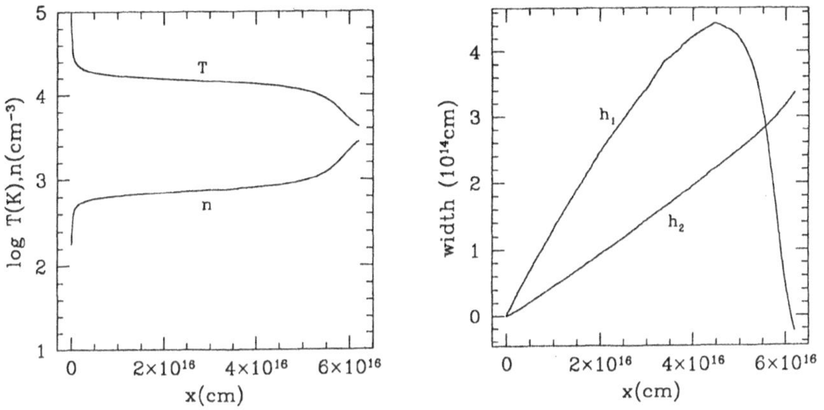

Fig. 2. Variation of physical parameters along the length of the mixing layer.

were vibrationally or rotationally cold. These authors also pointed out that the extrapolation of thermal rates outside of their temperature ranges is very often inappropriate and, since we encounter even higher temperatures, this caveat must also be applied to our results.

Mixing layers with $v_j < 100$ km s^{-1}have lower temperatures, which result in a lower ionization fraction, and molecular hydrogen is virtually undissociated. By contrast models with $v_j > 100$ km s^{-1}produce higher temperatures and ionization fractions. As v_j increases the H_2 molecules are progressively more dissociated, resulting in the formation of an extended, fully ionized region close to the source. Furthermore the width of the layer increases so that when the gas eventually cools to $\sim 10^4 K$ the entrainment does not efficiently replenish the molecules, resulting in the formation of a spatially extended, neutral region further downstream. Lower pressures have the effect of increasing the length scale of the mixing layer, so that the observed length of the mixing layer emission could become comparable to the total length of the HH jet.

We hope in the near future to extend these calculations through the collapse phase and examine the gas structure across the width of the layer.

5. High Velocity Molecular Outflow Emission

Our mixing layer models predict the existence of molecular emission at velocities similar to the velocity of HH jets (which are typically observed to have radial velocities ~ 100 km s^{-1}). Molecular emission at such high velocities has been observed in several outflows from young stars, in species found to be

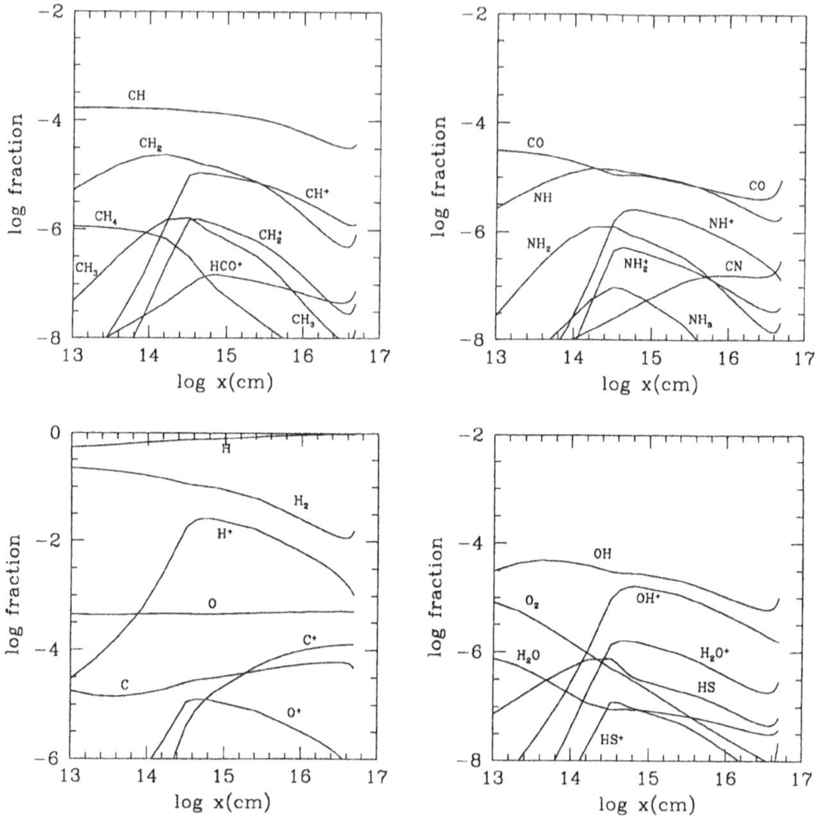

Fig. 3. Chemical abundances (relative to hydrogen nuclei) of some prominent species.

abundant in our models, including CO, CH, OH, HCO^+, and H_2 (see Taylor and Raga (1995) for a summary).

Possibly the best diagnostic of the presence of of a mixing layer would be the simultaneous detection along an edge of a jet of high velocity H_2, together with a $v = 1 - 0\,S(1)\,/\,v = 2 - 1\,S(1)$ line ratio that is incompatible with shock models.

References

Cantó, J.and Raga, A.C.: 1991, *Astrophys.J.* **372**, 646
Roberge, W. and Dalgarno, A.: 1982, *Astrophys.J.* **255**, 176
Taylor, S.D. and Raga, A.C.: 1995, *Astron.& Astrophys.* **296**, 823
Wagner, A.F. and Graff, M.M.: 1987, *Astrophys.J.* **317**, 423

SHOCK FOCUSING AND JET COLLIMATION IN YOUNG STARS

GARRELT MELLEMA
Department of Mathematics, UMIST, P.O. Box 88, Manchester M60 1QD, UK

and

ADAM FRANK
Department of Astronomy, University of Minnesota, 116 Church Street S.E., Minneapolis MN 55455, USA

Abstract. We study the collimation of a initially uncollimated outflow from a young star by its environment, a cloud core with a toroidal density distribution. It is shown that when cooling is unimportant the shape of the inner shock and the internal structure of the bubble will collimate the outflow and it is shown that in some cases cooling is unimportant. The effects of the focusing by an elliptical shock are studied. Numerical simulations show that when cooling is important, the collimation is actually much better and interesting combinations of energy– and momentum–conserving flow occur.

Key words: Stellar jets – Star formation – Hydrodynamics

1. Introduction

While the propagation of fully formed stellar jets has been well studied (see e.g. Raga 1995), the origin of these jets remains unclear. The current consensus tends to favour MHD processes such as disk wind models over pure gasdynamic collimation mechanisms such as DeLaval nozzles (Blandford & Rees 1974). Recent numerical studies of the formation of bipolar planetary nebulae have, however, revealed a new mechanism which can produce well-collimated supersonic jets through purely gasdynamic means (Icke et al. 1992). This process, which we call *Shock Focused Inertial Confinement* (SFIC) is based on the interaction of a central wind and a environment with a toroidal density distribution. The collimation occurs through a combination of non-linear, time-dependent processes that were either not treated or not fully appreciated in previous analytical studies (however see Cantó 1980, and Eichler 1983 for models which rely on similar physics). Because of the ubiquity of the winds interacting with tori in astrophysics, the SFIC collimation appears quite promising as means of producing stellar jets (see also Peter & Eichler 1995). Using conditions appropriate for young stellar objects Frank & Noriega-Crespo (1994) have recently carried out a preliminary study which demonstrated that the SFIC mechanism can produce well collimated outflows with properties consistent with observations of YSO bipolar outflows and jets.

Astrophysics and Space Science **233**: 145–153, 1995.

It is well known that a central wind interacting with a toroidal environment will produce a prolate bipolar bubble. What the SFIC mechanism adds to this picture is the gasdynamics occurring within the bubble. Numerical simulations show that in energy conserving bubbles the reverse shock which forms close to the wind source can also be prolate. The radially outflowing wind strikes this shock at an oblique angle and is refracted towards the polar axis initiating the jet collimation. The focusing of the wind by the reverse shock also produces a shear gradient between the shocked wind and the shocked ambient material leading to Kelvin-Helmholtz instabilities. Unstable material at the contact discontinuity is driven pole-ward by the ram pressure of the collimated wind producing a dense, cold chimney which helps inertialy confine the collimated flow producing a fully supersonic jet (see Fig. 1).

In this paper we examine the basic physics of shock focusing in more detail and attempt to determine its applicability to YSOs through estimations of the effects of cooling. We also extend the numerical work by including the effects of cooling. In the work we present here we focus on the interaction of spherical central winds embedded in aspherical density environments. We note however that processes we describe here should also be applicable to unfocused MHD disk winds.

2. Applicability to YSOs

The YSO simulations show that focusing at the reverse shock is the dominant process collimating SFIC jets. Thus if SFIC collimation is to work for YSOs the outer and reverse shocks must be relatively well separated, i.e. the wind-blown bubble must not be fully radiative. This implies that the shocked wind must have cooling timescales of the order of, or longer than the dynamical timescale for the bubble. Koo & McKee (1992) have formulated the evolution of wind blown-bubbles in a way which accounts for radiative cooling and we use parts of their analysis here to find what regions of YSO parameter space can lead to SFIC collimation.

We begin with the environment. The radial density gradient in a rotating collapsing cloud is assumed to follow a power law

$$\rho(r) = \rho_{01} r^{-k} \,. \tag{1}$$

The collapsing cloud core models of Ulrich (1976) and Terebey, Shu & Cassen (1984), have $k = 3/2$ and $\rho_{01} = \dot{M}_{\rm a}/(4\pi\sqrt{2GM_*})$, in which $\dot{M}_{\rm a}$ is the accretion rate and M_* is the mass of the star.

The bubble blown by the central wind will initially be fully radiative due to the large pre-shock densities in both the wind and ambient gas. The

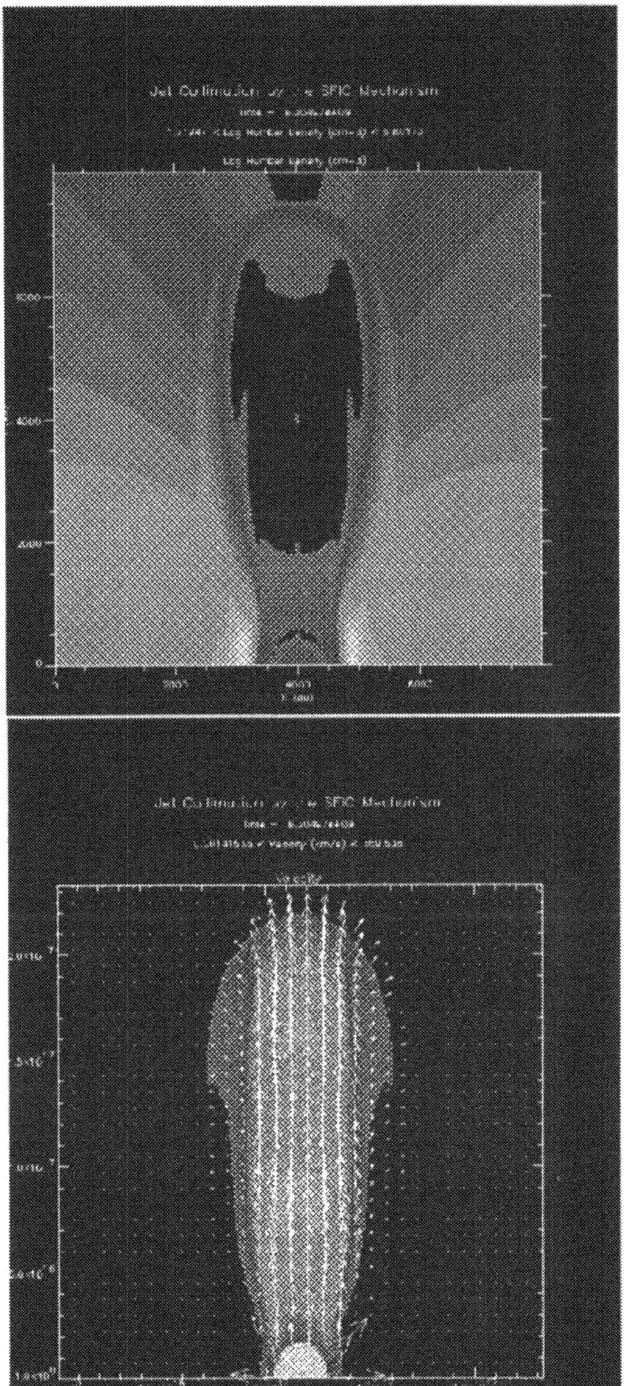

Fig. 1. SFIC Jet Collimation Model. Top: log of the density, bottom: log of velocity with velocity vectors. A spherical central wind expands into an accretion environment specified by a toroidal density distribution. The equator to pole density contrast is constant as a function of radius. The important initial conditions for the model are $\dot{M}_w = 10^{-7}\ M_\odot\ yr^{-1}$, $v_w = 550\ km\ s^{-1}$, $\dot{M}_a = 10^{-6}\ M_\odot\ yr^{-1}$, and q = equator to pole density contrast = 80. The flow is collimated primarily through the focusing by the prolate inner shock visible at the bottom center of the images. This simulation was carried out in cylindrical coordinates with the TVD method of Harten (1983) on a 256^2 grid.

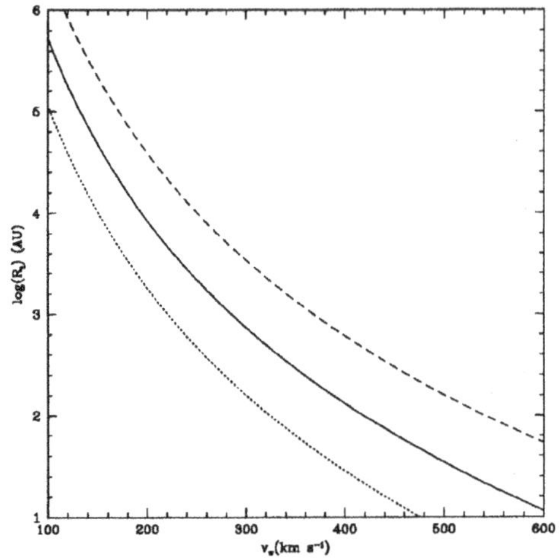

Fig. 2. Radius for transition to a fully energy conserving bubble as a function of wind velocity. When $R > R_t$ the inner shock can focus the wind into a jet. Solid line: $\dot{M}_w = 10^{-7}$ M$_\odot$ yr^{-1}, $\dot{M}_a = 10^{-6}$ M$_\odot$ yr^{-1}. Dotted line: $\dot{M}_w = 10^{-8}$ M$_\odot$ yr^{-1}, $\dot{M}_a = 10^{-6}$ M$_\odot$ yr^{-1}. Dashed line: $\dot{M}_w = 10^{-7}$ M$_\odot$ yr^{-1}, $\dot{M}_a = 10^{-5}$ M$_\odot$ yr^{-1}.. Note that these are upper limits to the transition radius.

expansion of a radiative, momentum-driven bubble into an environment given by Eq. (1) is

$$R_{s1} = \left(\frac{(3-k)\dot{M}_w v_w}{12\pi\rho_{01}}\right)^{\frac{1}{4-k}} \left(\frac{t}{a}\right)^{\frac{2}{4-k}} , \tag{2}$$

where $\dot{M}_w$ and v_w are the mass loss rate and velocity of the wind, R_{s1} is the radius to the outer shock, and $a = \sqrt{2/(12-3k)}$.

The cooling time t_c of the post-shock wind as a function of radius can be determined using the equation of continuity and the familiar results of Kahn (1976)

$$t_c = \frac{C v_s^3}{\rho_w(r)} \approx \frac{4\pi C v_w^4 r^2}{\dot{M}_w} , \tag{3}$$

where $C = 6 \times 10^{-35}$ g cm^{-6} s^4. In Eq. (3) we have assumed that the shock velocity v_s is approximately equal to the wind velocity v_w. Inverting Eq. (2) gives a dynamical time t_{dyn}. Setting $t_c = t_{dyn}$ allows us to solve for the transition radius where the bubble should enter the energy conserving phase. This transition radius is a function of the parameters of the wind and accretion flows,

$$R_t = \left(\frac{a}{4\pi C}\right)^{2/k} \left(\frac{12\pi\rho_{01}\dot{M}_w}{3-k}\right)^{1/k} v_w^{-9/k} \tag{4}$$

In Fig. 2 we show R_t versus v_w for different values of $\dot{M}_a$ and $\dot{M}_w$ and $k = 3/2$. We note that the bubble will become partially radiative (most of the wind mass has cooled but most of the bubble volume is filled with hot newly shocked wind, see Koo & McKee 1992) before it becomes energy conserving so Fig. 2 gives upper limits to R_t. This shows that for relatively fast, less massive winds the conditions for the SFIC mechanism are met.

3. Collimation at the Inner Shock

To determine the degree of focusing in the post inner shock flow we must solve the oblique shock jump conditions. Here we follow the analysis of Icke (1988). The angle (β) between the ellipse and radially directed wind will be a function of polar angle (θ) and is given by

$$\beta = \theta + \arctan\left(\frac{e^2}{\tan\theta}\right), \tag{5}$$

where e is the ellipticity of the shock, given by the ratio of equatorial over polar shock position: $e = R_{s2}(\mathrm{E})/R_{s2}(\mathrm{P})$

Because the inner shock is aspherical, the pre-shock flow variables (which we denote with subscript 0) will be also be a function of polar angle. Accounting for these variations we can write the post-shock variables (denoted with subscript 1) as

$$v_{1p}(\theta) = \left(\frac{\gamma - 1}{\gamma + 1}\right) v_0 \sin\beta(\theta), \qquad v_{1t}(\theta) = v_0 \cos\beta(\theta) \tag{6}$$

$$\rho_1(\theta) = \left(\frac{\gamma + 1}{\gamma - 1}\right) \rho_0(\theta) \tag{7}$$

$$s_1(\theta) = \left(\frac{\gamma - 1}{\gamma + 1}\right) s_0(\theta) \sqrt{\frac{2\gamma}{\gamma - 1} \mathcal{M}_0^2(\theta) \sin^2\beta(\theta) - 1}\,. \tag{8}$$

In Eqs. 9–11, v_{1p} and v_{1t} are the post-shock velocity parallel and tangential to the shock normal, s is the sound speed and $\mathcal{M}$ is the Mach number. The total angle of defection (χ) through the shock is given by

$$\tan\chi(\theta) = \frac{2\tan\beta(\theta)}{(\gamma + 1) + (\gamma - 1)\tan\beta(\theta)}\,. \tag{9}$$

In Fig. 3 we show curves of χ, ρ_1, speed v_1, and post-shock Mach number $\mathcal{M}_1$ as a function of polar angle for three values of the shock ellipticity e. Note that all points along the inner shock where $\chi > \theta$ have fully focused post-shock velocity vectors i.e. the flow at these points is directed towards the polar axis.

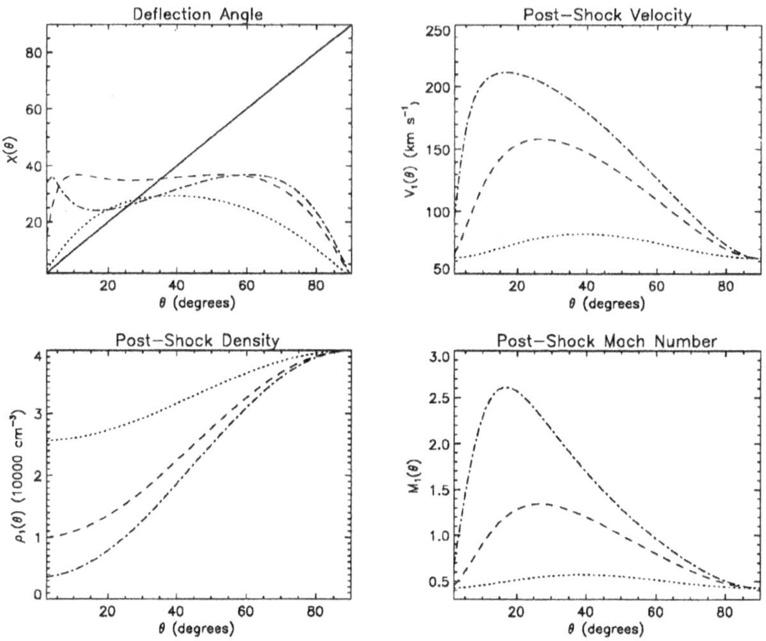

Fig. 3. Collimation at an elliptical inner shock. Shown are deflection angle and post-shock flow quantities for three shocks ($\mathcal{M}_0 = 10$, $v_0 = 250$ km s^{-1}, $n_0 = 10^4$ cm^{-3}) in a toroidal environment with different ellipticities (e). Dash-Dot $e = 0.3$; Dot $e = 0.5$; Dash $e = 0.8$. All deflection angles to the left of the solid line in the upper left figure yield post-shock gas parcels that are focused towards the polar axis.

Figure 3 demonstrates that significant focusing can occur for modest values of the ellipticity. The plot of $\mathcal{M}_1$ also shows that it is possible to obtain collimated supersonic flows through the inner shock in a way that is similar to the refocusing of free jets (Sanders 1983). Given the high Mach numbers of the YSO winds our results imply that fully developed supersonic jets can be formed purely through the action of the inner shock. We note also that the focusing by the inner shock becomes more pronounced as the adiabatic exponent (γ) decreases. Thus radiative cooling behind the shock will lead to better collimation.

4. Numerical Simulations with Cooling

Since cooling is likely to play an important role in many cases, we also performed a preliminary exploration of the SFIC mechanism under cooling conditions. In the numerical model a spherical central wind expands into a toroidal accretion environment given by the formulae in Frank & Noriega-Crespo (1994). The effects of gravity and rotation have been not been considered. The electron and ion densities are derived by calculating the time-dependent ionization fraction of hydrogen due to collision ionization and radiative recombination and these numbers are used to calculate the cooling using a standard cooling curve taken from Dalgarno & McCray (1972).

We show the result of a simulation with $v_w = 250$ km s^{-1}, $\dot{M}_w = 5 \times 10^{-8}$ M$_\odot$ yr^{-1}, and $\dot{M}_a = 5 \times 10^{-7}$ M$_\odot$ yr^{-1}. The other parameters are given in the figure caption (Fig. 4).

The results are complicated since the relevance of the cooling depends on the time and the position. Nevertheless some general trends can be seen. First of all the bubble is less wide than in the equivalent non-cooling case, leading to an intrinsically better collimation. But beside that the collimation is strongly enhanced by the rather extreme shape that the inner shock can attain in the case of significant cooling. When cooling is efficient the shape of the inner shock will follow the shape of the bubble, and the shock will therefore be much more aspherical than in the non-cooling case. This can be seen in the temperature plot, where the narrow interface of hot gas indicates the region between the inner shock and contact discontinuity. As was shown above a very aspherical shock will have serious consequences for the collimation. One of these effects is the formation of a dense, fast, narrow jet at the top of the bubble. This structure is most likely caused by the convergence of strong flows at the symmetry axis, where they collide and produce this structure, not unlike as in the mechanism described in Cantó (1980).

At later times, the top part of the bubble changes from momentum– to energy-conserving, as can be seen by the volume of hot gas at the top of the bubble. This is mainly caused by the lower densities in the wind at the top of the bubble. The position at which this will happen depends on the densities and wind velocity of course.

5. Conclusions

Based on numerical simulations and analytical derivations of cooling times, shock ellipticity, and post-shock flow focusing we conclude the collimation of outflows through the SFIC mechanism is a potentially powerful and universal means for producing the supersonic jets observed in star forming regions. In both cooling and non-cooling circumstances considerable collimation is achieved and the structures appear to be stable on the timescales of our simulations. Our calculations also show that the physics of these bubbles is neither fully momentum- or energy-conserving, which makes analytical treatment of the problem quite difficult.

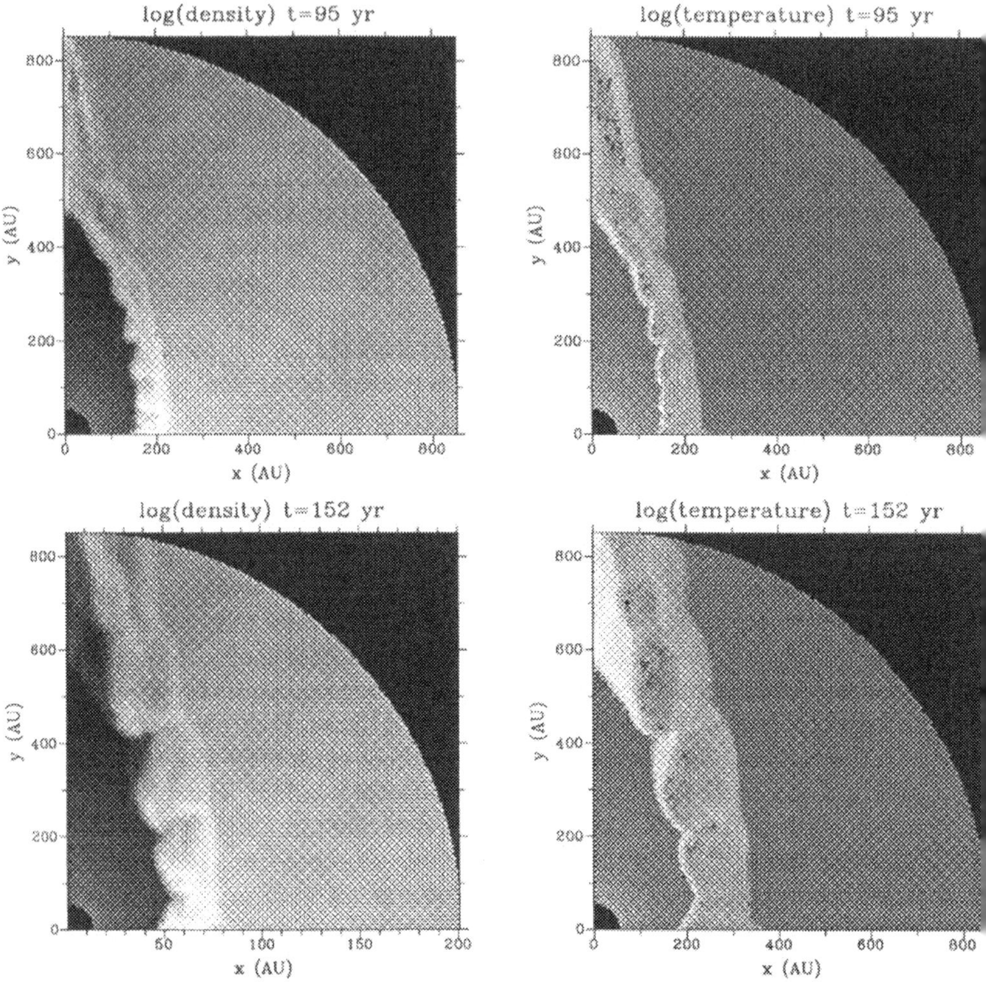

Fig. 4. Results of numerical simulations with cooling. Boxes on the left show the density, those on the right temperature. The important initial conditions for the model are $\dot{M}_w = 5\ 10^{-8}\ M_\odot\ yr^{-1}$, $v_w = 250$ km s^{-1}, $\dot{M}_a = 5\ 10^{-7}\ M_\odot\ yr^{-1}$, and q = equator to pole density contrast = 7.5. Simulation on a grid of 120^2 using the method of Roe (1981) as described in Mellema (1993). At $t = 95$ yrs, $\log\rho$ (cm^{-3}) ranges from 2.1 to 6.6, at $t = 153$ yrs from 2.0 to 6.4; the temperature ranges from 0.8 to 5.4 and 0.6 to 5.6

Numerous questions remain to be answered before we can be state definitively that the mechanism will work for YSOs. The effects of more realistic environments, rotation and gravitation are all expected to influence the results. In addition the ability of the processes describe here to collimate jets over long length and timescales ($L > 1$ pc, $t > 1000$ years) is not clear. We will address these issues in future investigations.

References

Blandford R.D., Rees, M.J.: 1974, *Mon. Not. R. astr. Soc.* **169**, 395
Cantó, J.: 1980, *Astron. Astrophys.* **86**, 327
Dalgarno, A. and McCray, R.: 1972, *Ann. Rev. Astron. Astrophys.* **10**, 375
Eichler, D.: 1983, *Astrophys. J.* **272**, 48
Frank, A. and Noriega-Crespo A.: 1994, *Astron. Astrophys.* **290**, 643
Harten, A: 1983, *J. Comput. Phys.* **49**, 357
Icke, V.: 1988, *Astron. Astrophys.* **202**, 177
Icke, V., Mellema, G., Balick B., Eulderink, F., Frank, A.: 1992, *Nat* **355**, 524
Kahn. F.D.: 1976, *Astron. Astrophys.* **50**, 145
Koo, B. and McKee C.F.: 1992, *Astrophys. J.* **388**, 93
Mellema, G.: 1993, *PhD thesis: Numerical models for the formation of aspherical planetary nebulae*, Leiden University
Peter, W. and Eichler, D.: 1995, *Astrophys. J.* **438**, 244
Raga, A.C.: 1995, 'Models of HH objects and molecular outflows' in S. Lizano, ed(s)., *Circumstellar disks, outflows and star formation*, in press,
Roe, P.L.: 1981, *J. Comput. Phys.* **43**, 357
Sanders, R.M.: 1983, *Astrophys. J.* **266**, 73
Terebey, S., Shu F.H., Cassen, P.: 1984, *Astrophys. J.* **286**, 529
Ulrich, R.K.: 1976, *Astrophys. J.* **210**, 377

DISINTEGRATION OF DUST AGGREGATES IN INTERSTELLAR SHOCKS AND THE LIFETIME OF DUST GRAINS IN THE ISM

C. DOMINIK
NASA Ames Research Center, Moffett Field, California
A.P. JONES
Space Sciences Laboratory, University of California, Berkeley
and
A.G.G.M. TIELENS
NASA Ames Research Center, Moffett Field, California

Abstract. We have studied to influence of grain porosity on the dust destruction mechanism in interstellar shocks. Our results show, that fluffy aggregates of dust grains can be broken up easily in shocks.

Key words: Dust Aggregates – Dust Destruction – Porous Grains

1. Introduction

Interstellar grains are involved in a complex lifecycle in the interstellar medium (ISM). Grains are produced in a variety of sources (red giant winds, supernovae, novae) and ejected into the ISM. Within the ISM, the grains are subject to several processes that alter the grain. A non-complete list is

- Accretion and evaporation of ice mantles
- Coagulation in dense regions
- Destruction by sputtering, shattering and evaporation in fast interstellar shocks.

The latter process is quite effective. Some degree of grain destruction is necessary in order to produce the observed element abundances in the interstellar gas. However, Jones *et al.* (1994) have shown, that the expected lifetimes of dust grains in the ISM are actually too short to explain the observed dust abundances. Therefore, the grains must be protected in some way. Besides refractory coatings on grains (Jones et al. 1994), grain destruction can also be substantially limited if the larger grains in the ISM are not solid blocks but aggregates of small grains. Protection here works in two different ways:

1. Porous aggregates of small grains have a relatively large surface-to-volume ratio which increases the efficiency of drag forces. Therefore, porous grains should be swept up by the gas much faster and both number and energy of grain-grain collisions will decrease.

Astrophysics and Space Science **233**: 155–159, 1995.

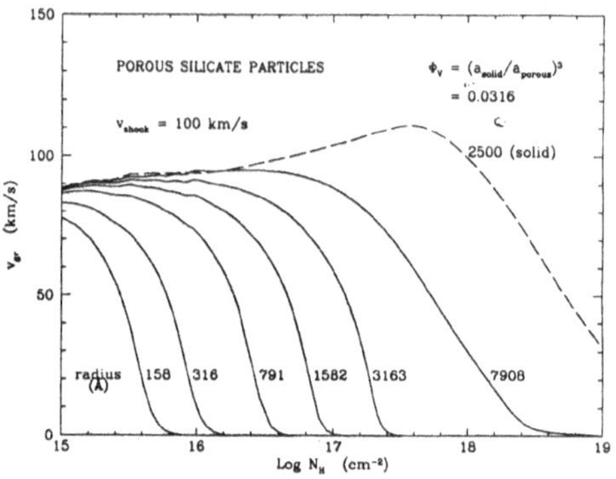

Fig. 1. Porous silicate grain post-shock velocities as a function of the shocked column (N_H), for various grain radii in a 100 km s^{-1} shock.

2. When a grain aggregate is hit by a small grain it might disintegrate into its constituent grains which are smaller and less susceptible to destruction. The depleted reservoir of big particles could be repopulated in dense regions by coagulation (e.g. Chokshi *et al.* 1993).

In the following we will discuss these two possibilities. We will do this by showing the velocities and collision probabilities for porous grains in a 100km $^{-1}$ shock. We also discuss numerical simulations of the collision between a small dust aggregate and a single grain.

2. Porous grains and grain velocity calculations

The motion of grains, relative to the gas, in the post-shock region is governed by a competition between the betatron acceleration and the drag deccelera-tion. The betatron acceleration of the grains occurs in the cooling gas where the compression produces an increasing B field which accererates the charged grains with respect to the gas. This acceleration is opposed by collisional and plasma drag forces which try to bring the grains to rest with respect to the gas.

In this calculation we study porous particles through the use of a volume filling factor

$$\Phi_V = \frac{a^3_{solid}}{a^3_{porous}} \tag{1}$$

and the mean density of the porous grains is then

$$\rho = \rho_{solid}\Phi_V. \tag{2}$$

We have adopted $\Phi_V = 10^{-1.5}$, which corresponds to highly porous grains with $\approx 97\%$ pores. The "inflated" grains are then treated in the same way as solid particles in calculating the post-shock velocity (Jones et al. 1994).

Figure 1 shows the post-shock velocities of porous silicate grains as a function of the shocked column. For homogeneous solid particles the grain destruction in a 100 km/s shock peaks at a shocked column of $\approx 10^{18}$ cm^{-2}. We show the post-shock velocity for a 2500Å solid (non-porous) grain for comparison (dashed line), which has the same mass as the 7908Å porous grain. Note the decreased peak velocity for the porous grain 7908Å grain compared to the solid grain of the same mass (2500Å), and also the more rapid "sweeping-up" of the porous grains by the gas (about an order of magnitude "sooner" in the shocked column N_H). Hence, compared to solid homogeneous grains, the destruction of all the porous grains considered here will be reduced by a factor of three (Jones et al. 1994), compared to solid particles.

In a 100 km/s shock only the largest grains (radii > 3800Å) experience more than one collision with other grains per shock.

3. Numerical Simulations of Aggregate-Particle Collisions

In the following we show results of a numerical simulation of collisions between an aggregate of dust grains and a fast single grain. We solve the full set of equations of motion for each single grain. The simulation includes

- Interaction of the grain by van der Waals forces and elastic repulsion (Johnson *et al.*, 1971).
- Energy dissipation by excitation of elastic waves (Chokshi *et al.* 1993).
- Energy dissipation by friction for both sliding and rolling (Dominik and Tielens 1995a,b).

The calculation have been carried through in two dimensions.

Figure 2 shows the results of collision simulations. The single grain comes in from the right hand side and hits the aggregate at the specified velocity. The figure shows the "final state" which is defined by the the fact that both sliding and rolling have stopped completely.

Very slow collisions (200 cm s^{-1}, first frame) do not even deform the aggregate. The grains just sticks to the aggregate. For intermediate collision velocities ($\approx$1000 cm s^{-1}, third frame), some compression takes place. At higher velocities (2000 cm s^{-1}, fourth frame) the aggregate first breaks into a few pieces, until it finally completely disintegrates at collision velocities of about 10000 cm s^{-1} (last frame). It is also important to note that the incoming grain does not punch a hole through the cluster. Since the grain sizes are all the same, energy is distributed very effectively between the grains and all contacts break.

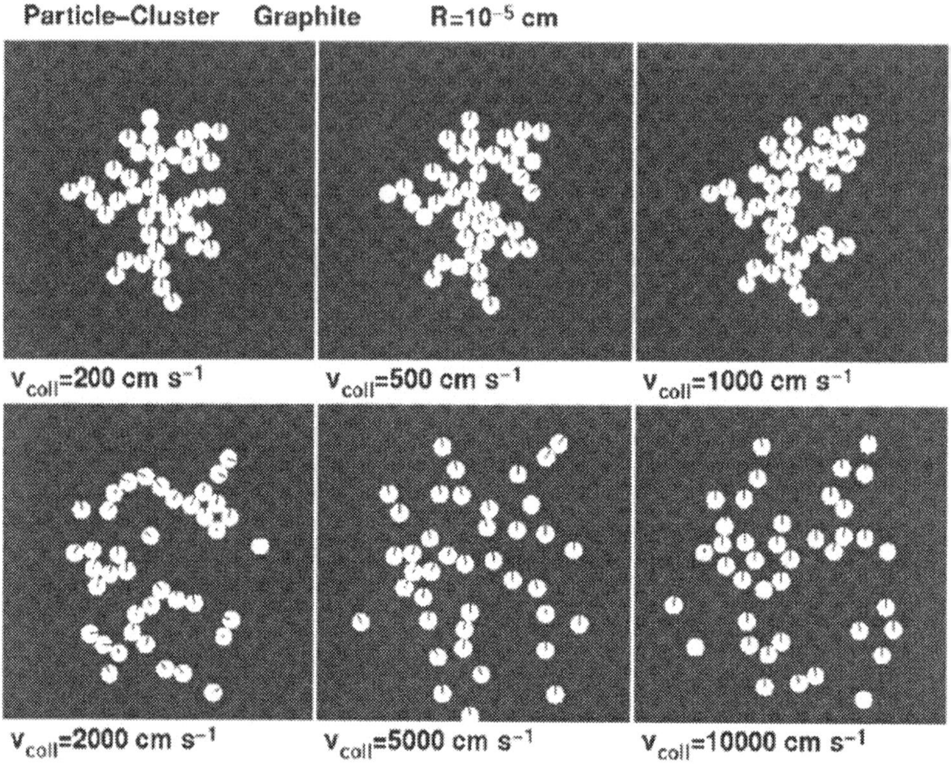

Fig. 2. Outcomes of collisions between a single grain and an aggregate at different velocities.

4. Conclusions

1. Due to the reduced velocities alone, porous grain destruction is reduced by a factor of three (Jones et al. 1994), compared to solid particles, in interstellar shock waves. The resulting grain life-times are enhanced by a factor of three, and the equilibrium fraction of the condensible elements determined to be in stardust is of order 25%.
2. Only the largest porous grains (radii > 3800Å) will experience at least one collision per 100 km s^{-1} shock.
3. If the size of the impacting grain is comparable with the size of the aggregate sub-particles, energy redistribution is so effective that virtually all contacts within the aggregate will be broken.
4. The energy required to disintegrate a 40-grain aggregate can be provided by a 5000-10000cm s^{-1} collision. For bigger aggregates this energy should scale roughly with the number of grains involved. The critical velocity will therefore increase as the square root of the aggregate size. Since grain

velocities in interstellar shocks are 2 orders of magnitude larger than the velocity used above, much bigger aggregates could still be destroyed.

References

Chokshi, A., Tielens, A.G.G.M. and Hollenbach, D.J.: 1993, *Astrophys. J.* **407**, 806

Dominik, C. and Tielens, A.G.G.M.: 1995, *Phil. Mag. A* **in press**,

Dominik, C. and Tielens, A.G.G.M.: 1995, *Phil. Mag. A* **in preparation**,

Johnson, K.L., Kendall, K. and Roberts, A.D.: 1971, *Proc. R. Soc. Lond.* **342**, 301

Jones, A.P., Tielens, A.G.G.M., Hollenbach, D.J. and McKee, C.F.: 1994, *Astrophys. J.* **433**, 797

Jones, A.P. and Tielens, A.G.G.M.: 1994, '' in The Cold Universe, XIIIth Moriond Astrophysics Meeting, ed(s)., *T.Montmerle, C.J.Lada, I.F.Mirabel, and J.Tran Thanh Van*, Editions Frontieres: Gif-sur Yvette, p. 35

McKee, C.F., Hollenbach, D.J., Seab, C.G. and Tielens, A.G.G.M.: 1987, *Astrophys. J.* **318**, 674

Tielens, A.G.G.M.,McKee, C.F., Seab, C.G. and Hollenbach, D.J.: 1994, *Astrophys. J.* **431**, 321

MOLECULAR DIAGNOSTICS OF DIFFUSIVE BOUNDARY LAYERS

J. M. C. RAWLINGS
Department of Physics and Astronomy, University College London, Gower Street, London WC1E 6BT

and

T. W. HARTQUIST
Max-Planck-Institut für extraterrestrische Physik, 85740 Garching bei München, Germany

Abstract. We have assessed the role of diffusion in determining chemical abundances in molecular interface regions. Chemical models have been developed which include the appropriate diffusion terms and that are appropriate to a narrow diffusion region (~0.01pc) that may exist at the interface between a dark core and a hot, shocked T-Tauri wind. We have assumed pressure balance throughout and have calculated the chemical abundances as functions of time and position through the interface. The results show that significant enhancements of detectable molecules/transitions are expected (e.g. CO J=6→5, OH and CH). Using a realistic value of the diffusion coefficient a diffusive region of dimension 0.01pc may be established within about 10^4 years. In general it seems likely that diffusion processes are highly significant on these and smaller lengthscales.

Key words: Stars – Chemistry – Diffusion

1. Introduction

Mass, momentum and energy transfer processes in boundary layers between hot tenuous flowing plasma and the clumps embedded within them are important for the global structures and evolutionary behaviour of nearly all diffuse astrophysical media. Examples include star forming regions, supernovae, the intergalactic medium during the era of galaxy formation and starburst galaxies and other active galaxies (e.g. McKee and Ostriker 1977; Harquist and Dyson 1993). However, these processes are not subject to first principles theoretical investigation; rather, they must be studied semiempirically. Amongst the boundary layers that are most likely to be resolved spatially and to be studied with the greatest variety of diagnostics are those that exist between T-Tauri stellar winds and the dark molecular cores that are the progenitors of star formation.

Charnley et al. (1990) and Nejad and Hartquist (1994) have performed some investigations of the chemistry in such dark core – T-Tauri wind interface regions, but the models employed were very simple with ion source terms added to the chemical rate equations to describe the mixing of ions into the neutral gas; the mixing of neutral gas into the ionized material was not considered. In order to introduce a plausible self-consistent treatment of the

Astrophysics and Space Science **233**: 161–164, 1995.

mutual effect of the mixing of ions into neutral gas and molecules into ionized gas we adopt a diffusion model of the boundary layer between the TMC-1 dark core and the wind of the T-Tauri star nearest to it. We assume that either ambipolar diffusion or magnetic reconnection occurs in the turbulent boundary layers in order for diffusive mixing to occur (Hartquist and Dyson, 1988).

In the absence of diffusion the time-dependence of the abundances of the various chemical species is typically governed by equations of the type:

$$\frac{dn_i}{dt} = \sum_A \sum_B k_{AB} n_A n_B \pm \sum_i \beta_i n_i$$

where n_i is the abundance of species i, k_{AB} and β_i are the rate co-efficients for bi-molecular reactions and photoreactions respectively.

If diffusion is included then extra terms are generated by the well-known diffusion equation:

$$\frac{\delta n_i}{\delta t} = \frac{\delta}{\delta x}\left(\kappa \frac{\delta n_i}{\delta x}\right)$$

2. The Chemical Models

As a first step towards developing descriptive chemical models of diffusive interfaces we have considered a narrow diffusion region ($\sim$0.01 pc) that may exist at the interface between a dark cloud (of radius $\sim$0.1 pc, T=20 K) and a hot, shocked wind (T=10^4 K). We have assumed pressure balance throughout (with nT=10^7 cm^{-3}K, which may be appropriate for the densest fragments in TMC1). Such a crude method clearly does not accurately describe the temperature structure through the diffusive zone but at least with this approach there is an implied coupling between the density scaleheight and the turbulent lengthscale. We assume that quasi-equilibrium conditions exist immediately prior to initiating the diffusion calculations. We have used a reduced chemistry involving 30 species in about 260 chemical reactions. The chemistry has been limited to the elements H, He, C, O and a representative low ionization potential metal (Na). We used a standard ordinary differential equation integration package (GEAR) to calculation the abundances as functions of time and position. The radial grid consisted of some 100 points, integrated separately and coupled via a cubic spline fitting procedure. This had the advantage of naturally generating the spatial second derivatives from which the diffusion terms were calculated. We assume that each atomic, or molecular species can be treated as a separate fluid and that the diffusion coefficient (κ) is the same for all species. Hartquist and Dyson (1988) argued that the effective turbulent viscosity is self-regulatory such that the effective

Reynolds number is of order 1000 and the thickness of the boundary layer is typically a few percent of the cloud size. From this we may deduce that κ lies in the range 10^{19}–10^{21}. A 'standard' value of 10^{20} is adopted. The model parameters are summarised in Table 1.

TABLE I

Parameters for the static interface model

Pressure	10^7 cm^{-3} K
Temperature (Outer region)	10^4 K
Temperature (inner region)	20 K
A_v (Outer boundary)	1.0 mag.
Diffusion coefficient (κ)	10^{20}
$R_{interface}$	0.01 pc

3. Results

Examples of the results obtained (for the molecular species CO and O_2) are shown in figure 1. The results are presented as the log of the number density as a function of position. The initial contact discontinuity occurs at $R = 1.5\times 10^{16}$ cm. The region on the left of this position is the hot, shocked wind; whilst that on the right is the cold, dark cloud. Curves are plotted for different output times and show the progress of the diffusion region.

These results demonstrate some very interesting features:

- The turbulent diffusion zone is established very quickly. The timescale goes like (R^2/κ) where R is the size of the interface region. For R=0.01 pc and κ=10^{20} this timescale is about 2000 years
- Very strong (detectable) enhancements of many molecular species occurs in the interface. CO J:6→5 emission (requiring T=100 K, n=10^5 cm^{-3}) should be particularly noticeable.
- The inclusion of diffusion terms is essential in 'high resolution' chemical modelling where enhanced observational techniques push us closer to the diffusion limit.

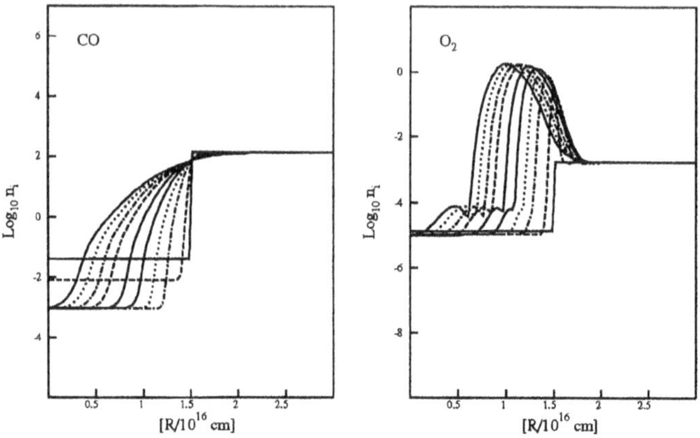

Fig. 1. Abundances as a function of position and time. The curves correspond to the output times: 0, 6.3, 81, 195, 369, 629, 966, 1260, 1590 and 1920 years.

4. Future Developments

We are currently developing the models so as to describe the flow of material through the ablation fronts associated with low mass star-forming regions. It is in these regions that we expect turbulent mixing to be most significant. The flow-time of the hot (10^4 K) gas past the cloud is about 10^4 years. At the outer boundary of the diffusive zone we assume that the diffusion of hot, shocked gas balances convection away from the region thus preventing further growth of the boundary layer. In this case we are looking for a quasi-steady-state solution for a wind as it passes through the ablation front. Preliminary results again indicate that very significant enhancements are expected for observable species, such as CO, CH and OH.

References

Charnley, S.B., Dyson, J.E., Hartquist, T.W. & Williams, D.A.: 1990, *Mon. Not. R. astr. Soc.* **243**, 405
Hartquist, T.W. & Dyson, J.E.: 1988, *Astrophys. Sp. Sci.* **144**, 615
Hartquist, T.W. & Dyson, J.E.: 1993, *Q. J. R. astr. Soc.* **34**, 57
McKee, C.F. & Ostriker, J.P.: 1977, *Astrophys. J.* **218**, 148
Nejad, L.A.M. & Hartquist, T.W.: 1994, *Astrophys. Sp. Sci.* **220**, 253

HH 30 : THE INTERACTION OF A HERBIG-HARO JET WITH A STRATIFIED MEDIUM ?

A. C. RAGA*
Mathematics Dep., UMIST, P.O. Box 88, Manchester M60 1QD
and
J. CANTÓ
Instituto de Astronomía, UNAM, Ap. 70-264, 04510 México, D. F.

Abstract. This paper describes a comparison of observations of the HH 30 jet/counterjet system and theoretical models of jets propagating in a strongly stratified medium. We find that the observed westward bending of the HH 30 jet and counterjet can be explained as the result of a plane-parallel pressure stratification of the surrounding environment. This model predicts specific properties for the kinematics of the outflow, that could be straightforwardly checked with future spectroscopic and proper motion studies of HH 30.

Key words: Hydrodynamics – ISM – Jets and Outflows

1. Introduction : The HH 30 Jet

The HH 30 jet has been studied in detail in a number of papers by Mundt and collaborators (Mundt, Brugel and Bührke 1987; Mundt, et al. 1990; Mundt, Ray and Raga 1991). López et al. (1995) have recently shown that the HH 30 jet and the counterjet both curve westward, resulting in a morphology that, though less extreme, is somewhat reminiscent of the morphologies of "head/tail" radio jets. This curvature of the HH 30 jet/counterjet system is clearly seen in figure 1.

This striking morphology could be the result of a movement of the outflow source through the surrounding environment. Balsara and Norman (1992) have presented numerical simulations which show that such a motion of the source does result in a "swept-back" morphology that resembles "head/tail" radio jets. López et al. (1995) indeed favour such a scenario for explaining the curvature in the plane of the sky of the HH 30 jet and counterjet.

An alternative scenario (which appears to be favoured by more recent, and still unpublished observations) is that the environment of the HH 30 jet has a large-scale pressure gradient which bends the beams of the jet and the counterjet. We discuss this scenario in detail in the following section.

* Now at the Instituto de Astronomía, UNAM

Astrophysics and Space Science **233**: 165–168, 1995.

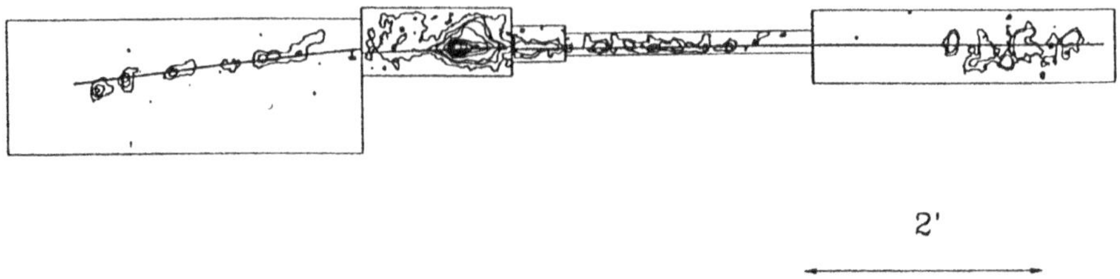

Fig. 1. [S II] 6717+31 image showing the morphology of the HH 30 jet/counterjet system. The peak intensity corresponds to the approximate position of the source. Three straight lines have been drawn to show the approximate direction of the outflow close to the source, as well as the directions of the jet and counterjet farther away from the source. This figure has been taken from López et al. (1995).

2. A Model for HH 30

Cantó and Raga (1995) have presented an analytic formalism for solving the problem of a radiative jet travelling in an environment with a plane-parallel, exponential pressure stratification, and we can use their results to compute a model for HH 30. In figure 2, we show the locus of the beams of a jet/counterjet system with an initial (isothermal) Mach number $M_0 = 16$, moving at an initial angle $\theta_0 = 30°$ with respect to the isobars of the environmental pressure stratification. It is clear from this figure that it is indeed possible to obtain a curvature that is qualitatively similar to the one of the HH 30 outflow, over a distance of several environmental scale heights.

An interesting result is that the beam that is initially travelling in the direction of increasing pressure suffers a larger deflection than the oppositely directed jet beam. We would therefore associate the HH 30 counterjet with the beam that is initially travelling into the pressure stratification (as it is more strongly curved than the HH 30 jet, see figure 1).

Figure 1 also shows the modulus of the flow velocity along the beams of the jet and counterjet. It is interesting to note that while the jet (i. e., the beam that is initially directed down the pressure gradient) has an approximately constant velocity, the more strongly curved counterjet appreciably slows down with increasing distance from the source. This prediction could

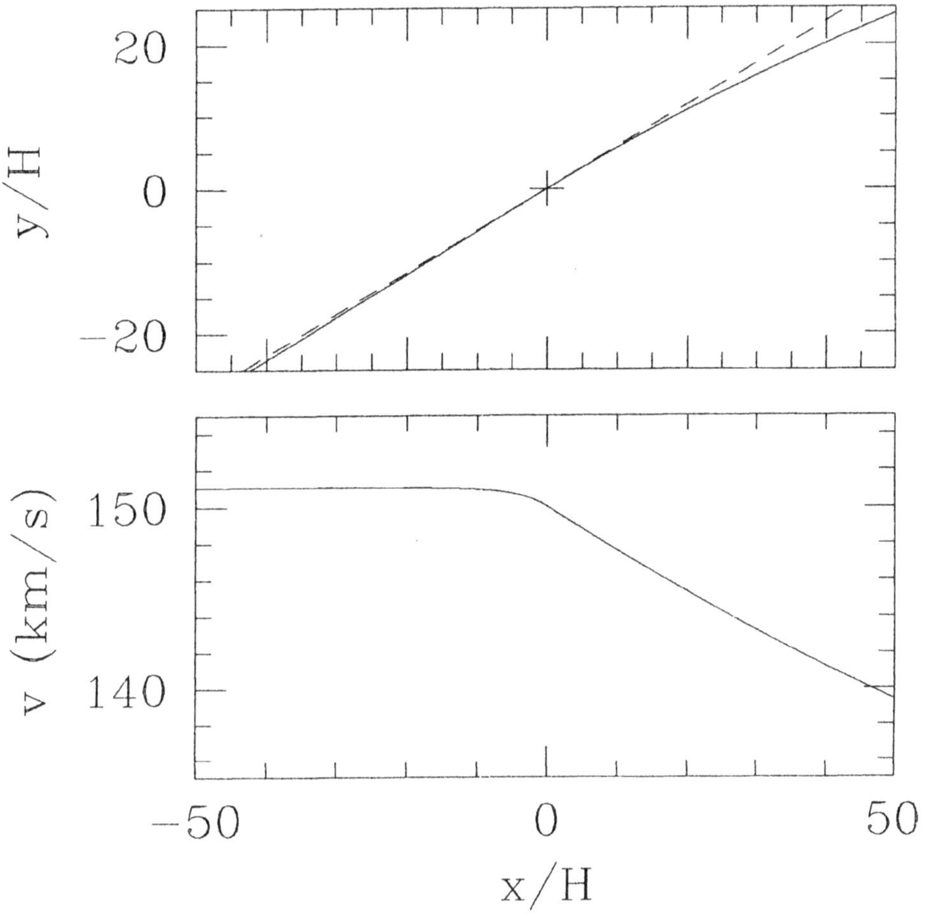

Fig. 2. *Top* : locus of a jet/counterjet system ejected from $(x, y) = (0, 0)$ (the position of the source being indicated by a cross) with an initial (isothermal) Mach number $M_0 = 16$ (resulting from an initial velocity $v_0 = 150$ km/s and temperature $T_0 = 10^4$ K) at an angle of 30° with respect to the x-axis. It is assumed that the environment has a pressure stratification $P(y) = P_0 \exp(y/h)$. *Bottom* : modulus of the velocity along the beams of the jet and the counterjet for the model described above.

be checked with future proper motion and spectroscoipc observations of the HH 30 outflow.

3. Summary

The HH 30 jet/counterjet system has a curved, arch-like morphology that resembles "head/tail" radio jets. We have presented a simple model that explains this curvature as the result of a large-scale pressure gradient in the surrounding environment.

Our model also gives a prediction of the expected distribution for the flow velocity along the jet and counterjet. This prediction can be straightforwardly used to either confirm or reject this model when radial velocity and proper motion measurements of the HH 30 outflow become available.

References

Balsara, D. S. and Norman, M. L.: 1992, *Astrophys. J.* **393**, 631
Cantó, J. and Raga, A. C.: 1995, *Astrophys. J.* , submitted
López, R., Raga, A. C., Riera, A., Anglada, G. and Estalella, R.: 1995, *Mon. Not. R. astr. Soc.* **274**, L19
Mundt, R., Brugel, E. W. and Bührke, T.: 1987, *Astrophys. J.* **319**, 275
Mundt, R., Ray, T. P., Bührke, T., Raga, A. C. and Solf, J.: 1990, *Astron. Astrophys.* **232**, 37
Mundt, R., Ray, T. P. and Raga, A. C.: 1991, *Astron. Astrophys.* **252**, 740

LOW VELOCITY SHOCK-CLOUD ENCOUNTERS I.

Two Dimensional Hydrodynamical Model

ANDRÁS HORVÁTH JR.
Loránd Eötvös University, Budapest, Hungary
Széchenyi István College, Győr, Hungary

and

L. VIKTOR TÓTH
Loránd Eötvös University, Budapest, Hungary

Abstract. Shocks propagating in the interstellar medium (ISM) play an important role in the life of molecular clouds. Through a theoretical study of interaction between clouds and shocks we can understand, for example, the density distribution of observed molecular clouds and the first steps of star formation. The only way to study of interaction in detail is via a numerical hydrodynamical simulation. In this paper we present the first results of a hydrocode which is able to follow the processes after the collision between the cloud and shock front.

Our main theoretical result is that the chemical processes (e.g. H_2 dissociation) can affect the dynamical processes significantly. Global parameters of the cloud are calculated for the comparision of the simulation and the observations.

Key words: molecular clouds – shock fronts – numerical simulation

1. The Numerical Model

In our calculations we take into account the forces of the gas pressure, the radiative cooling in optically thin approximation; gravitational forces are neglected in the present version. In the model we can use two kinds of equation of state, namely that of an ideal gas, and a H_2–H–He mixture (Monaghan and Lattanzio 1992). In the high-density range of the cloud ($\rho > 10$ H-atoms cm^{-3}) we use a cooling function which includes H_2 and CO cooling (Hollenbach and McKee 1979). The numerical model is based on the ZEUS hydrocode (Stone and Norman 1992). Cylindrical symmetry (2D simulation) is assumed. In our calculations the initial cloud was spherically symmetric and embedded in pressure equilibrium within a homogeneous low-density, warm ISM. For the initial model of the shock we used the Rankine-Hugoniot conditions. In the present version we neglected radiative cooling in the low-density range.

The parameters of the model are the following: The density ratio of the cloud and the ambient medium (χ), the Mach-number of the shock front (M), the initial density ($\rho_{\rm cl}$), radius ($R_{\rm cl}$) and temperature ($T_{\rm cl}$) of the cloud. The time is measured in "cloud crossing time" units: $T_{\rm cc} = 2R_{\rm cl}/v_{\rm shock}$

Astrophysics and Space Science **233**: 169–173, 1995.

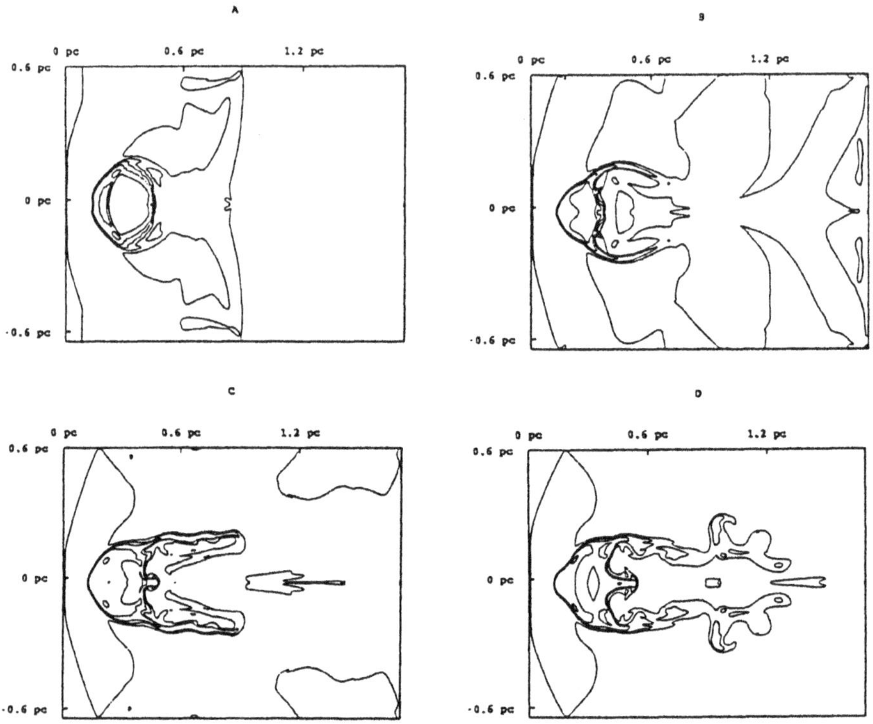

Fig. 1. Calculation with ideal gas dynamics ($M = 2.6$, $\chi = 80$, $\rho_{\rm cl} = 80\, m_{\rm H}\,{\rm cm}^{-3}$, $R_{\rm cl} = 0.2\,{\rm pc}$, $T_{\rm cl} = 100\,{\rm K}$).

The calculations were performed on a RISC/6000 workstation. We used a 320×120 equidistant grid. The Courant number was equal to 0.4 .

2. Results

The validity of the model was checked by test calculations. Comparision with analytical solutions in some test cases and comparision with the previous simulations was performed. We could reproduce the results of Bedogni and Woodward (1992) to within a small difference with ideal gas dynamics.

In Fig. 1 and Fig. 2 we show the effect of changing the equation of state. The timestep between pictures is $\Delta T = 2.0 T_{\rm cc}$. The figures are density contour maps with contours of 1, 3.16, 10, 31.6, ... in $m_{\rm H}\,{\rm cm}^{-3}$ units.

We can establish that the equation of state significantly influences the evolution of the cloud. In the ideal gas case, a tube-like part appears and

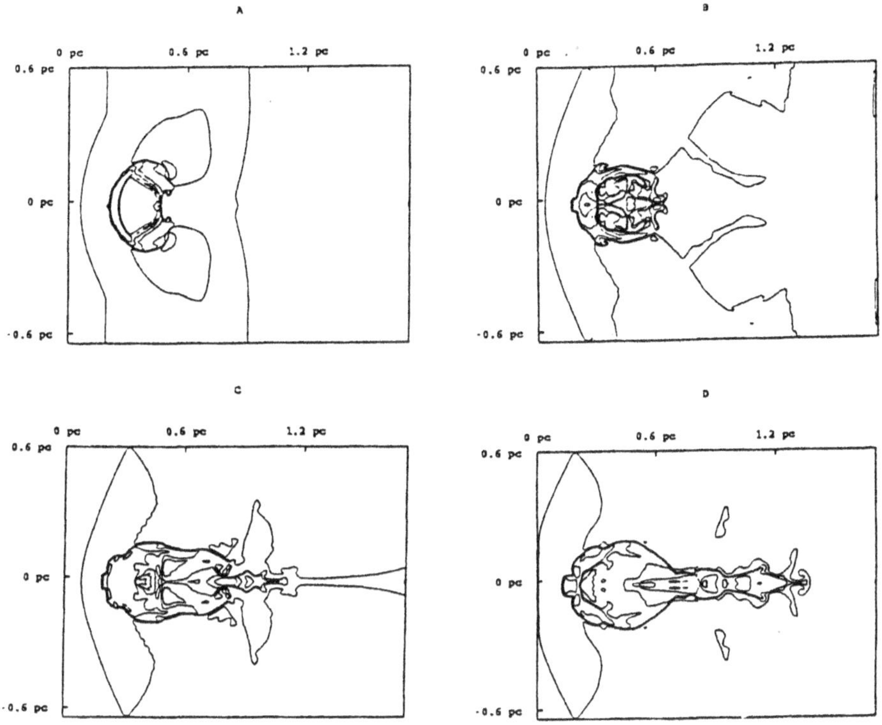

Fig. 2. The same as Fig. 1. but with a H–He–H_2 mixture.

grows while in the non-ideal case this part forms a tail along the axis of symmetry. Another difference is that the maximum density is higher by a factor of 6 in the second case. This higher density can affect, for example, star formation.

The point of our calculations is to compare the numerical model with observations. For this reason we divided the cloud into 3 density regions: 10–100 m_H cm^{-3}, 100–1000 m_H cm^{-3} and 1000–10000 m_H cm^{-3}. The X coordinate and the velocity of center of mass of different regions are presented in Fig. 3 and Fig. 4. (The signs of different regions are $\diamond$, $+$ and a $\square$ respectively.) We hope that it is possible to calculate, for example, the Mach-number if we compare these kinds of graph with the large-scale parameters of clouds.

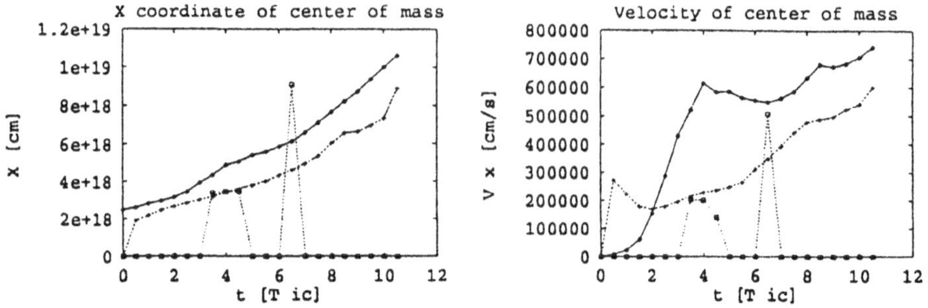

Fig. 3. X coordinate and velocity of center of mass ($M = 5$, $\chi = 80$, $\rho_{\rm cl} = 80\, m_{\rm H}\, {\rm cm}^{-3}$, $R_{\rm cl} = 0.6\,{\rm pc}$, $T_{\rm cl} = 100\,{\rm K}$).

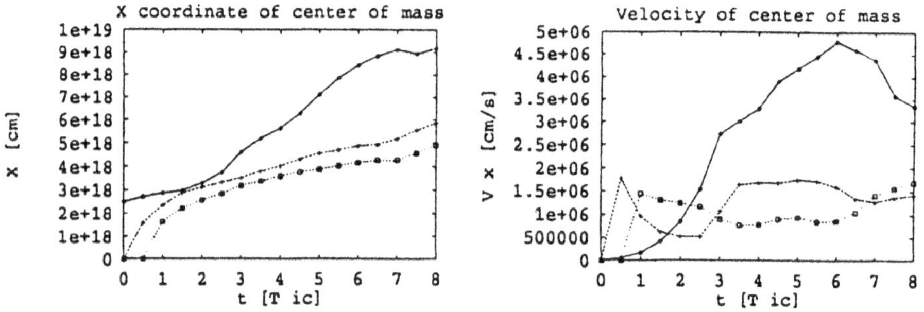

Fig. 4. The same as Fig. 3. but $M = 10$.

Acknowledgements

Many thanks to Udo Ziegler (University of Würzburg) for the first version of the hydrocode and to the "Universitas Győr Foundation" for the financial support. This research was partly funded by the Hungarian Research Foundation, OKTA-F4239.

References

Bedogni, R. and Woodward, P.R.: 1992, *Astron. Astrophys.* **231**, 481
Hollenbach, D. and McKee, C.: 1979, *Astrophys. J. Suppl.* **41**, 555
Monaghan, J.J. and Lattanzio, J.C.: 1989, *A Simulation of the Collapse and Fragmentation of Cooling Molecular Clouds*, Lawrence Livermore National Laboratory

Stone, J.M. and Norman, M.L.: 1992, *Astrophys. J. Suppl.* **80**, 819

LOW VELOCITY SHOCK-CLOUD ENCOUNTERS II.

Observations and Interpretation

L. VIKTOR TÓTH AND ANDRÁS HORVÁTH JR. *
Loránd Eötvös University, Budapest, Hungary
and
LAURI K. HAIKALA
Helsinki University Observatory, Finland

Abstract. We report on a comparative study of nearby shocked clouds with and without star formation, based on IRAS, HI(21cm), CO(1-0) NH_3 (and other molecular line) observations. The dark clouds L1780 (no star formation) and L1251 (high SFE) are discussed here. Their density and velocity structure are compared with the predictions of the HD model of Horváth & Tóth (1995), Paper I.

Key words: Molecular Clouds – Interstellar Shocks

1. Introduction

According to our present knowledge of the SN frequency and the filling factor of cold dense gas in the Galactic disk, clouds can not avoid shock collision(s) during their lifetime (see e.g. Heathcote & Brand 1983). Shocked diffuse clouds may became gravitationaly bound and even form stars with a high star formation efficiency (SFE). The shocked low and intermediate mass clouds should be important locations of low and intermediate mass star formation. The shock should have a low velocity (or small Mach number) not to destroy the small clouds. Two examples of the slow shock – cloud collision phenomenon are discussed here.

2. L1780: A Shocked Cloud – Relaxed

The L1778/L1780 dark cloud is a compact high latitude cloud probably associated with the Loop I shells. The cloud has a significant surface brightness above the sky level. This prototypical "bright dark nebula" has a well defined boundary at its northern and western sides, and is more diffuse on the south-eastern side. A tail, pointing eastwards from the cloud, is also outlined in the surface brightness distribution. The core, tail structure tail is also seen on IRAS images (see Figs. 10 and 11 in Tóth et al. 1995 (THLM)).

* A. Horváth's present address: Széchenyi István College, Győr, Hungary

Astrophysics and Space Science **233**: 175–179, 1995.

^{12}CO, ^{13}CO and $C^{18}O$ (1→0) (THLM) observations revealed a cometary cloud structure: a dense core asymmetrically surrounded by a diffuse outer layer. The mass of the cloud calculated from the CO observations is $21.6M_\odot$. The compact ^{13}CO core has a mass of 8.3 $M_\odot$ and is found to be in virial equilibrium. THLM suggested that the structure and the morphology of L1780 have been influenced by two large scale environmental effects: (1) the asymmetric interstellar UV radiation field and (2) a shock front, both caused by the high mass stars of the Sco – Cen OB association.

2.1. Evolutionary scenario of L1780

The outermost Loop I shells were formed from the gas swept up by the shock front of a supernova or supernovae in the Upper Centaurus-Lupus subgroup of Sco – Cen (de Geus 1991). At the position of L1780, 100pc away from the supernova, the shock front was too slow (shock speed 10-15 km s^{-1}) to destroy the pre-L1780 cloud fragment. Even a small ($\approx 50M_\odot$) cloud may survive a 10 km s^{-1} shock collision, which forms a cometary cloud from the preexisting spherical cloud. The time τ elapsed since the collision can be estimated. Three Hα emission stars were found recently by Martin & Kun (1995) 1.2 degrees west of L1780 (2.3 pc at 110 pc distance), i.e. towards the centre of Upper-Centaurus Lupus (UCL). Assuming that the formation of these stars was triggered by the same shock as L1780, the shock was at the position of the Hα objects at about 2×10^6 years ago. (We assume a few times 10^5yrs for the shock induced core formation and an age of 1.5×10^6yrs for the PMS stars generated there). The pre-L1780 cloud was shocked a minimum 1.5×10^6 years ago (including 0.2×10^6 years traveling time of a $v = 10$ km s^{-1} shock). There are indications (e.g. on IRAS 100μm intensity maps) that the shock front has already reached the dark clouds L134, L169, L183, L184, about 11 pc distant from L1780, and has ignited core formation there. An estimate for τ from this would be 1.6×10^6yrs, assuming 0.6×10^6yrs for reaching the maximum compression by the shock. From both estimates the time τ is about the minimum relaxing time for small shocked clouds, enough to form a tail and to start to produce a warm, partly ionized envelope. Heathcote & Brand (1983) estimated $\approx 1.5\times10^6$yrs to form a tail and reach pressure equilibrium. The UV radiation from the massive stars of Sco – Cen increased the dissociation rate of the molecules and heated up the gas at the galactic-plane-facing "bright side" of the L1780 cloud. The "shadow" side kept its sharp boundary and its gas content remained molecular. The HI halo was formed on the bright side. These correspond to the warm HI (WNM) envelope expected according to the ISM model of McKee & Ostriker (1977).

3. L1251

The medium latitude dark cloud L1251 has been the subject of a number of optical, radio and infrared studies. This is partly due to its favorable location and partly due to its high star forming efficiency ($\approx 10\%$, Kun & Prusti 1993, K&P). The indicators of active star formation in L1251 are: embedded FIR point sources, PMS stars, CO outflows and an H_2O maser (K&P, Sato et al. 1994, Tóth & Walmsley, 1993 (TWa)).

The three interconnected dark clouds L1251, L1247 and L1243 are situated on the flank of the Cepheus Flare giant molecular cloud and of these, L1251 has the largest opacity and area on POSS prints. The opaque "head" of this elongated cloud extends into the hot void between the Cepheus Flare and Cassiopeia GMCs suggesting that the compression of the head is attributable to the SNR which caused the Cep-Cas "hole" (Grenier et al. 1989). The distribution of the dense molecular gas in L1251 was traced by NH_3 inversion lines. The calculated $n(H_2)$ molecular hydrogen number density distribution is shown in Fig. 4 of Tóth & Walmsley 1995 (TWb). TWb detected six gravitationaly bound ammonia cores and two condensations with typical sizes (at half power) of 1.5-3' (0.15-0.3 pc at 350 pc distance) and with a maximum H_2 density of 9×10^4 cm^{-3} and an average T_{kin} of 10 K. The masses range from 5 to 30 $M_\odot$, $\sim$40% (200$M_\odot$) of the total molecular mass is in dense cores which are surrounded by the less dense envelope seen in ^{13}CO maps of Sato et al. (1994).

The core group at the "head" of L1251 shows a velocity gradient of 1.3 km s^{-1} pc^{-1}. (in the NE-SW direction, see Fig. 5 of TWb). The average velocity of the "tail" NH_3 core group (at the western end of the cloud) is similar to that of the head, but the velocity dispersion is smaller. The observed radial velocity changes less than 0.5 km s^{-1} pc^{-1} going from W to E. The velocity of the solitary core in the northern part of L1251 deviates most (-1 km s^{-1}) from the average. We assume that the observed velocity dispersion reflects the dispersion in the v_y velocity component which is perpendicular to the main axis of the cloud. This axis is assumed to be parallel to the propagation direction of the Cep-Cas SNR shock.

3.1. Evolutionary scenario

The distribution of the class I (Lada 1991) point sources in L1251 is well correlated with the distribution of molecular gas (Sato et al. 1994, TW). However, the center of mass of the Hα emission stars is 1 ° ($\approx$6 pc at 350 pc distance) offset from the center of L1251. The present cometary shape of the cloud was formed by a shock – cloud collision. We suggest that the PMS stars were formed in the pre-shocked L1251 before the time of the shock collision. The displacement of these objects from the molecular cloud is due

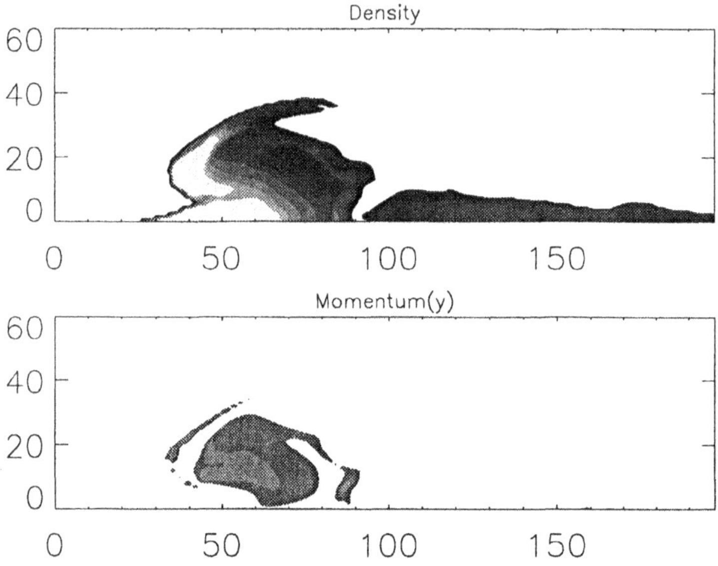

Fig. 1. (a) The density, and (b) the momentum distribution predicted by our numerical model (Paper I). The "head" and "tail" density peaks both contain a considerable mass but the mv_y momentum appears positive (expansion) only at the head region. The tail shows only compression. This configuration appears at a wide range of preshock conditions when $v_{shock} < 40$km s^{-1}.

to the shock. Assuming a shock velocity of ≈ 10km s^{-1}, 10^5yrs is required to produce a $\approx 1°$ offset. Simulation of the shocked cloud is shown on our Fig. 1 as calculated by our model (see Paper I).

Acknowledgements

This research was partly found by the grants: OTKA F-4239 and F-4342.

References

Balázs, L. G., Eislöffel, J., Holl, A., Kelemen, J., Kun, M.: 1992, *Astron. Astrophys.* **255**, 281

de Geus, E.J.: 1991 *The Formation and Evolution of Star Clusters*, K. Janes, Ed(s), A.S.P. Conf. Ser., Vol 13, p 40

Grenier, I. A., Lebrun, F., Arnaud, M, Dame, T. M., Thaddeus, P.: 1989, *Astrophys. J.* **347**, 231

Heathcote, S. and Brand, P. W. J. L.: 1983, *Mon. Not. R. astr. Soc.* **203**, 67

Horváth, A. and Tóth, L. V.: 1995, this volume (Paper I)

Kun, M. and Prusti, T.: 1993, *Astron. Astrophys.* **272**, 235

Martin and Kun, M.: 1995, private comm.

Lada C.J.: 1991, 'The Physics of Star Formation and Early Stellar Evolution' in C.J.Lada and N.Kylafis, ed(s)., *Kluwer*, Dordrecht, 329
McKee, C.F., and Ostriker, J.P.: 1977, *Astrophys. J.* **218**, 148
Sato, F. et al.: 1994, *Astrophys. J.* **435**, 279
Tóth, L. V. and Walmsley, C. M.: 1993, *Andromeda* **10**, 26
Tóth, L. V. and Walmsley, C. M.: 1995, in prep.
Tóth, L. V., Haikala, L. K., Liljeström, T. and Mattila, K.: 1995, *Astron. Astrophys.* **295**, 755

SHOCKS OF HIGH VELOCITY CLOUDS AND THE GALACTIC DISK

3D Numerical Simulations

A. HETEM JR. *
CEA Saclay - Service d'Astrophysique
91191 Gif-sur-Yvette Cedex, France

Abstract. The ISM shock waves due to the collisions of a high velocity cloud (HVC) and the galactic disk can generate turbulence and star formation processes. As a particular feature, a fraction of gas is ejected out of the disk, and this could be the origin of some molecular complexes like the Chamaeleon and *rho* Ophiuchi clouds. One can observe some hydrodynamical phases during the shock, when the cloud presents a coalescence stage, followed by re-expansion and deep disk penetration. We present results of evolving shocks done by 3D hydrodynamical simulations of a falling high velocity cloud on the galactic disk with conditions more realistic than preceding studies, with equations which include molecular cooling.

Key words: ISM – Galactic Structure – Hydrodynamics – Shocks

1. Introduction

The origin of giant molecular clouds (GMCs) situated more than 100 pc below the galactic plane (e.g. Orion and Monoceros clouds) is an interesting object of study. The compression and gravitational instability induced by the impact of a high-velocity cloud (HVC) over the galactic disk is a process that can be applied to explain the formation of some complexes located far from the galactic plane.

Some HVC observational data and discussion are presented by Mirabel (1989). In 1988, Franco et al. (hereafter F88) presented results from numerical simulation of these phenomena and concluded that some main features of the Orion and Monoceros cloud complexes are due to a shock in the disk which occurred about 20 million years ago. This kind of interaction produces cloud coalescence, re-expansion and full disk penetration. Most of these simulations used a 2D code to simulate a HVC shocking with a galactic disk.

Recently, Lépine & Duvert (1994, hereafter LD) presented a revisited version for this problem with emphasis on the star formation due to the HVC collision, and they argued that this process has an important role in the creation of separated groups of OB stars which tend to be aligned in a sequence of ages, in opposition to the ideas first proposed by Elmegreen & Lada (1977, hereafter EL). The falling HVC mechanism also explains the presence of some

* Work supported by CNPq grant process 201793/92-0

groups of young T Tauri stars positioned far from their original molecular clouds or completely isolated T Tauri stars, like TW Hya (Gregorio-Hetem et al. 1992). LD also show that the component of the young stars momentum in the Z direction can establish an orbit that crosses the disk every $\sim 3\ 10^7$ years, and can bring the star far from the galactic plane. LD also presented some simulations made by a simple hydrodynamics code.

2. The model

The equations for the gas dynamics of a polytropic gas under a gravitational potential, neglecting magnetic fields, are the equations of continuity, momentum, energy and an equation of state:

$$\dot{\rho} + \nabla \cdot (\rho \mathbf{v}) = 0 \tag{1}$$

$$\rho \dot{\mathbf{v}} + \rho(\mathbf{v} \cdot \nabla)\mathbf{v} + \nabla P = 0 \tag{2}$$

$$\rho \dot{u} + \rho(\mathbf{v} \cdot \nabla)u + P\nabla \cdot \mathbf{v} + \Lambda = 0 \tag{3}$$

$$P = (\gamma - 1)\rho u \tag{4}$$

where the symbols are density ρ, velocity $\mathbf{v}$, pressure P, and energy density u. The symbols Λ and γ represent the energy loss function and the ratio of specific heats.

We want the gas to be *preferentially accretive*, or in other words if there is a local increase in density it remains stable for a given time, which is an important property for the generation of stars. A way to do this is to subtract the energy excess when there is the tendency to expansion, giving the cooling function Λ. Consider the radiative loss of energy due to molecular cooling. The molecule formation rate tends to grow with the increase of density at low temperatures, and so does the energy loss rate. In general, the approximation $\Lambda \propto \rho^\alpha$, with $\alpha > 1$ is valid for $T < 50$ K (e.g. tables and graphics in Hollenbach & McKee 1979). We consider the high densities drive star formation and act as a regulation mechanism triggered by a critical density. Consider the scales involved in the simulation. Taking into account all the movements the gas may do during the shock, mainly after crossing the galactic disk, one should consider a volume of about 800^3 pc^3, to simulate a complex like the Orion GMC. For computational reasons, the number of gas elements should not be larger than $5\ 10^5$, limiting the size of each element to about 10^3 pc^3. As a second simplification, we assume the pressure gradients are only observable at scales smaller than the size of a gas element, as for star formation or sound waves. These two assumptions can avoid the use of an artificial viscosity and conduction terms.

A third assumption is that the only gravitational potential involved in the shock and its evolution is due to the galactic disk, $G(z)$. Based on these three assumptions, one can re-write expressions (2) and (3) as

$$\dot{\mathrm{v}} = -G(z), \tag{5}$$

$$\rho\dot{u} = -\Lambda, \tag{6}$$

and, naming the constant pressure $P = P_0$, the equation of state becomes

$$u = P_0/(\gamma - 1)\rho. \tag{7}$$

These new expressions together with expression (1) represent a gas under a external potential, and highly compressible due to energy loss via cooling processes. The system of equations is also a generalized form for the expressions presented in LD.

In this work the star formation is limited to a critical density that, once reached inside an element, drives the generation of a *star formation region* (SFR) – not of an isolated star. The star formation process is based on three parameters: the critical density for star formation, $\rho_\star$, the process efficiency, $\alpha_\star$, and the time needed for this gas to collapse, $t_\star$ (see the formalism in EL). The original position and velocity of the SFR are those of the centre of its parental gas element.

The HVC at $t = 0$ is simulated by a spherical distribution with radius R_c of constant density and velocity. We do not represent all the galactic disk but a rectangular slab of the moving gas. Initially, the HVC is just above the disk slab. The disk slab is the region around the HVC collision, which is a distance $R_{\mathbf{gal}}$ from from the Galactic Center (GC), and has a differential disk velocity $V_d = f(r)$ and scale high h. The disk gas movement is defined by the Oort constants. The disk gas density is initially taken to be a Gaussian distribution perpendicular to the galactic plane, expressed by the solution for the condition of hydrostatic equilibrium with the gravitational potential $\nabla_z P = -\rho G(z)$.

These expressions are integrated in a Eulerian-Cartesian grid by a direct differences method, with the time step constrained by the CFL condition.

3. Conclusions

We simulated a shock with a fast cloud (100 km s^{-1}). This more realistic scenario includes movement in the initial conditions for the disk gas, and also the disk refreshing in the boundaries as the gas "enters" the scenario.

As expected, during the simulation the hole created by the infalling gas does not keep its cylindrical shape, but suffers the action from the disk gas that acts as a "wind", pushing the shock structures to the left. The GMC formed 200 pc below the disk (after 10 Myr) also has its velocity affected by this wind and has a x velocity component smaller than in the preceding case. On the other hand, the new SFRs have high values for their z velocity component, and can achieve high Z. These objects are not affected by the hydrodynamic forces and follow orbits determined by the gravitational potential only. The velocity field of the shock waves is a time evolving mix of the HVC and disk gas fields. After 6 Myr, the main fraction of the shocked gas is moving in the $-Z$ direction, but in the disk there are shock waves that enlarge and push the hole. Most of the flow that will generate the GMC is still getting out of the disk. At 12 Myr the z component of the velocity in the gas out of the disk is smaller, but not zero. The hole in the disk is half filled with low density, low velocity gas in the inner region, but higher layers present perturbations due to the shock waves. The future GMC gas presents an almost uniform movement.

Our results are in very good agreement with the prediction of F88 for the Orion GMC. We also compared the resulting data with IRAS maps of the Orion region, obtaining coincident positions for the GMC cores and the galactic disk hole. If one considers the position and shape of many of the most important high z GMCs out of the galactic plane, we conclude this kind of shock occurs in greater number and frequency than is currently supposed.

More realistic HVC and disk models can be done. Studies on the geometry of molecular clouds (Hetem & Lépine 1993) show these kind of objects can not be represented by simple low-resolution grids. The same can be said about the gas in the galactic disk. It should be desirable to enhance the star-gas interactions, as well as gas heating and ionization. Also, the inclusion of super-novae explosions, due to the massive and localized introduction of energy and momentum in the gas, may improve the results.

References

Elmegreen, B.G. and Lada, C.J.: 1977, *Astrophy. J.* **214**, 725

Franco, J., Tenorio-Tagle, G., Bodenheimer, P., Rozyczka, M. and Mirabel, I.F.: 1988, *Astrophy. J.* **333**, 826

Gregorio-Hetem, J., Lépine, J.R.D., Quast, G.R., Torres, C.A.O. and de la Reza, R.: 1992, *Astron. J.* **103**, 549

Hetem Jr., A. and Lépine, J.R.D.: 1993, *Astron. Astrophys.* **270**, 451

Hollenbach, D. and McKee, C.F.: 1979, *Astrophy. J. Suppl.* **41**, 555

Lépine, J.R.D. and Duvert, G.: 1994, *Astron. Astrophys.* **286**, 60

Mirabel, I.F.: 1989, 'Tenorio-Tagle, G., Moles, M. and Melnick, J.' in Structure and Dynamics of the Interstellar Medium, ed(s)., *Kluwer*, Dordrecht, 396

NON-THERMAL EXCITATION OF H_2 IN MOLECULAR CLOUDS

I. O'BRIEN
Dublin Institute for Advanced Studies, School of Cosmic Physics
5 Merrion Square, Dublin 2, Ireland

Abstract. High-energy collisional cascades in H_2 initiated by collisions with streaming ions in magnetically-moderated shocks can lead to large non-thermal variations in excitation level populations. These effects are significant even for relatively low (15 $\mathrm{km\,s^{-1}}$) ion-neutral streaming velocities. Amended cooling functions taking this effect into account could affect shock dynamics and help explain observed line ratios.

Key words: Molecular Clouds – Continuous shocks

1. Introduction

Magnetically-moderated "C-shocks" (Draine 1980) in molecular clouds have been extensively studied using magnetohydrodynamic (MHD) techniques (see Draine & McKee (1993) for a review) with energy and momentum transfer due to friction between ion-neutral streaming ("ambipolar diffusion") taken to be a continuum process. Brand et al. (1988) found significant non-thermal excitation populations of H_2 in OMC-1. Chang & Martin (1991) examined non-thermal effects due to the radiation depleting upper levels. High energy ion-neutral collisions, however, will lead to highly non-thermal excitation populations in the neutrals, at least for the first few generations of the resulting cascades. In the densities that prevail for C-shocks, where collision times are of the order of the time for forbidden transitions to occur (Draine et al. 1983), this will lead to changes in the relative emission line strengths and overall emission. These cooling changes may also influence the dynamics of the shock.

A Monte-Carlo simulation of the collisional cascades following high-energy ion-neutral collisions was carried out to study these non-LTE effects. The gas is considered to be optically thin (Roberge & Dalgarno 1989), and the effects of grains, electrons and species other than the dominant species, molecular hydrogen, are ignored; at these densities hydrogen cooling should dominate (Smith 1989).

2. The Model

The critical transition rates needed for a simulation of this type are unavailable or of poor quality, particularly for temperatures above a few hundred

Astrophysics and Space Science **233**: 185–188, 1995.

K, corresponding to collision velocities of a few $km\,s^{-1}$. The cross–sections used to derive the transition rates in Danby et al. (1987) were used (Flower, D., private communication) for J–transitions, with reasonable extrapolation, and those from Lepp & Shull (1983) are used for mixed V,J–transitions. Microscopic reversibility was used to invert the available downward rates for upward transitions.

The rate at which a particle's kinetic energy is dispersed by collisions and, therefore, the number of collisions over which non–thermal effects can be expected to be significant, is governed by the differential cross–sections. Work is ongoing to generate these cross–sections (Schaefer, J., private communication; Martin, P.G., private communication) but nothing comprehensive will be available for at least another few years, so we have used S–wave scattering throughout. This has the effect of maximising this energy dispersion and, therefore, minimising non–LTE effects.

The main effect of the ions in this model is just to transfer momentum to the neutrals; the exact nature of the ion is not critical and an "average ion" of mass 30 m_H may be used (de Jong et al. 1980). Molecular excitation due to ion–neutral collisions has been ignored, despite these being the collisions with the highest available energy, as they depend on the nature of the ion, which is not well–known in many environments. Due to this limitation, level populations presented here, particularly for high excitation levels, should be considered lower bounds. Collisional dissociation has also been ignored.

The cascade following an initial ion–neutral collision is simulated for a number of generations. At each stage in the cascade new collision partners are selected for existing particles. As cascades will be only a few mean free paths in length, over which the bulk properties of the system remain constant, it is reasonable to select these partners from a sample molecular population in an assumed thermal equilibrium state.

3. Results

In Fig. 1 the excitations due to a single ion–neutral collision are plotted against the cascade generation for given neutral density ρ_n, temperature T and ion–neutral streaming velocity v_{in}. The excitation populations grow rapidly during the first few generations as highly non–thermal particles undergo collisions, after which the excess kinetic energy has been well distributed. The falloff in these populations is governed by the relative importance of collisional deexcitation and radiation, with radiation dominating in this system: the radiation rate was about 50% higher than for a thermal system with the same energy and significant non–thermal excitation was present after the ten generations considered, with this increase more pronounced for cooler systems.

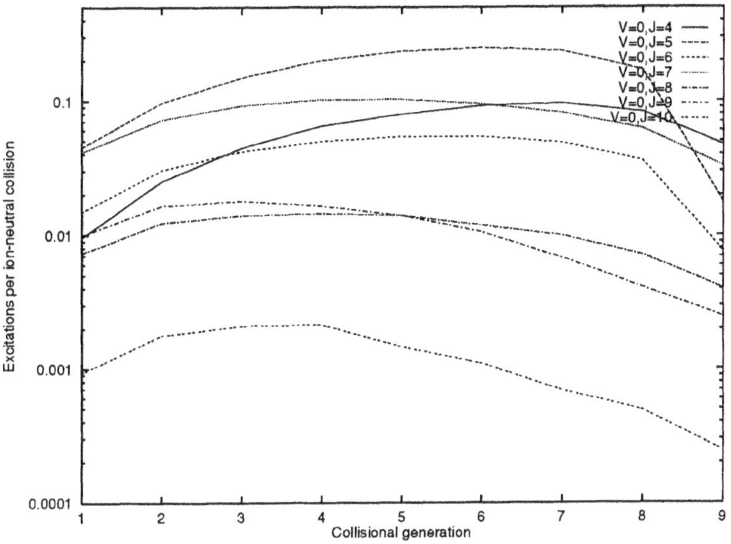

Fig. 1. The evolution of a selection of internal excitation levels during a collisional cascade. Initial high excitation levels decrease due to radiation and collisional deexcitation, while excitation is reduced as the non-thermal kinetic energy is distributed. The system had $\rho = 100\ \mathrm{cm}^{-3}$, $v_{\mathrm{in}} = 15\ \mathrm{km\,s}^{-1}$, T=500 K.

Fig. 2 shows the excitation levels of the system derived from emitted radiation compared to that which would be expected for a system at 750 K. Rotational progressions for each V have been shown separately. The relative underabundance of low-J levels for V>0 is due to their short lifetimes against radiation, and are due to the assumed initial Boltzmann population distribution which, in gases at these densities, is almost certainly invalid. The form of the results is very similar to that observed by Brand et al. (1988), albeit at a lower temperature: excitation temperatures derived from the ratios of these levels could give widely varying results for the system temperature.

4. Conclusions and Further Work

This analysis shows strong non–LTE effects in this system. Unlike previous studies of non–thermal effects in C–shocks (e.g. Chang & Martin 1991) we find enhanced populations for many internal states and significant radiation from high internal states even for low temperatures. It may help to explain some of the observed line ratios in OMC–1. The results presented assume a steady state system and so the results are only an indication of what a calculation over a full shock could look like. The enhanced emission from this model could make higher velocity non–dissociative C–shocks possible, and

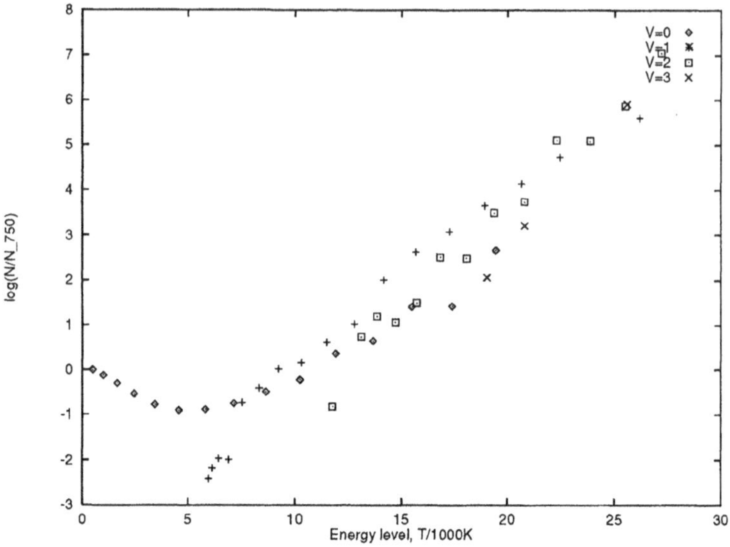

Fig. 2. Internal state populations derived from emitted radiation for a system with $n_H = 10^2 \mathrm{cm}^{-3}$ at 500 K undergoing collisions from ions at 15 km s^{-1}. The derived populations have been normalised against the expected populations at 750 K, and show significant excitation.

could provide an alternative to some of the proposed non–magnetic models for OMC–1.

The first step in this process is to cover the model's main parameter spaces: neutral density, neutral temperature and ion–neutral velocity. This will allow the potential dynamical effects on shocks to be estimated through the use of a modified cooling function.

References

Brand, P.W.J.L., Moorehouse, A., Burton, M.G., Bird, M., Wade, R., 1988, *Astrophys. J.*, **334**, L103
Chang, C.A., Martin, P.G., 1991, *Astrophys. J.*, **378**, 202
Danby, G., Flower, D.R., Montiero, T.S., 1987, *Mon. Not. R. astr. Soc.*, **226**, 739
de Jong, T., Delgarno, A., Boland, W., 1980, *Astron. Astrophys.*, **91**, 68
Draine, B.T., 1980, *Astrophys. J.*, **241**, 1021
Draine, B.T., McKee, C.F., 1993, *Ann. Rev. Astron. Astrophys.*, **31**, 373
Draine, B.T., Roberge, W.G., Delgarno, A., 1983, *Astrophys. J.*, **264**, 485
Lepp, S., Shull, M., 1983, *Astrophys. J.*, **270**, 578
Roberge, W.G., Draine, B.T., 1990, *Astrophys. J.*, **350**, 700
Smith, M.D., 1989, *Mon. Not. R. astr. Soc.*, **238**, 235

PHASE TRANSITIONS IN THE ISM
A SOURCE OF DISSIPATIVE BEHAVIOUR

A. C. QUILLEN
Ohio State University, Department of Astronomy, 174 West 18th Avenue, Columbus, OH 43210
Email: quillen@payne.mps.ohio-state.edu

and

C. B. QUILLEN
Division of Applied Mathematics, Brown University, Providence, RI 02912

Abstract. By considering a simple fluid model, we investigate the role of phase transitions in the ISM on the galaxy- scale gas dynamics. Cooling and heating timescales in the ISM are typically shorter than typical galactic rotation timescales, so the individual phases in the ISM can be assumed to be in temperature equilibrium with the radiation field. Using this approximation we can construct an equation of state which depends upon the average density and mass fractions in the individual phases. Previous studies suggest that there is an equilibrium phase fraction as a function of pressure. We incorporate evolution towards this equilibrium state as a relaxation term with a time to obtain equilibrium τ. We derive a condition in terms of a critical Mach number when one dimensional shocks should be continuous. For small values of the relaxation time τ we show that the relaxation term acts like a viscosity. We show with one dimensional simulations that increasing τ causes shocks to become smoother. This work suggests that phase changes can strongly effect the gas dynamics of the ISM across spiral arms and bars.

Key words: Galactic structure – Gas dynamics

1. Introduction

Recent studies of spiral galaxies have found that the column density of HI across spiral arms peaks at a different location when compared to the H_2 (traced in CO) column density (Rand & Tilanus 1990). These observations, coupled with the recent work of Elmegreen (1993) who modelled the dependence of the molecular to atomic mass fraction in clouds as a function of pressure and the UV radiation field, suggest that molecular hydrogen can be dissociated and produced by pressure and radiation field changes across a spiral arm or galactic bar. In this paper, by considering a simple fluid model, we investigate the role of phase transitions in the interstellar medium (ISM) on the gas dynamics. Our long term goal is to construct multiphase models of the ISM that can be tested through observations of gas at different temperatures and densities in spiral and barred galaxies where changes in the pressure and radiation field are expected as a function of position in the galaxy.

Astrophysics and Space Science **233**: 189–193, 1995.

The ISM of the Galaxy is inhomogeneous, consisting of gas clouds and regions at a range of temperatures and densities. Since the sound crossing time in most phases of the ISM is smaller than other timescales such as the cooling timescale, it is generally assumed that the pressure in each phase is equal to the mean interstellar pressure. Most of the models proposed for the ISM have assumed that the average pressure in the galactic plane is static. However, the theories of spiral density waves and gas response in barred galaxies predict pressure variations as a function of position in a galaxy. This provides an observational setting in which we can explore the effect of pressure changes in the ISM.

2. Continuum Equations

Here we show that cooling timescales are generally faster than rotational timescales in a galaxy. This justifies using a first approximation in which each phase is considered to be in thermal equilibrium with the radiation field. In the cold gas $T \sim 100$K, the cooling rate per hydrogen nucleon is $\sim 10^{-25}$ erg s^{-1}. This implies a cooling timescale of $t_{cool} \sim$ kT/(dE/dt)$\sim 10^4$ years which is fast compared to galactic scales. Likewise, in the ionized medium with $T \sim 8000$ K, the cooling for the same pressure is about the same, so that $kT/(dE/dt) \sim 4 \times 10^5$ years. For comparison, one galactic rotation period in a galaxy with circular rotation speed v_c at a radius r is

$$T = \frac{2\pi r}{v_c} \sim 3 \times 10^7 \text{yr} \left(\frac{200 \text{ km s}^{-1}}{v_c} \right) \left(\frac{r}{1 \text{ kpc}} \right).$$

Variations in pressure and density caused by a spiral density wave or barred potential should extend over a solid angle which corresponds to a time that is larger than few percent of the rotation period. Therefore, except in the inner regions of galaxies, heating and cooling timescales are generally faster than the gas response to a barred potential or a spiral density wave. This implies that in each phase, the state of the gas can be described as a function of the radiation field and the local average pressure.

Using the approximation that each phase is in thermal equilibrium with the radiation field, we do not require equations to describe the thermal state of the fluid in each phase. We adopt here a formalism where the variables are averages over size scales larger than the size of individual clouds. We assume here that the mean velocity v of all phases is the same so that all phases move together. Mass conservation is then given by

$$\frac{\partial \rho}{\partial t} + \nabla \cdot (\rho \mathrm{v}) = 0$$

where ρ is the average density summed over a scale greater than that of individual clouds. Conservation of momentum or the Euler's equation is

$$\frac{\partial \mathbf{v}}{\partial t} + (\mathbf{v} \cdot \nabla)\mathbf{v} = -\frac{\nabla P}{\rho}$$

where P is the pressure, which is a function of the phase mix as well as ρ.

Previous studies suggest that at a given pressure there exists an equilibrium phase mix (e.g. Begelman & McKee 1990, and Elmegreen 1993). If we assume that the medium requires a relaxation timescale τ to achieve equilibrium, which we expect is greater than the cooling timescale in the individual phases, then we can describe the evolution towards equilibrium as a relaxation process. For f, the mass fraction of one phase of a two phase medium, we can describe the evolution towards equilibrium as

$$\frac{\partial(\rho f)}{\partial t} + \nabla \cdot (\rho f \mathbf{v}) = \frac{\rho\,(f_{eqm} - f)}{\tau}$$

where τ is the timescale required to obtain equilibrium and we have written the equilibrium phase fraction as f_{eqm}. We expect f_{eqm} to be a function of local variables such as the pressure and the radiation field. We note that relaxation terms can strongly effect the dynamics of the Euler equation and can produce dissipation-like terms in this equation.

We note that our set of equations is similar to the equations that describe relaxation effects in gases, where the internal energy may lag behind the equilibrium value. For example vibrational and rotational degrees of freedom may take longer to reach equilibrium than kinetic or translational ones in a molecular gas. We can view the set of previous equations in terms of two systems. If we consider timescales of order τ then we require the full set of equations listed above. The characteristics for this system are $v, v \pm c$ for v the velocity and c the sound speed where $c^2 = \partial P(\rho, f)/\partial \rho$. If we average the fluid flow over timescales larger than τ then we can consider the phase fraction to be at the equilibrium value. This is system is described by only two equations, however the phase fraction is then equal at all times to the equilibrium value and the characteristics are $v \pm c'$, with sound speed c' where $c'^2 = c^2 - d^2$ and

$$d^2 = -\frac{\partial P}{\partial f}\frac{df_{eqm}}{dP}\frac{\partial P}{\partial \rho}/(1 - \frac{df_{eqm}}{dP}\frac{\partial P}{\partial f}).$$

Viewed from the averaged system (averaging over timescales larger than τ) which takes the flow between two uniform states in equilibrium, the shock will be continuous in micro system if $v_2 - c_2 < U < u_1 + c_1$ for U the velocity of the shock. Subscripts here refer to quantities before and after the shock. A frozen shock followed by a region of relaxation will occur when $U > u_1 + c_1$ or

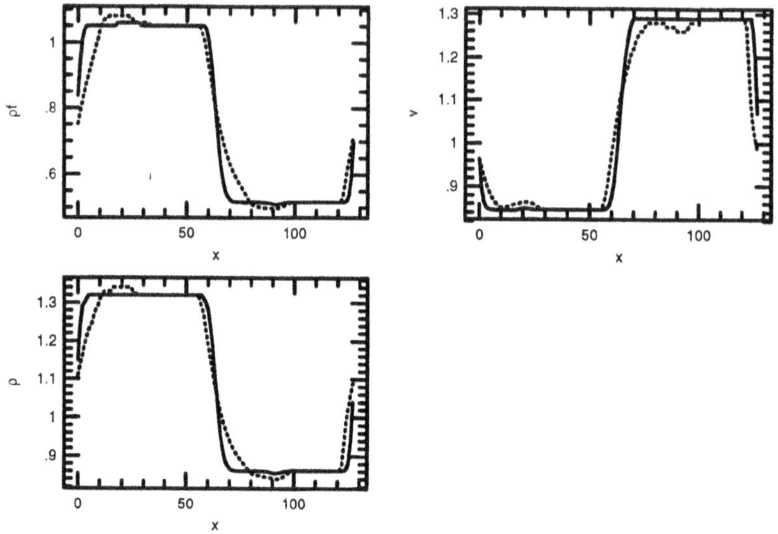

Fig. 1. Comparison of two 1-dimensional $M < M_{crit}$ shocks with different values of τ, the relaxation time. Note that the one with larger τ (dotted line) has a smoother profile, consistent with the dissipative behavior caused by the phase changes.

when the Mach number $M \equiv (U - u_1)/c_1' > c/c'$. The critical Mach number $M_{crit} \equiv c/c'$.

Sometimes a dispersion relation can be used to gain some insight into the stability and dissipational behavior of a set of dynamical equations. Here we show the dispersion relation for for small τ which is derived by considered first order perturbations. We obtain the following dispersion relation

$$\left(-\omega^2 + 2kv_0\omega + k^2\left(c^2 + v_0^2\right)\right)\left(-i\omega + ikv_0 + \frac{1}{\tau} - \frac{1}{\tau}\frac{\partial f_{eqm}}{\partial f}\right) + \frac{k^2}{\tau}\frac{\partial P}{\partial f}\frac{\partial f_{eqm}}{\partial \rho} = 0.$$

For small τ we can expand the dispersion relation in powers of τ and to first order in τ we find that

$$\omega = kv_0 \pm kc + \frac{ik^2\tau}{2}\frac{\partial P}{\partial f}\frac{df_{eqm}}{dP}\frac{\partial P}{\partial \rho}.$$

The imaginary component of ω should be less than zero, and the solutions are damped. If we incorporate a kinematic viscosity ν into the Euler's equation and neglect phase changes, then we find to first order in ν that

$$\omega = kv_0 \pm kc - \frac{ik^2\nu}{2}.$$

It is clear from comparison of these two previous equations that the relaxation term behaves similar to a viscosity for small τ. If the ratio of the dissipation terms

$$\left(\frac{\tau}{\nu}\right)\frac{\partial P}{\partial f}\frac{df_{eqm}}{dP}\frac{\partial P}{\partial \rho} \gg 1$$

then the dominant form of dissipation is due to phase changes for a system that has both phase changes and viscosity. This inequality in order of magnitude is equivalent to $\tau c^2/\nu \gg 1$ which we expect is satisfied when τ is greater than the mean collision time of clouds.

By incorporating phase changes in a simplistic way into a continuum fluid equation, we find that phase changes can significantly effect the gas dynamics of the ISM. The sound speed for the system with phase equilibrium should be lower than the sound speed of a system with a frozen phase mix. Phase changes have a dissipative effect on the gas dynamics, and for small values of τ the relaxation timescale, the phase change acts similar to a viscosity. This dissipative effect causes simulated one-dimensional shocks with larger values of τ to be smoother than those with small values of τ.

References

Begelman, M. C. and McKee, C. F.: 1990, *Astrophys. J.* **358**, 375
Elmegreen, B. G.: 1993, *Astrophys. J.* **411**, 170
Rand, R. J., & Tilanus, R. P. J.: 1990, '' in H.A. Thronson Jr. and J.M. Shull, ed(s)., *The Interstellar Medium in Galaxies*, Kluwer Academic Publishers: Dordrecht, p. 525

ALL'S KNOT QUIET ON THE RECOMBINATION FRONT

M. P. REDMAN, R. J. R. WILLIAMS and J. E. DYSON
Department of Physics and Astronomy,
University of Manchester,
Oxford Road, Manchester M13 9PL, UK

Abstract. Molecular clouds are clumpy on length scales down to the limits of observational resolution. At least some ultracompact HII regions (UCHIIR) may result from the interaction of a young early type star and this type of cloud. The clumps can act as reservoirs of ionized gas distributed within the HII region. These models reproduce the relatively long lifetimes implied by the population statistics of UCHIIR. We present line profile and emission measure plots based on the simplest case where the flow remains supersonic through to a recombination front. The morphology agrees with the shell-like UCHIIR as classified by Churchwell. The predicted line profiles are broad and double peaked with a separation of about $50\,\mathrm{km\,s^{-1}}$ for the example given.

Key words: Hydrodynamics – Stars: mass loss – ISM: Structure – HII regions

1. Introduction

Ultracompact HII regions are important since they are associated with the later stages of massive star formation. They are heavily obscured at optical wavelengths and so most information on them comes from infrared and high frequency radio data. At these wavelengths they are amongst the brightest point sources in the galaxy. Churchwell (1990) reviews typical parameters for UCHIIR. They are small ($\lesssim 0.03\,\mathrm{pc}$), have high emission measures ($\langle n_e{}^2\rangle L \gtrsim 10^7\,\mathrm{cm^{-6}\,pc}$) and are thus dense ($\langle n_e{}^2\rangle^{1/2} \gtrsim 10^4\,\mathrm{cm^{-3}}$). Churchwell defines 5 main morphological types: cometary ($\sim$ 20 per cent), core-halo ($\sim$ 16 per cent), shell ($\sim$ 4 per cent), irregular or multiply peaked ($\sim$ 17 per cent) and spherical or unresolved ($\sim$ 43 per cent). The cometary regions seem well explained as steady-state partially ionized structures behind bow shocks which are being driven into molecular cloud material by the winds of moving stars. Although it is likely that some non-cometary types may be explained by a viewing angle effect (Mac Low et al. 1991), other explanations are probably necessary.

Dyson (1994) has suggested that some UCHIIR may be naturally explained in terms of the interaction of an OB star with a clumpy surrounding medium. It has long been appreciated that molecular clouds are clumpy on scales down to the limit of present spatial resolution. These clumps can undergo mass loss either by photoionization or through ablation by the wind (Dyson 1968; Hartquist et al. 1986). We here examine the particular case of an initally hypersonic stellar wind being decelerated by mass loading to supersonic velocities. Line profiles and emission measure distributions indicate that this

Astrophysics and Space Science **233**: 195–198, 1995.

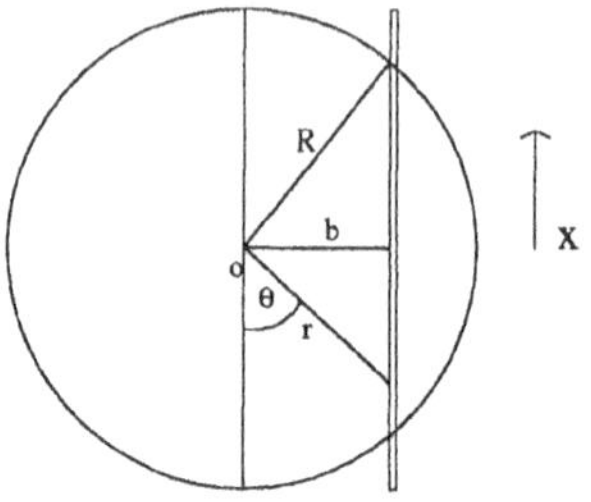

Fig. 1. The geometry of the UCHIIR

simple model may account for the shell-like UCHIIR classified by Churchwell and collaborators in their radio surveys of these objects (Kurtz, Churchwell & Wood 1994; Wood & Churchwell 1989).

2. The Flow

Hartquist et al. (1986) have discussed hydrodynamic ablation by a mass loaded wind. They find that if the flow is supersonic, then mass is injected at a rate independent of the Mach number of the flow. We will therefore take the mass injection rate to be constant throughout the ionized flow region and also assume it is injected at zero velocity relative to the star.

The steady state mass conservation condition is, neglecting the stellar wind gas,

$$4\pi r^2 \rho u = \frac{4}{3}\pi r^3 \dot{q}, \tag{1}$$

where $\dot{q}$ is the volume mass injection rate and ρ and u respectively the density and velocity. Since the flow is supersonic, the momentum equation is

$$4\pi r^2 u^2 \rho = \dot{\mu}_* \; (\equiv \dot{M}_* V_*), \tag{2}$$

where $\dot{M}_*$ and V_* are, respectively, the stellar mass loss rate and stellar wind velocity. Hence the velocity and density distributions are simply

$$u = \frac{3\dot{\mu}_*}{4\pi \dot{q} r^3} \quad \text{and} \quad \rho = \frac{4\pi \dot{q}^2 r^4}{9\dot{\mu}_*}. \tag{3}$$

3. The Line Profiles

The geometry is shown in Figure 1. The intensity of a line at velocity w

is

$$I_w = 2\int_0^{x_{\max}} \left(\frac{j_\lambda}{4\pi}\right) \frac{1}{c_i\sqrt{2\pi}} \exp\left\{ \frac{-(w - u\cos\theta)^2}{2c_i^2} \right\} \,dx, \tag{4}$$

where c_i^2 ($\equiv kT/\mu m_H$) is the isothermal sound speed and j_λ is the emission coefficient for the line being observed, which we assume is proportional to ρ^2, and dx the element of path length.

The widths of the profiles are determined by the Mach number (M_R) at the recombination front (radius R) which is given by

$$M_R \equiv \frac{u_R}{c_i} = \frac{3\dot{\mu}_*}{4\pi R^3 \dot{q} c_i}, \tag{5}$$

where u_R is the flow velocity at the recombination front. The radius of the recombination front is given by the usual Strömgren relationship. Hence (Dyson et al. 1995)

$$M_R = 0.78 S_{48}^{-1/4} R_{17}^{-1/4} \dot{\mu}_{28}^{1/2} f^{-1/4}. \tag{6}$$

Here $S_{48} \equiv S_*/10^{48}\mathrm{s}^{-1}$; $\dot{\mu}_{28} \equiv \dot{\mu}_*/10^{28}\mathrm{g\ cm\ s}^{-2}$; $R_{17} \equiv R/10^{17}\mathrm{cm}$ and f is the clump filling factor of high density gas. Williams, Hartquist & Dyson (1995) have investigated the deceleration of mass loaded winds in detail and as long as the Mach number stays above about 2 up to the recombination front, global shocks will not occur in the flow. With the filling factor taken to be $f = 0.02$ and $S_{48} = R_{17} = \dot{\mu}_{28} = 1$, we find $M_R = 2.1$.

Two line profiles are displayed in Figure 2 at two different fractional offsets, z, from the central star. They show two broad lines, separated in velocity by about $50\,\mathrm{km\,s^{-1}}$. In Figure 3 we show the emission measure as a function of z. Clearly, this simple model can reproduce the shell appearance of some UCHIIR. To check our velocity structures, high spatial *and* spectral resolution radio recombination line observations of unresolved and shell-like UCHIIR are now needed. If these distinctive features are seen it would lend strong support to this model. More details are given by Dyson et al. (1995).

Work is in progress to make this simple model more detailed by including, for example, radiative transfer and dust in and around the region. The full range of other models outlined by Dyson (1994) is also being investigated in parallel with more detailed work on the structure and behaviour of recombination fronts.

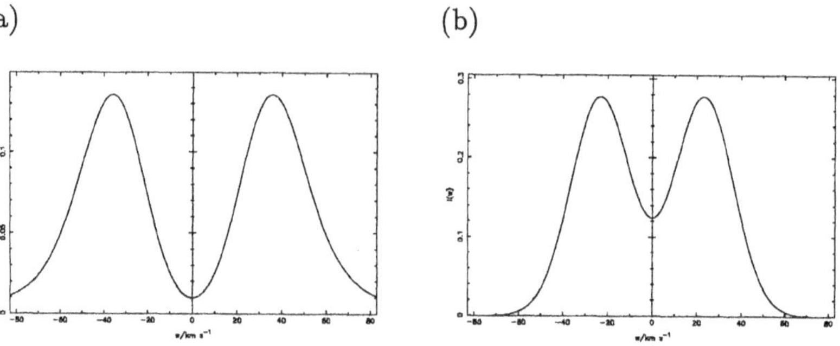

Fig. 2. Line profiles at (a) $z = 0.1$ and (b) $z = 0.7$, normalized to unit total emission across the nebula, for $M_R = 2.2$.

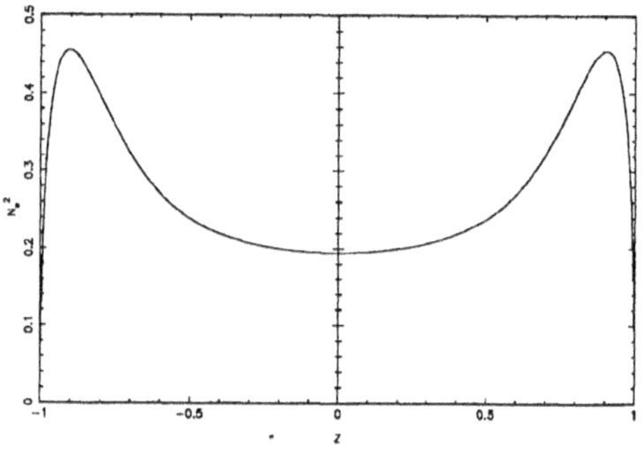

Fig. 3. Emission measure as a function of z.

References

Churchwell E.: 1990, *Astron. Astrophys. Rev.* **2**, 79
Dyson J. E.: 1968, *Astrophys. Space Sci.* **2**, 461
Dyson J. E.: 1994, in Ray T. P., Beckwith S. V. W., ed(s)., Lecture Notes in Physics 431. Springer-Verlag: Berlin, 92
Dyson J. E., Williams R. J. R. and Redman M. P.: 1995, *Mon.Not.R.Astr.Soc.* submitted
Hartquist T. W., Dyson J. E., Pettini M. and Smith L. J.: 1986, *Mon.Not.R.Astr.Soc.* **211**, 715
Kurtz S., Churchwell E. and Wood D. O. S.: 1994, *Astrophys.J.Suppl.* **91**, 659
Mac Low M. M., Van Buren D., Wood D. O. S. and Churchwell E.: 1991, *Astrophys.J.* **369**, 395
Williams R. J. R., Hartquist T. W. and Dyson J. E.: 1995, *Astrophys.J.* **446**, 759
Wood D. O. S. and Churchwell E.: 1989, *Astrophys.J.Suppl.* **69**, 831

FLOWS AND SHOCKS IN ACTIVE GALAXIES AND THEIR NUCLEI

J. E. DYSON and R. J. R. WILLIAMS
Department of Physics and Astronomy, University of Manchester, UK

and

J. J. PERRY
Institute of Astronomy, University of Cambridge, UK

Abstract. Observations of active galactic nuclei imply that shocks must be an essential and important part of their structure. We outline the basic observations, and discuss those features which must be addressed by any physical model of active nuclei. These features, in particular the observed spectrum of strong emission and absorption lines, lead naturally to the conclusion that shocks are present. The velocity widths of these lines, which range from hundreds to many thousands of kilometres per second, are most readily explained by models in which shocks play an important role in the generation of cool gas. The extreme parameters of the shocks in and around active nuclei provide a unique application for the physics discussed in this meeting.

Key words: Shock Waves – ISM: Kinematics and Dynamics – Galaxies: Active

1. Introduction

The multifaceted phenomenon known as *Activity in Galactic Nuclei* remains mysterious even after a third of a century of study. Most astrophysicists agree that the theoretical understanding of the formation, structure and evolution of these nuclei is incomplete (some would even say primitive). We believe that some of the most significant gaps in our understanding of these elusive objects can be filled only by a proper consideration of hydrodynamical phenomena in the dense nuclei of galaxies, and that therefore much of the physics discussed at this meeting is central to the development of theories of Active Galactic Nuclei (AGN).

'Active' (according to the Oxford Dictionary "..... abounding in action; energetic") is itself ill-defined when applied to galaxies: originally synonymous with 'non-stellar' or 'non-thermal', it has evolved to mean 'an output of energy unsustainable over the Hubble time'. Currently, as ever more detailed high resolution observations of galactic nuclei accumulate, objects once considered to be quite different are becoming seen as members of a very extended family—or as Bernard Pagel emphasised in 1984 (in a paraphrase of George Orwell's famous remark) "all galactic nuclei are active, but some are more active than others".

The gross characteristic of the rest-frame spectra of AGN, shared by almost all members of the class, is that the energy radiated spans up to

Astrophysics and Space Science **233**: 199–214, 1995.

10 decades in frequency (unlike, for instance, stellar black-body spectra), and is remarkably uniform in power per decade of frequency from the IR to the X-ray. Prominent emission lines, with line widths between hundreds and thousands of km s^{-1}, are superposed on the underlying continuum and are considered a defining feature of active galaxies (except for blazars). They are a rich source of detailed physical information about the very centre of the galaxy where the lines are emitted. The relative constancy of their equivalent widths tie them tightly to the production of the central continuum and require that the energy production process and the line emission must be considered together. This then leads, unequivocally, to the realisation that shocks must play a central role in the physics of AGN.

The apparent magnitudes of QSOs and AGNs, when corrected for distance—and assuming isotropic emission—imply that the integrated luminosities are $\gtrsim 10^{42} - 10^{49}$ ergs (equivalent to $\gtrsim 10^{8} - 10^{15} L_{\odot}$). There are short time scale variations in the continuum emission: in low luminosity objects the X-ray fluxes often vary substantially within hours, and sometimes even minutes. *If* the variability is due to a single central object, then that source must, in some objects at least, be smaller than a light-hour, or $\lesssim 10^{14}$ cm. The variability appears to be correlated to luminosity, with the lowest luminosity objects varying the most rapidly, although the relationship has yet to be quantified; the largest inferred sizes of the central continuum source are $\lesssim 10^{17}$ cm. One of the central problems in AGN research is to explain the generation of the extraordinary output of energy from such small volumes in the nuclei of galaxies.

Some of the earliest views regarding the nature of AGN invoked stars, particularly through their evolution to the supernova stage (e.g. Shklovskii 1960; Field 1964; Colgate 1967). However, when severe difficulties arose regarding the necessary efficiency of energy generation and the observed rapid continuum variability, there was a change of opinion in favour of long-lasting gravitational processes (Lynden-Bell 1969). Accretion of matter onto a supermassive black hole (BH) via the intermediate agency of an accretion disc became, and has remained ever since, the paradigm for the powerhouse driving AGN luminosities. However, very little observational proof of the existence of supermassive BHs in galactic nuclei exists to date. A competing proposal is that the variability arises in compact supernova remnant 'flares' in a starburst (Terlevich et al 1993; Filippenko 1992).

The maintenance of energy generation by accretion onto a BH requires a large enough reservoir of accretable material to sustain the source over its lifetime. It appears likely, at least for low luminosity objects, that the interstellar medium and/or mass loss from stars in the galaxy which houses the AGN can provide enough mass to fuel the central engine. This view is supported by the (rather sparse) evidence for extranuclear fuelling (Heckman 1992). In this context, extranuclear implies a region not significantly affected by the central engine—as will be discussed later, such regions may be rare. However, the high mass-consumption rates (up to hundreds and even thousands of solar masses of material per year) required to explain the energy generation in high-luminosity objects cannot be readily explained in this manner, and the existence of a dense nuclear star cluster surrounding the BH seems to be necessary (Williams & Perry 1994).

In order to construct physical models of the broad emission line regions (BELR) and derive densities, temperatures and excitation conditions, emission lines from many ions must be measured simultaneously. Yet, in the spectral range available for any one object, too few lines are observed: therefore throughout the history of AGN modelling it has been necessary to construct "composite" spectra assuming that there is a continuity between objects of different z and absolute luminosity, so that their spectra can be pieced together. If, for example, there are systematic differences between high and low luminosity objects, these will be difficult to determine on the basis of the analysis of composite spectra.

Despite these caveats, a number of very important general features have been established from spectral analysis. The line spectra are well explained by assuming that the excitation mechanism is photo-ionization by the broadband "underlying" continuum; the thermal and ionisation structure determined by the interaction of dense ($n_e \gtrsim 10^9$ cm^{-3}) gas with the continuum. (The continuum here refers both to the central non-thermal radiation and to an enhanced component of UV and soft X-ray radiation generally ascribed to radiation from an accretion disc surface). There are two main, physically distinct, components to the emission line spectra: the low ionization lines (LIL) and high ionization lines (HIL), consisting primarily of, respectively, the Balmer series and lines from singly ionized species (e.g. C II, S II, Fe II) and of Ly α and multiply ionized species, e.g. C IV, N V, O VI. The HIL is systematically blue-shifted with respect to the LIL, which is generally interpreted as meaning that the HIL are formed in gas having a systematic component of bulk radial motion (probably outflow, possibly inflow), and that there is some material obscuring redshifted emission. The full width at half maximum (FWHM) of the emission lines appears to be correlated with luminosity, which implies that the velocity of the emission line gas is larger in more energetic objects. The lines vary, apparently in response to variations in the ionizing continuum, on time scales ranging from several days in low luminosity objects to more than several years in high luminosity QSOs.

Rather paradoxically, the sheer success of such photoionisation models has been a major hindrance to the development of actual physical models of the BELR (Perry 1992). This is because adjustments to the thermal and ionisation equilibria take place on time scales which are effectively instantaneous. We then can develop a detailed understanding of the local properties of the clouds (densities, temperatures, ionising incident fluxes, etc.) from line ratios, without addressing the important issues of the origin and evolution of individual emitting regions. Observational data provide scant direct information on these issues. Yet an understanding of the BELR is essential, since the photoionisation models firmly place it in the innermost nuclear regions. A central issue is how to account hydrodynamically for gas which is moving hypersonically at Mach numbers anywhere from 10 to more than 100 (suggesting that shocks are involved), and which occupies only a small fraction of the spatial region where it is found.

Broad absorption lines (the BALs) are present in a small percentage (10% or so) of high luminosity objects, with velocities ranging up to a tenth the speed of light; their occurrence suggests that violent dynamic effects take place in the nuclear regions. These absorption lines are invariably blue-shifted relative to the corresponding emission lines. The BALs represent one of the most puzzling aspects of all AGN phenomena and it is no exaggeration to say that progress to date in understanding them has been minimal. However, it is most likely that shock phenomena is involved.

The very existence of dynamically active BEL and BAL regions coupled with the fuelling process imply that high speed flows exist in the nuclear regions and that shock phenomena will occur. Moreover, mass-loss in stars is ubiquitous throughout the HR diagram. In particular, high speed mass-loss occurs in winds from early type stars and in supernova explosions. Clearly, such mass-loss would produce shocks in any surrounding medium, such as the gas involved in the fuelling.

Shocks in active galaxies and their nuclei have several unique features, compared to other interstellar shocks. They will have extreme parameters: velocities of thousands of km s^{-1}, high incident radiation luminosities, and small filling factors. Diagnostics of the line emitting gas—line variability and absorption—require that the shocks process gas on observable timescales. Since line emission is usually biased towards the densest gas at the highest pressure, it is the properties of the post-shock gas which must be modelled successfully in order to account for the observations.

The influence of an AGN is not confined to the nucleus itself, but is felt throughout—and even outside—the host galaxy. Although not discussed here, the widespread existence of jets, on scales ranging from nuclear to intergalactic, has implications for many sorts of shock activity. Direct observational evidence for outflow from the nuclear regions does not exist, and although the BALs are generally assumed to arise in a hypersonic global wind, the distinction between outflow or the effects of outflow is often not made. As we discuss below, the BALs may be indicative of quite complex interactions involving both local and global phenomena.

If such hypersonic winds do exist and are powerful enough—and the available, rather sparse, evidence suggests this may be the case—then they will certainly have an effect on the general interstellar medium (ISM) in the host galaxy. Very crudely one might imagine interactions similar in nature to the well studied ones between massive stars and the ISM (e.g. Dyson 1994) where shocks play a major role, though those in AGN are more energetic and on much larger scales. The analogue of the swept up stellar-wind driven shell may contribute to the formation of another type of absorption line, namely the narrow absorption lines (NALs) seen in the spectra of many QSOs (Dyson, Falle and Perry, 1979). The NALs have widths of the order of the sound velocity in the absorbing gas (about 20 km s^{-1}) although the velocity of this gas relative to the nucleus of the host galaxy is several thousand km s^{-1}.

Less overly dramatic—but equally important—phenomena are associated with the observed narrow emission lines. Narrow emission lines are observed in most AGN spectra; they have widths less than 1000 km s^{-1}, although their wings may be extensive, ranging up to 2000 km s^{-1}. From the intensity ratios of the forbidden lines, electron densities of $n_e \approx 10^3$ to 10^4 cm^{-3} are deduced. The narrow line gas must lie in a region (the NLR) of the order of 10^3 times as far from the nucleus as the BELR, if they are both photoionized by the same central continuum flux. The narrow emission lines thus 'map' the outer regions of the galactic centres, intermediate between the BELR and the full galaxy. High resolution mapping of the NLR in several Seyfert galaxies (e.g. Unger et al 1986) has shown very strong evidence for the association of the optical NLR emission and radio emitting plasma which plausibly may have its origin in the nuclear regions. That, together with the fact that the velocity widths in the NLR are usually significantly higher than could be expected from a gravitational origin, is suggestive that in some way, the NLR is directly linked with the escape of plasma from the nuclear regions. Again shocks must be generated. In addition to the classic NLR, further kpc-scale regions of narrow line emission have been discovered in narrow band imaging of Seyfert Galaxies.

In this review, we concentrate on four major aspects of AGN phenomena where shocks are important; namely, the BELR, the NLR, the BALs and NALs.

2. The Broad Emission Line Region

We here explore the role and properties of the BELR within the context of a model postulating the symbiosis of a black hole and a nuclear starburst cluster first advanced by Dyson & Perry (1982) and developed in much greater detail by Perry & Dyson (1985-henceforth PD), Collin-Souffrin et al (1987) and Williams & Perry (1994).

Very simple considerations show why shocks are inevitable in such a system of dense stellar cluster and BH (Perry 1992). If a BH exists at the centre of galaxy, the Keplerian velocities at distance r_{pc} from a central black hole of mass M_8 solar masses is $V_k \approx 660(M_8/r_{pc})$ km s^{-1}. The sound speed in the surrounding medium depends on its temperature; this will almost certainly be the Compton temperature of the non-stellar continuum, $T_c = T_7 10^7$ K (T_7 is typically about unity). Thus $V_s \approx 380\ T_7^{1/2}$ km s^{-1}. All stars within $r \approx 3(M_8/T_7)$ pc therefore move through the local ISM supersonically. It then follows that all mass injection into the ISM, whether through stellar winds, supernovae, or as a result of tidal stripping or stellar collisions, must involve shocks. This would be true even in the improbable event that the ISM were completely static. However, the ISM originates from stellar mass-loss processes (Williams & Perry 1994) and the associated energy and momentum input ensures that typical velocities are roughly three times the sound speed, i.e. the velocities are those which are characteristic of thermal winds.

The main role of a shock is to convert kinetic energy into thermal energy: the post-shock temperature and pressure are increased and in the strong shock limit, the post-shock temperature is $T_s \approx 10^8 \omega^2$ K, where the shock velocity ω is expressed in units of 3000 km s^{-1}. Note that a very large range in velocity is expected, since it is a vector sum of the Keplerian velocity, the local ISM velocity and the velocity associated with the injection process.

To appreciate the significance of these shocks for the formation of the BELR, we first briefly consider the physical state of the emitting gas. The properties of optically thin gas in thermal and ionisation equilibrium with a non-thermal continuum radiation source are essentially determined by a single parameter, the ionisation parameter, over a broad range of densities characteristic of the BELR. A physically meaningful way of defining this is as the ratio (Ξ) of the local ionising radiation pressure to the gas pressure. Since the radiation pressure depends on the radiation flux, if Ξ can be determined directly from the observed line spectra, it determines the distance of the irradiated gas from the radiation source. High values of Ξ (i.e. low pressures) correspond to gas where Compton heating balances Compton cooling and the equilibrium temperature is close to the Compton temperature. Low values (i.e. high pressures) correspond to gas whose temperature is determined by the balance between photoionisation heating and recombination cooling; characteristically its temperature is about $2\ 10^4$ K. This latter state is established to be in very good agreement with that deduced from emission line data on the BELR.

When gas with a high value of Ξ enters a strong shock, the increase in temperature and pressure takes the gas out of equilibrium with the radiation field and it cools, initially by Compton cooling, then by bremsstrahlung, and finally by line emission. Eventually it cools back into equilibrium with the radiation field. If the shock is physically large, and the pressure remains more or less constant as the gas cools, the cooled gas has a much lower value of Ξ

than the pre-shock gas. The ionisation parameter can be decreased by up to a factor of $40\omega^2/T_7$—which can be very large.

Such radiative shocks are the basis of the mechanism for the formation of BELR proposed by Dyson & Perry (1982) and PD. Its advantages can be stated simply. Firstly, the gas cooling behind the shock front is thermally unstable once its temperature falls below that at which Compton cooling (which is thermally stable) dominates. The cooled gas will form small clouds whose existence is confirmed by the very small filling factors deduced from emission line fluxes. Secondly, the clouds will retain some component of their pre-shock velocity and thus no acceleration mechanism is needed.

This latter point is coupled with the question: can small clouds survive in the interstellar medium of the AGN for any significant time? We know that no two-phase equilibrium for the global ISM exists within the active nucleus for Compton temperatures, T_c, $\lesssim 10^7$; indeed limits on the density of the hot ISM phase show that it has a pressure far lower than that of the line emitting gas. Furthermore, unconfined clouds expand and dissipate in times of the order of their sound crossing time, at least an order of magnitude less than the flow time in the broad line region. How then can unconfined clouds exist long enough to explain the existence of the observed broad lines? The simplest answer, which escaped notice until pointed out by PD, is that, provided clouds are continuously being created, there is no need for them to be long-lived, i.e. evanescent clouds replace permanent clouds.

The properties of the line emitting clouds depend directly on the structure of the radiative shocks. The postshock pressure gradients and flow velocities depend on the size and energy and momentum input by the obstacle—the star, stellar wind, or supernova. Small shocks around stars have flow times short compared to the cooling time, and emit only cooling radiation. Larger shocks, where the flow time from the immediate post-shock entry to the sonic point (or line) in the flow is longer than the cooling time can maintain the gas at roughly constant pressure while it cools. PD noted that the bow-shocks set up when fast flowing ISM encountered more or less fully expanded supernova remnants in the nuclear stellar cluster are large enough to provide such constant pressure cooling regions. A comparison of the flow time between the post-shock stagnation zone and the sonic line and the cooling times gave the remarkable result that cool cloud (i.e. BEL) formation would occur *everywhere interior* to some radius which agreed extremely well with the characteristic BELR radius deduced from the ionisation parameter Ξ. In fact, cooled gas will be formed also in the earlier stages of the supernova expansion, behind closed shocks in the ISM produced by still-expanding supernova remnants.

A very powerful argument for this general model comes from an examination of the requirements on supernova rates dictated by the need to provide the mass and covering factor observed for the BELR. Williams & Perry (1994) obtained the important result that the rates needed were in harmony with those in a nuclear stellar cluster which was simultaneously providing mass at a rate sufficient to fuel the central engine.

In contrast to high luminosity objects where the cooling shocks are expected to be formed by supernovae, in low luminosity objects (e.g. Seyfert galaxies) supernovae are probably not relevant to the production of the BELR, since a single supernova there can blow away the entire BELR—and may provide an explanation for the change from Seyfert 1 to Seyfert 2 type spectra observed on short time scales in some objects (Seyfert 2s do not have strong BELs). It is much more likely that, in Seyfert nuclei, the cooling takes place essentially as described above, but in stellar wind shock interactions, in the wind gas from OB or WR stars (Perry, Williams & Dyson—in preparation).

Very detailed modelling of the structure behind a wide range of shocks spanning the range of conditions from the outer to the innermost regions of the BELR has been carried out recently by Innes & Perry (in preparation). They calculate line ratios for the emission from the shocked gas and compare them, where appropriate, with observational data. The agreement of the model results with the data is striking and provides strong support for shock-confined BELR models.

3. The Broad Absorption Lines

The BALs are an enigmatic feature in the spectra of roughly 10% of radio quiet QSOs. They are regarded as the most definite evidence for mass loss intrinsic to the QSO. The frequency of observation can be interpreted as indicating either that the BAL phenomenon occurs in a special class (10% or so) of QSOs in which the absorbing gas has a covering factor of 100%, or that the absorbing gas has a covering factor of about 10% and BALs are present in all QSOs. The latter interpretation is currently favoured. BAL profiles have a variety of forms, ranging from almost featureless to highly structured, and either attached or detached from the adjacent emission lines. The ions observed indicate gas temperatures of about 10^4 K and evidence from the lack of associated forbidden line emission suggests electron densities above 10^{6-7} cm^{-3}. This, together with column density estimates, then shows that the absorbing gas must be distributed in thin sheets across the line of sight.

There is rather weak evidence to indicate that some connection can be made between the BELR and the BALs. Firstly, if the absorbing gas is in photoionisation equilibrium with the central continuum, its ionisation parameter must be comparable with that in the BELR. Secondly, estimates of the distance of the BAL gas from the central source put it contiguous but largely external to the BELR. Finally, correlations exist between BAL line profile properties and those of associated emission lines. However the BALs do have very significantly larger velocity widths than those in the BELR.

Most theoretical attempts to explain this phenomenon have been in terms of global flow models, for example the acceleration of clouds by winds (Weymann et al 1982) or radiation pressure (Arav, Li & Begelman 1994). There are severe hydrodynamic difficulties with either mechanism. It is well known that ram pressure acceleration of clouds disrupts them before their velocity has increased much above that of the shock driven into them by the initial wind impact. Radiation driving, whilst efficient in principle, requires a mechanism for cloud confinement which does not simultaneously produce an appreciable drag force on the clouds.

Perry & Dyson (1991) noted that there was an interesting coincidence between the typical BAL velocity widths and that of the highest velocity ejecta seen in supernovae. They therefore argued that a natural connection between the BELR and the BALR arises from the supernovae integral to the BELR model of PD, and that in fact the BALs are evidence for local, as opposed to global, high velocity flows, although these latter do play a role. A supernova must produce absorbing material before it slows down appreciably. In this 'Phase I', a fast forward shock is driven into the local ISM: a reverse shock moves into the ejecta. The forward shock moves outwards at the speed of the ejecta whilst the reverse shock accelerates back towards the explosion centre (e.g. Bode & Kahn 1985). Thus the ISM is always shocked to a high temperature. The temperature of newly shocked ejecta is initially low but increases with time as the reverse shock accelerates. In order that this gas produce observable BALs, the cooling time in shocked gas must be less than the slowing down time and the fast moving cool gas must persist long enough to be observed. This basically demands that it persist at high velocity for a period greater than a few years.

Simple modelling of the Phase I evolution (Perry & Dyson 1992) shows that to satisfy these constraints, supernovae must explode into interstellar material which has a density somewhat lower than that required by the PD models for BEL generation, i.e. somewhat outside the BELR. The only gas which is cool enough to produce absorption is the 10% or so of ejecta which is shocked only to a low temperature when the two shocks initially move very closely together. This model reproduces well both the column densities of absorbing gas (apart from the very much higher ones associated with the rare Mg II systems) and the timescales of a few years associated with changes in line profile structures. As has been known for many years (e.g. Gull 1975), the ejecta-shocked-ISM interface is Rayleigh-Taylor unstable and the resulting fragmentation and accompanying destruction of fragments may contribute to the velocity structure in the line profiles.

This model currently has a number of unresolved problems: the most immediately obvious is that a supernova would produce both blue and red-shifted absorption troughs. Provided that the ISM flow is directed more or less radially outwards in the vicinity of the explosion, the supernova expansion is asymmetric and the shell section moving towards the nuclear regions—that which would produce redshifted absorption—slows down more rapidly and reaches a smaller radius than the shell section moving towards the observer. Although no detailed statistics have been investigated, simple models for asymmetric expansion (Wright, unpublished) show that this is not implausible. Moreover, the cooled gas will experience an outward force due to the radiation pressure exerted by the continuum radiation field which will enhance such effects.

One major problem with this simple idea may lie in the details of the global flows in the nuclear regions. Williams (1994) calculated aspherical radiatively-driven models for these flows. The adopted structure was that of a central massive black hole and an accretion disk. Around this region, a central star cluster injects matter into the ISM. In these models, inflow along the plane of the accretion disk coexists with strong outflows along the axis perpendicular to the disk. The flow therefore has a pronounced bipolar structure. Clearly supernovae occurring in inflow regions will have entirely the wrong behaviour as far as the suppression of red-shifted absorption is concerned. Nevertheless, supernovae have at least the merit of not requiring a contentious acceleration mechanism.

The complex structures of these global flows throw up other shock phenomena. Close to the BH, some fraction of an accretion flow along the disk plane can be diverted into an axial wind (Williams 1994). The computed global solutions reflect the interaction between this outflow and the surrounding region where the flow is mass-loaded by stellar mass-loss (cf. Williams, Hartquist & Dyson, 1995). If this central wind is stalled by the mass-loading, a varying shock forms in the wind. For rather stronger winds, a series of explosive events drives nuclear mass-loss. For the strongest winds, the flow reaches a steady equilibrium state.

In all these cases, strong shocks form in the global flow. In the one case, they are generated by the explosions; in the other, steady shock structures form around the stagnation point as is, in general, required if a strongly aspherical flow is to become fully supersonic at large radii (Williams & Dyson 1994).

4. The Narrow Line Region

It has long been known (Shields & Oke 1975) that in the NLR, the emitting gas is distributed in an assembly of small clouds, as in the BELR. The emission spectrum is very well accounted for by photoionisation theory which, in a manner analogous to the BELR, divorces the thermal and ionisation balance from the dynamics and origin of the emitting gas. In recent years, studies of the NLR in Seyfert galaxies have shown that close links exist between the optical NLR emission and the nuclear non-thermal radio emission. Correlations have been found between the λ21-cm radio power and both the [O III] λ5007 luminosity and line width. The radio emitting plasma appears to be in approximate pressure equilibrium with the thermal gas producing the optical emission. Spatial correlations have also been found between the radio and optical components within the NLR. Of particular interest are cases where the optical structure can be resolved and where direct spatial association of the optical and radio structures is seen (e.g. NGC 5929—Whittle et al 1986).

These associations led Pedlar, Dyson & Unger (1985) to propose a model for the NLR in which the optical emission is produced by the interaction of the radio components with the ambient galactic ISM. They proposed that spherically symmetric radio-emitting bubbles of plasma ('plasmons') expanding into the ISM, compressed and accelerated it in a manner entirely analogous to that occurring when supernova ejecta expand into the ISM. The NLR in this model results from shocked interstellar matter which has cooled back into equilibrium with the nuclear radiation field. In order for optical emission to result, the cooling time in shocked gas has to be less than the expansion time of the plasmon. Again this is analogous to the expansion of supernova remnants, and demands that the plasmon expansion velocity has dropped below a critical value of about 350 km s^{-1} (which is very insensitive to the physical parameters, e.g. ISM density, plasmon energy content etc.). Pedlar et al (1985) showed that this simple model reproduced reasonably well the observed filling factors, densities and emission measures of the NLR thermal gas.

The radio components could represent the working surfaces of jets on, say, giant molecular clouds, or it could be that plasmons are emitted directly from galactic nuclei. This latter possibility had already received some theoretical support from Smith et al (1983), who suggested that plasma ejection from the central regions of low-power AGN takes place in the form of bubbles. whereas in high-power AGN it is more likely to take the form of jets. The original work of Pedlar et al (1985) assumes that the expansion velocity of plasmons is much greater than any linear velocity they might have. However, where the NLR is resolved, such as in NGC 5929, the NLR has a non-spherical morphology suggesting that they may better be described at the other extreme where the plasmon linear velocity is much greater than its expansion velocity.

The alternative model explored by Taylor et al (1989) and Taylor, Dyson & Axon (1992–henceforth TDA) therefore assumes that plasmons are ejected supersonically from the nucleus and that they drive bowshocks into the galactic ISM. Shocked gas which cools into equilibrium with the nuclear UV continuum forms the NLR components, exactly as in the expanding-plasmon model.

Apart from the inclusion of an external continuum radiation field, this model has many similarities to bowshock models of Herbig-Haro (HH) objects (e.g. Hartigan, Raymond & Hartmann 1987) but with one very important difference. A general assumption of the HH models is that the shocked gas cools more or less instantaneously, i.e. the cooling length is much less than the scale size of the shocked zone. However, this is not a valid assumption to make when modelling the NLR since the relatively high velocities into the shocks ($V_s \sim 1000$ km s^{-1}) and the low pre-shock densities (~ 1 cm^{-3}) result in very large cooling lengths. The cooling time of an individual packet of gas depends upon where it enters the bowshock and the models of TDA took this into account.

The finite cooling time combined with the linear motion of the plasmon produces the observed displacement between the radio components and the regions of significant optical emission. Two particularly clear examples are the cases of NGC 5929 and Mkn 78 where the optical sources appear to lie closer to the nucleus than the radio sources, because they have a smaller observed separation than do the radio components. Taylor et al (1989) successfully reproduced the separation between the radio and the optical components of NGC 5929 in terms of this finite cooling length model.

A particularly interesting feature of the model is that the cooling length, defined as the distance from the apex of the shock to where significant optical emission appears, is rather insensitive to the shock velocity (going roughly as only V_s) in contrast to the case of cooling behind a plane shock where the cooling length is roughly proportional to V_s^4 (for Kahn's (1976) $T^{-1/2}$ cooling law). In the TDA model, the concept of a single cooling length must be replaced by that of an effective cooling length, corresponding to the minimum point on the cooling curve; it is a function of where the gas enters the shock and the shape of the shock. The displacement of the optically emitting region from the shock apex is found to be surprisingly insensitive to V_s. For example, for a quadratic shock shape, this effective cooling length is proportional only to $V_s^{16/13}$, and the line profiles calculated depend only on the viewing angle and the value of V_s. This model is very simplified, in that the structure of the shocked region is calculated on the assumptions that the post-shocked gas is well coupled as far as momentum is concerned, but that individual gas elements cool independently of each other, and that the shocked region is optically thin to the central ionising radiation. Nonetheless, the model predicts, significantly, that the wide diversity of NLR profiles and velocity structures can be understood simply as a two-parameter family, dependent on only viewing angle and plasmon velocity (TDA).

5. Larger Scale Phenomena

There is ambiguous observational evidence for extremely energetic outflows from AGN. The high velocity blue-shifted BALs, as noted above, are usually cited as direct evidence for non-relativistic nuclear outflows. A possibly related phenomenon is that of the galactic superwinds (e.g. Heckman, Armus & Miley 1990) which seem to be associated with the presence of nuclear starbursts. Evidence for these winds comes from large scale ($\geq$ kpc) optical line and X-ray emission perpendicular to galactic discs; line ratios consistent with shock phenomena; double peaked optical emission line profiles extending over kpc which are consistent with expanding bubbles or bipolar flows, with line splitting $\sim$ 200–600 km s^{-1}.

There is a variety of ways in which winds could be driven out of the nuclear regions. Compton heated winds from accretion discs have received considerable attention. These are essentially thermal winds whose terminal velocity will be about three times the sound velocity in gas heated to the Compton temperature, i.e. $\sim$ 1000 km s^{-1}. Starburst winds are driven by the thermalisation of stellar ejecta, i.e. supernovae and stellar winds. Williams (1994) studied radiatively driven winds which have somewhat higher terminal velocities, up to a few thousand km s^{-1}.

Whatever the wind driving mechanism, the impact of these winds on the galactic ISM has many similarities to the impact of OB star winds on galactic ISM. A strong shock is driven into ambient gas whilst a reverse shock decelerates the wind. The implications of wind activity are profound. For example, they inject chemically enriched material, energy and momentum into the intergalactic medium (e.g. Heckman, Lehnert & Armus 1993). They have also been suggested as the means of accelerating ISM clouds to typical NLR velocities (Smith 1993), although there are very severe hydrodynamic difficulties with this due to the tendency for the acceleration process to disrupt the accelerated clouds. Here we will concentrate on the suggestion made originally by Dyson et al (1979) that wind driven shells may be responsible for the class of narrow absorption lines which have velocities within a few 1000 km s^{-1} of the associated emission lines.

Statistical studies of the frequency and distribution of narrow metal absorption line systems (e.g. Anderson et al 1987; Weymann, Carswell & Smith 1981; Perry, Burbidge & Burbidge 1978) provide evidence that absorption with $z_{abs} \approx z_{em}$ shows a statistical excess in radio-loud QSOs and is therefore likely to be intrinsic to the objects. (Most absorbing systems are produced in foreground intervenors, e.g. haloes of galaxies.)

The best studied of these associated systems is the complex in 3C 191 (Williams et al 1975) which contains at least two absorption components. The main system has a velocity of $0.0027c$ relative to the emission line redshift (which is, as usual, taken as defining the rest frame). Studies of the absorption spectrum suggest that the absorbing gas is located at a distance of about 10 kpc from the ionising continuum source, that it has a number density $\sim 10^3$ cm^{-3} and that the gas temperature is $\sim 2 10^4$ K (Williams et al 1975). Dyson et al (1979) noted that the gas density is roughly that expected in gas which has cooled to about the estimated temperature behind a shock moving at 0.0027c into galactic ISM with a number density of unity. They therefore used the standard theory of stellar wind driven bubbles (e.g. Dyson 1981) and showed that the absorbing gas could be the shocked ISM in the wind-driven shell.

The more general problem of the interaction of a nuclear wind and radiation field with galactic ISM was studied by Dyson, Falle & Perry (1980) and Falle, Perry & Dyson (1981—henceforth FPD). The initial stages of the interaction were discussed by Dyson et al (1979; 1980); FPD showed that the inclusion of the nuclear radiation field has important hydrodynamic consequences for the later evolution of the driven shell, because of radiation pressure exerted on the gas in the shell. Provided that the radiative driving is sufficient, the wind driven shell accelerates. A positive pressure gradient develops in the cooled gas in the shell as the downstream piston pressure (i.e. the shocked wind pressure) decays. The radiative acceleration in photoionised gas is, under the circumstances appropriate here, proportional to the gas density (Röser 1979), and the development of the pressure gradient is thereby amplified. A velocity gradient is simultaneously set up in the shell and, in the shock frame, the gas is moving fastest at the piston. Provided this velocity is subsonic, as in normal stellar wind driven flows, the shock and piston are causally connected. However, when radiation driving is included, the velocity becomes sonic and the shock and piston disconnect. At large enough times after disconnection, the flow becomes that of an accelerating radiative shock wave which sweeps through the ambient medium. Absorption line formation effectively takes place between the outer shock and the sonic point in the postshock gas, and the velocity dispersion is about equal to the sound speed there, as indicated by the absorption line widths.

There are restrictions on the velocities which can be reached. Firstly, the optical depth to the continuum steadily increases and eventually the radiative acceleration begins to fall. This may not be too serious a restriction since Perry & Dyson (1990) suggested that heating due to turbulent dissipation of energy in the shell might raise the gas temperature high enough ($\sim 2 - 3 10^4$ K) to decrease the neutral hydrogen content and thereby reduce the estimated optical depths. A more serious problem is that for radiative driving to be operative, the cooling time in the post-shock gas must be less than the dynamical timescale. They estimated that it was unlikely that the velocity of the absorbing gas could become greater than about $0.01c$ relative to the rest frame. As noted earlier, 3C 191 has multiple absorbing systems, and this is also seen in other sources. There are two possibilities for generating these. Firstly, gas subjected to radiative driving which is proportional to the gas density, is subject to radiatively driven instabilities (Mestel, Moore & Perry 1976). Secondly, it is generally agreed that radiative shocks with power-law cooling functions ($L \sim T^{\gamma}$) in the cooled gas exhibit an oscillatory instability provided $\gamma \approx 1$ and the shock velocity is greater than ~ 140 km s^{-1} (Innes, Giddings & Falle 1987; Gaetz, Edgar & Chevalier 1988). The velocity condition is easily satisfied here and either mechanism may contribute to the appearance of multiple systems.

6. Conclusions

Activity in galactic nuclei appears to be the result of a symbiosis of an accreting black hole at the centre of a dense starburst stellar cluster. The young stars in the cluster lose mass, creating a fast moving, relatively dense interstellar medium which is heated by the radiation produced by accretion of the interstellar medium itself by the black hole. Shocks occur at the interfaces of the stellar mass loss envelopes and the interstellar medium created by the cluster. These shocks are radiative, and the resulting cool, post shock gas is photoionized by the central continuum radiation; the fast moving high pressure photoionised gas radiates the broad emission line spectrum characteristic of AGN. Similar radiative shocks further out in the galaxy radiate the narrow emission lines, and when the shocks lie on the line of sight to the nucleus they create both broad and narrow absorption features in the spectra. Clearly, the physical processes which are the subject of this meeting are important well outside their traditional remit, and may provide the missing clues in our search for an understanding of AGN.

References

Anderson, S.F., Weymann, R.J., Foltz, C.B. & Chaffee,F.H.: 1987, *Astron. J.* **94**, 278
Arav, N., Li, Z.-Y. & Begelman, M.C.: 1994, *Astrophys. J.* **432**, 62
Bode, M.F. & Kahn, F.D.: 1985, *Mon. Not. R. astr. Soc.* **217**, 205
Colgate, S.A.: 1967, *Astrophys. J.* **150**, 163
Collin-Souffrin, S., Dyson, J.E., McDowell, T.C. & Perry, J.J.: 1987, *Mon. Not. R. astr. Soc.* **232**, 537
Dyson, J.E.: 1981, in F.D. Kahn, ed(s)., *Investigating the Universe,* D. Reidel Publ. Co.: Dordrecht, 125
Dyson, J.E.: 1994, in T.P. Ray & S.V.M.Beckwith, ed(s)., *Star Formation and Techniques in Infrared and mm-Wave Astronomy,* Springer-Verlag: Berlin, 93
Dyson, J.E. & Perry, J.J.: 1982, *Proc. 3rd European IUE Conference, Madrid,* 593
Dyson, J.E., Falle, S.A.E.G. & Perry, J.J.: 1979, *Nature* **277**, 118
Dyson, J.E., Falle, S.A.E.G. & Perry, J.J.: 1980, *Mon. Not. R. astr. Soc.* **191**, 785
Falle, S.A.E.G., Perry, J.J. & Dyson, J.E.: 1981, *Mon. Not. R. astr. Soc.* **195**, 397
Field, G.B.: 1964, *Astrophys. J.* **140**, 143
Filippenko, A.: 1992, in W.J. Duschl & S.J. Wegner, ed(s)., *Physics of Active Galaxies,* Spring-Verlag: Berlin, 345

Gaetz, T.J., Edgar, R.J. & Chevalier, R.A.: 1988, *Astrophys. J.* **329**, 927
Gull, S.F.: 1975, *Mon. Not. R. astr. Soc.* **171**, 237
Hartigan, P., Raymond, J. & Hartmann, L.: 1987, *Astrophys. J.* **316**, 323
Heckman, T.M.: 1992, in S.S. Holt & C.M. Urry, ed(s)., *Testing the AGN Paradigm*, AIP Press: New York, 593
Heckman, T., Armus, L. & Miley, G.: 1990, *Astrophys. J. Suppl.* **74**, 883
Heckman, T., Lehnert, M. & Armus, L.: 1993, in J.M. Shull & H.A. Thronson, ed(s)., *The Environment and Evolution of Galaxies*, Kluwer Academic Publ.: Dordrecht, 455
Innes, D.E., Giddings, J.R. & Falle, S.A.E.G.: 1987, *Mon. Not. R. astr. Soc.* **227**, 1021
Kahn, F.D.: 1976, *Astron. Astrophys.* **50**, 145
Lynden-Bell, D.: 1969, *Nature* **223**, 690
Mestel, L., Moore, D.W. & Perry, J.J.: 1976, *Astron. Astrophys.* **52**, 203
Pagel, B.F.: 1985, in J.E. Dyson, ed(s)., *Active Galactic Nuclei*, Manchester Univ. Press: Manchester, 373
Pedlar, A., Dyson, J.E. & Unger, S.W.: 1985, *Mon. Not. R. astr. Soc.* **214**, 463
Perry, J.J.: 1992, in B. Rocca-Volmerange, B. Guideroni, M. Dennefeld & J. Tran Thanh Van, ed(s)., *First Light in the Universe: Stars or QSOs?*, Editions Frontières: Gif-sur-Yvette, 225
Perry, J.J., Burbidge, E.M. & Burbidge, G.R.: 1978, *Publ. Astron. Soc. Pacific* **90**, 337
Perry, J.J. & Dyson, J.E.: 1985, *Mon. Not. R. astr. Soc.* **213**, 665
Perry, J.J. & Dyson, J.E.: 1990, *Astrophys. J.* **213**, 362
Perry, J.J. & Dyson, J.E.: 1992, in S.S.Holt, S.G. Neff & C.M. Urry, ed(s)., *Testing the AGN Paradigm*, AIP Press: New York, 353
Röser, H.-J.: 1979, *Astron. Astrophys.* **80**, 179
Shields, G.A. & Oke, J.B.: 1975, *Astrophys. J.* **197**, 5
Shklovskii, I.S.: 1960, *Sov. Astron. A.J.* **4**, 885
Smith, M.D., Smarr, L., Norman, M.L. & Wilson, R.J.: 1983, *Astrophys. J.* **263**, 432
Smith, S.J.: 1993, *Astrophys. J.* **411**, 570
Taylor, D., Dyson, J.E., Axon, D.J. & Pedlar, A.: 1989, *Mon. Not. R. astr. Soc.* **240**, 487
Taylor, D., Dyson, J.E. & Axon, D.J.: 1992, *Mon. Not. R. astr. Soc.* **255**, 351
Terlevich, R., Tenorio-Tagle, G., Franco, J., Boyle, B., Rozyczka, M. & Melnick, J.: 1993, in B. Rocca-Volmerange, B. Guideroni, M. Dennefeld & J. Tran Thanh Van, ed(s)., *First Light in the Universe: Stars or QSOs?*, Editions Frontières: Gif-sur-Yvette, 261
Unger, S.W., Pedlar, A., Booler, R.V. & Hanson, B.A.: 1986, *Mon. Not. R. astr. Soc.* **219**, 387
Weymann, R.J., Carswell, R.F. & Smith, M.G.: 1981, *Ann. Rev. Astron. Astrophys.* **19**, 41
Weymann, R.J., Schiano, A.V.R., Scott, J.S. & Christiansen, W.A.: 1982, *Astrophys. J.* **262**, 497
Whittle, M., Haniff, C.A., Ward, M.J., Meurs, E.J.A., Pedlar, A., Unger, S.W., Axon, D.J. & Harrison, B.A.: 1986, *Mon. Not. R. astr. Soc.* **222**, 189
Williams, R.E., Strittmatter, P.A., Carswell, R.F. & Craine, E.R.: 1975, *Astrophys. J.* **202**, 296
Williams, R.J.R.: 1994, Ph.D. Thesis, Cambridge University.
Williams, R.J.R. & Dyson, J.E.: 1994, *Mon. Not. R. astr. Soc.* **270**, L52
Williams, R.J.R, Hartquist, T.W. & Dyson, J.E.: 1995, *Astrophys. J.* **446**, 759
Williams, R.J.R. & Perry, J.J.: 1994, *Mon. Not. R. astr. Soc.* **260**, 437

PHOTOIONISING SHOCKS IN SNRS AND AGN

MICHAEL A. DOPITA
Mount Stromlo and Siding Spring Observatory,
Institute of Advanced Studies, The Australian National University,
Private Bag, Weston Creek P.O., ACT 2611, Australia

Abstract. A high velocity radiative shock, or one moving into high-metallicity gas, provides an efficient means to generate a strong local UV photon field. The optical emission from the shock and precursor region is dominated by the photoionised gas, rather than by the cooling region, and the total optical + UV emission scales as the mechanical energy flux through the shock. In this paper, such models are applied to oxygen-rich supernova remnants and AGN. For AGN, the degree of magnetic support in the post-shock gas is an important parameter. LINER and cooling flow spectra can be understood as resulting from high velocity shocks without precursors, while Seyfert 1.5-2 galaxy emission line ratios result from high velocity shocks with their photoionised precursor HII regions. This model explains the problem of the high electron temperatures observed in both classes of object.

Key words: - AGN – SNRs – Liners – Shock Waves

1. Introduction

A fast, radiative shock in interstellar space is a powerful source of ionising photons. These photons, are produced in the hot, post-shock plasma as it cools, and can diffuse both upstream and downstream. Those photons diffusing upstream may produce an extensive precursor HII region, and those finding their way downstream will profoundly influence the ionisation and temperature structure of the recombination region of shock. Since much of the available enthalpy is radiated in this photon field, the photoionisation effects produced by the diffuse UV field may even come to dominate the optical emission from the shock itself.

The first reference to the photoionisation effects of precursors to shocks was in Daltabuit and Cox (1972) in the context of $\approx$ 1000 km s^{-1} cloud-cloud collisions in the broad-line regions of AGN. For lower velocity radiative shocks (v < 150 km s^{-1}), the shock models of Raymond (1976) and Dopita et al. (1984a) allowed for photoionisation both up- and down-stream. However, since the photoionising fields are weak in these low velocity shocks, the emission from the precursor region is negligible, and the photons serve merely to pre-ionise the gas coming into the shock. Nonetheless, this change in ionisation state radically affects the emission line spectrum of the shock and so sensitive shock diagnostics can be made on the basis of the emission line spectrum. Shocks with velocities lower than about 100 km s^{-1} are unable to pre-ionise the precursor region, and these produce an enhanced two-photon

Astrophysics and Space Science **233**: 215–230, 1995.

continuum through collisional excitation of the neutral hydrogen, an effect recognised by Dopita, Binette and Schwartz (1982).

As the shock velocity increases, the shape of the cooling curve ensures that the cooling plasma cools in a thermally unstable fashion. Innes, Giddings and Falle (1987a,b) claimed that this invalidated all shock diagnostics. However further work by Innes (1992) has established rather that the plasma undergoes limit cycle oscillations, and that the time averaged mean spectrum is more like the steady-flow solution. In fully three dimensional modelling, it is clear that secondary shocks driven by the development of pressure differences between adjacent cooling regions may have an important contribution to make to the resultant optical/UV spectrum, but as yet such models have not been made. An early attempt to investigate the effects of radiative transfer in thermally unstable cooling flows or radiative shocks was by Binette, Dopita and Tuohy (1985). The hot plasma was modelled on the assumption of collisional ionisation equilibrium (CIE) cooling using the Raymond and Smith (1977) code, and the effect of thermal instabilities was approximated by assuming the formation of photoionised dense cloudlets in pressure equilibrium with the hot plasma. The opposite extreme, of plane parallel sheets cooling in thermal stability was also treated. Hybrid models for fast shocks with an externally imposed photoionising field have been extensively discussed in a series of papers by Viegas-Aldrovandi and Contini (1989 and references therein); and Contini (this conference).

A fully self-consistent model for high velocity shocks cooling in a thermally stable fashion had to await the development of codes capable of dealing with the non-CIE cooling, and including X-ray emission processes and radiative transfer. Recently, this was accomplished in the code MAPPINGS II (Sutherland and Dopita 1993) and similar results have been produced by Schmutzler and Tscharnuter(1993). These models form the basis of the results presented in this paper.

As stated above, a radiative shock is one which radiates all or most of the available enthalpy. At high velocities, the predominant emission mechanisms are electron thermal bremstrahlung and line emission. The non-equilibrium ionisation in the post-shock plasma enhances line emission enormously over the equilibrium case, particularly for the hydrogen-like and helium-like ionic species of intermediate mass elements such as oxygen and neon for which the collisional excitation rates are far higher than the ionisation rates. Thus, these elements undergo many collisional excitations of their principal resonance lines before they are ionised. The output spectra for a number of shock velocities is shown in Figure 1. Note that the bulk of the radiation occurs in the EUV region, with an exponential bremstrahlung tail extending to thermal X-ray wavelengths.

Much of the work described in this paper is from the thesis of my ex-student Ralph Sutherland (1993), published in Sutherland and Dopita (1993,

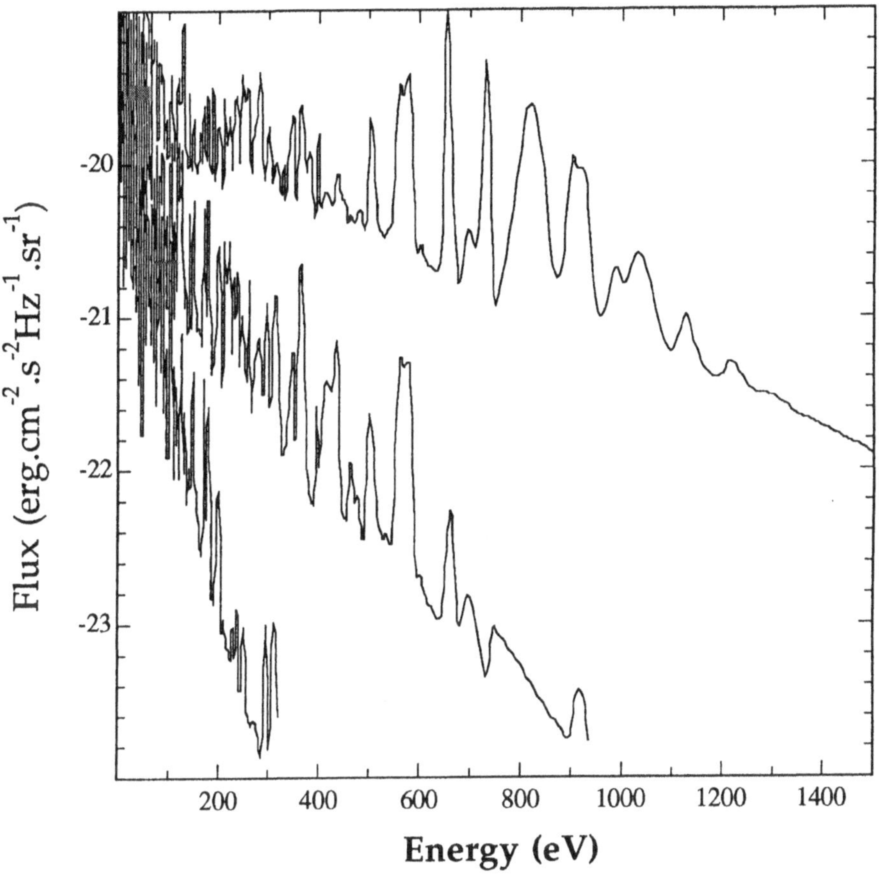

Fig. 1. The EUV/soft X-Ray fluxes generated by fully radiative shocks with velocities 200 km s^{-1} (lower curve), 300 km s^{-1} (middle) and 500 km s^{-1} (upper).

1995a,b) and in Sutherland, Bicknell and Dopita (1993). The material on fast shocks in AGN including the effects of magnetic support is from Dopita and Sutherland (1995; in preperation).

2. Supernovae Remnants

2.1. Supernova 1987A

SN 1987A has already provided one of the more spectacular examples of a photoionising shock. At the time of shock breakout through the photosphere of the star, an intense flash of EUV photons was produced. About 70 days after the supernova explosion, this flash was seen to reach and to photoionise the dense waist of the bipolar nebula originally thrown out by the precursor red giant, and subsequently shaped by the faster blue giant wind which immediately preceded the supernova event. The photoionisation was initially detected through the narrow UV emission lines which were excited (Lundqvist and Fransson 1987; Fransson et al 1989), and, as the ring itself gradually cooled and recombined (Lundqvist and Fransson 1991) it was imaged both from ground-based observations (Crotts, Kunkel and McCarthy, Wampler et al 1990) and in a series of spectacular observations with HST (Jacobsen et al. 1991; Plait et al. 1994).

An even more spectacular event is predicted by Luo, McCray and Slavin (1994). According to these authors, this will occur around AD1999, when the expanding ejecta of SN 1987A will finally impact on the dense equatorial ring. The collision is expected to drive a radiative shock into this ring with a velocity of 200-400 km s^{-1}. The UV cooling lines produced in the shock are expected to achieve luminosities as high as 800 $L_{\odot}$. The powerful EUV field produced will both pre-ionise the material in the ring, producing strong, narrow [O III] and Balmer lines. Furthermore, the photons which escape back towards the ejecta will re-ionise this material, producing visible emission of broad (FWHM $\approx$ 104 km s^{-1}) Balmer lines. All in all it should be a spectacular display, and one that will provide a powerful test of of the fast shock theory.

2.2. Oxygen Rich Supernova Remnants

The young supernova remnants (SNRs) provide a key to test theories of nucleosynthesis in massive stars, since only in these objects do we find optically-thin material from the stellar interiors exposed to direct investigation. This small class of SNRs shows fast moving ($>$ 1000 km s^{-1}) knots of material emitting in lines of oxygen, neon, and other heavy elements, but apparently contain no trace of hydrogen or helium, consistent with their origin from within the helium-burnt layers of a massive ($>$ 12 $M_{\odot}$) progenitor star. In our Galaxy, Cas A is the prototype of this class (see Chevalier and Kirshner 1979). The other members are the galactic objects G292+1.8 and Puppis A (Goss et al. 1979; Winkler and Kirshner 1985), three remnants in the Magellanic Clouds (Lasker 1978; Dopita et al. 1981; Kirshner et al. 1989), and two unresolved objects in more distant galaxies M83 and NGC 4449 (Long et al. 1981, Blair et al. 1983).

A fundamental barrier to a quantitative interpretation of the spectra from these objects has been our inability to produce a plausible model for the excitation of these knots. It is clear that the knots are formed by an early instability in the supernova ejecta. Evidence for such instabilities has been found in the case of SN 1987A. The dense knots of oxygen-rich material are excited at the time that they pass through the reverse shock which is propagated back into the low-density phase of the ejecta as this interacts with the surrounding ISM. Logically, therefore, the knots are excited by slow shocks running into them, being driven by the increase in external pressure. However, the structure of these shocks is quite different from those which occur in material of normal composition. In particular, Itoh (1981a, b) and Dopita, Binette and Tuohy (1984b) demonstrated that the very high metallicity causes the cooling time to remain shorter than the recombination time until the shocked plasma has cooled to a few hundred degrees K. As a consequence, the optical [O III] lines are predicted to be bright, but lines of species of lower ionisation are weak, unless the shock velocity is less than 100 km s^{-1}. However, in these remnants, lines of both [O I] and of [O II] are observed to be bright. Itoh (1981b) argued that the powerful EUV field generated in the cooling zone could escape upstream to produce an extensive warm photoionised precursor region. This idea has the difficulty that the extent of this region would have to be be greater than the observed size of the knots. Furthermore, if oxygen rich, the temperature of this region would high enough to excite optical forbidden lines only if it was composed of pure oxygen. The efficient cooling in the fine-structure lines of any other elements such as carbon, expected to be present, would force the photoionisation equilibrium temperature in this zone to lie close to 300 K, suppressing the optical cooling. At this point, it might be appropriate to remark that evidence for a photoionised precursor in the 'normal' composition gas ahead of the main blast wave has been adduced in two oxygen-rich SNRs, one in the SMC (Dopita and Tuohy, 1984), and the other in the LMC (Morse et. al, 1995, in prep).

For the photoionised precursors in the oxygen-rich material, Dopita (1987) suggested a non-equilibrium solution. The buildup of the strong precursor radiation field occurs over the cooling timescale of the shocked plasma, allowing an ionisation front to be propagated into the cloud. For an R-Type ionisation front driven by a strong photoionising field, the heating effect of the photon field dominates at the leading edge of the front, and the gas is strongly superheated, producing optical forbidden lines of a range of ionisation states. As the gas approaches its photoionisation equilibrium, the temperature of the plasma approaches its equilibrium value of a few hundred degrees, and the optical lines are suppressed in this low temperature, high ionisation plasma.

This idea was fully developed in Sutherland's thesis (1993); Sutherland and Dopita (1995b). Sutherland showed that the cloud shock by itself could drive the R-Type ionisation front, rather than the bow-shock emission which had been invoked by Dopita (1987). The buildup of the EUV emission is sufficiently rapid to allow an R-Type ionisation to be detached from the shock in 1/10 of the cooling timescale; about 3/n years, where n is the pre-shock density in units of cm^{-3}. The optical emission is produced in a brief period during and after the cloud shock becomes radiative, and as the thin superheated zone close to the ionisation front sweeps through the cloud. The ionisation front velocity is comparable with the dynamical timescale, the time need for the cloud to fully enter the reverse shock, and this timescale is itself a few times shorter than the cloud crushing timescale. For this reason, the lifetime of an individual knot is short, and theory shows that, for cloud densities $\approx$ 100 cm^{-3}, it is comparable with the 25 year e-folding lifetimes observed for the Cas A knots by Kamper and van den Bergh (1976). The combination of the (steady-flow) cloud shock of 100-200 km s^{-1} and its R-Type precursor gives a good description to the various ratios that can be formed from the [O I] 6300Å, [O II] 3727,9Å and 7316,24Å, and the [O III] 4363Å and 5007Å lines. However, UV diagnostics from observations such as those by Blair et.al (1989) are required in order to obtain useful abundance data, especially for elements such as C, Mg and Si. We look forward to HST to provide such data in Cycle 4 and 5.

3. Narrow Line Phenomena in AGN

All classes of AGN show some evidence of 'narrow Line' emission in the [O III] lines. Seyferts of Classes 1 and 2 (Khachikian and Weedman, 1974) show a rich spectrum of narrow lines, as do the narrow line radio galaxies (NLRGs). The universal characteristic of these narrow line regions is a very large [O III] $\lambda 5007/H_\beta$ ratio ($\approx$ 7-20), the presence of HeII and strong lines of [O I], [N II] and [S II]. Although the lines are termed 'narrow', this is something of a misnomer, since the typical velocity widths are $200 < \mathrm{FWHM} < 500$ km s^{-1}, and velocities in a few objects are seen even to exceed 1000 km s^{-1} (Veilleux, 1991a,b).

There is a wide variety of circumstantial evidence that shocks are important in these objects. Close correlations between the forbidden line regions and the radio continuum regions are found in the sample of nearby Seyferts studied in the extensive survey by Tsvetanov, Fosbury and Tadhunter (1994). In particular, there is often an excellent alignment with inner radio jets (Unger et al. 1987; Wilson and Tsvetanov 1994). Higher redshift and more luminous sources likewise show strong alignments between the NLR and the radio (McCarthy et al. 1987: Baum et al. 1988; Meisenheimer et al. 1994).

Evidence of balance between the thermal pressure in the NLR gas and the relativistic and magnetic pressure in the lobes was found by De Bruyn and Wilson (1978) and Unger et al. (1987). Correlations of line flux with radio luminosity have been found by Baum and Heckman (1989; [N II] + H_α), Forbes and Ward (1993; [Fe II]), and Morganti et al. (1994; [O III]). A dynamical correlation between [O III] line width and the radio luminosity was found for high luminosity sources by Wilson and Willis (1980) and Whittle (1992, 1994). Evidence for global outflow and for line splitting and broadening of several hundred km s^{-1} has been found in many objects (see review by Whittle, 1989). To give a few particular examples, 3C120 (Meurs et al. 1989), NGC 1068 (Antonucci and Miller 1985; Cecil and Bland 1989), NGC 3516 (Mulchaey et al. 1992; Veilleux et al. 1993), N2992 (Tsvetanov, Dopita and Allen 1994, in prep), and NGC5252 (Morse et al., in press). A relationship between the surface brightness and level of excitation of the NLER was discussed by Fosbury (1989) and Prieto et al. (1989).

The so called 'unified models' of AGN (Barthel 1989) rely fairly heavily on the observation of cones of highly ionised narrow-line gas extending from the nucleus (Pogge, 1988a,b,1989; Tadhunter and Tsvetanov; Fosbury 1989). In the standard picture, the nuclear ionising UV radiation is assumed to be directionally beamed to illuminate the extranuclear gas (Morganti et al. 1991,1992). Despite a deal of success in describing the general features of the emission line spectrum, the weakness of such models is that a number of parameters remain arbitrary. In particular, the model fails to give an explanation for the relatively small range of ionisation parameter over a very large radial scale range, and it fails to address the question of the observed velocity dispersion of the gas. On the other hand, the observed velocity dispersion and the correlations described in the previous paragraphs might naturally find an explanation in a model based upon a direct mechanical interaction between the ambient medium and outflowing jet.

The other major class of narrow-line AGNs is that of the low-ionisation nuclear emission-line regions, LINERs (Heckman 1980). These fall into a distinct region of excitation space when classified according to their emission line ratios (Baldwin, Phillips and Terlevich 1982; Veilleux and Osterbrock 1987). In particular, the [O II] lines are stronger than [O III] λ5007, and the [O III] $\lambda 5007/H_\beta$ ratio is less than about 3. In the red, their spectra are quite similar to those of Seyfert Galaxies. LINERs continue the trend of decreasing velocity dispersion and Balmer luminosities towards lower excitation. A survey of 'normal' elliptical galaxies by Phillips et al. (1986) showed that low-level LINER activity is found in an appreciable fraction of these. Elliptical LINER galaxies cover a range of radio properties, including radio quiet objects, weak nuclear sources, and Fanaroff-Riley (FR) Class I sources. However, LINERs are never found in the high-power FR Class II radio sources. In terms of the correlation between the optical and the radio luminosities, a similar trend emerges. There is a good correlation for the FR II class, but this disappears smoothly at the FR I-II transition. These correlations suggest that the optical emission makes a transition from predominantly jet-excited to some other excitation mechanism at the FR I-II transition.

LINERS show a variety of internal motions. Generally speaking, the lowest excitation objects show a more ordered rotation, whereas the higher excitation AGN are more chaotic systems (Baum, Heckman and van Breugel 1992). This strongly suggests that mechanical energy input and excitation are correlated. This is consistent with the expectations of shock model for the excitation, first suggested in the case of NGC 1052 by Fosbury et al. (1978).

LINERs, or objects having LINER-like spectra, are also found in the nuclei of spiral galaxies lacking nuclear starbursts (Keel 1983a,b), in as many as 50% of elliptical galaxy nuclei (Phillips et al. 1986), and in the 'cooling flows' associated with first-ranked elliptical galaxies in clusters (Heckman 1981; Cowie et al 1983; Johnstone, Fabian and Nulsen 1987; Heckman et al 1989). Again, typical velocity dispersions observed in all these objects are a few hundred km s^{-1}.

In the last fifteen years, the hypothesis that all these classes of narrow line regions (NLRs) are excited by a hard power-law spectrum of UV photons has gained almost universal acceptance. In such a model, the Seyfert and QSO NLRs would be excited by a fairly flat UV power law, or else a truncated power law having an ionisation parameter (U) of order 10^{-2} (Koski 1978; Stasinska 1984; Veilleux and Osterbrock 1987; Osterbrock, Tran and Veilleux 1992; Baum, Heckman and van Breugel 1992). On the other hand if LINERs are photoionised, the ionising spectrum is similar to the NLRs, but the ionisation parameter is low, $U \approx 10^{-4}$ (Ferland and Netzer 1983, Ho et al. 1993). The source of these photons is assumed to be the nucleus, but the exact means whereby the non-thermal UV spectrum is generated remains obscure. Generally what is done is to simply join the observed soft X-ray to the observed EUV point, or to extrapolate the EUV slope with an exponential cutoff to match the soft X-ray. Neither of these procedures properly addresses the physical processes producing the emission, and there remains a serious shortfall in the estimated number of ionising photons based on an extrapolation of the observed UV spectrum (Binette, Parker and Fosbury 1993).

In the following section, we postulate that all these classes of object result from the input of mechanical energy through fast shocks, rather than on photoionisation from a central source. The source of the mechanical energy may be either turbulent motions of clouds in the potential of the galaxy, or direct energy deposition from an outflowing jet of relativistic electrons. In the hydrodynamic calculations of Bicknell (1990; 1994), a mildly supersonic jet confined by a denser medium will strongly interact with the surrounding medium through the operation of the Kelvin-Helmoltz instability at the boundary. This generates slow shocks in the surrounding medium, and turbulent vortices pull off dense clouds of gas embedded in a lower-density phase which is partially entrained in the jet. The turbulent collisions of clouds generate fast shocks, and slower shocks are driven into them.The strength of the photoionising field produced by the cooling plasma ensures that the optical spectra of this class of shocks is dominated by photoionisation effects rather than by the cooling plasma. It should be remarked that this idea is by no means new. Osterbrock (1971) suggested shocks as a means to set up the multi-phase structure of AGN, and Daltabuit and Cox (1972) pointed out how such high velocity shocks in a turbulent medium could of themselves provide an ample supply of ionising photons.

4. Models for Fast Radiative Shocks in AGN

The emission line spectra for a (low density, solar abundance) grid of fast shocks has been computed using the general purpose modelling code MAPPINGS II, described in Sutherland and Dopita (1994). These models include all the relevant physics required to compute non-equilibrium cooling and ionisation, and the code includes a very large atomic data set enabling us to treat all ionisation stages of cosmically abundant elements up to fully stripped nickel. This said, the models presented here contain a number of important approximations. First, although the models are fully time dependent and explicitly include non-equilibrium ionisation and cooling, they assume a steady and one-dimensional flow. Since the plasma is known to pass through regimes of thermally unstable cooling (see also Falle, this conference), this is a serious limitation of these models. Second, the diffuse field in the shock is treated in the downstream-only approximation; the upstream photon field is assumed to escape freely from the shock structure, where it can photoionise the gas moving into the shock. This approximation will have relatively little effect on the computations, since the opacity of the hot plasma is very low in the region where most of the UV flux is generated. Third, the precursor photoionised region is assumed to have the ionisation structure of an equilibrium HII region. In practice, this is not a bad approximation for shocks having v $>$ 190 km s^{-1}, since these can drive an ionisation front into the precursor material at a velocity in excess of the shock velocity, allowing the formation of an equilibrium HII region. Pre-shock equilibrium temperatures and ionisation parameters are given in Table 1:

The total shock luminosity will scale as the mechanical energy flux through it; ρV_s^3. However, the fraction of this luminosity that appears as Balmer lines

TABLE I
The effective ionisation parameter of upstream radiation field and the photoionisation equilibrium temperature in the gas entering the shock as a function of shock velocity.

V_s (km s^{-1})	200	250	300	350	400	450	500
QT (km s^{-1})	207	398	619	872	1234	1642	2025
T_e (K)	7220	9640	11200	11620	13340	16570	18150

decreases with velocity. Simple scaling relations (for the radiative shock only) are:

$$L_T = 2.28 \times 10^{-3} \left[\frac{V_s}{100kms^{-1}}\right]^{3.0} \left[\frac{n}{cm^{-3}}\right] \quad erg\ cm^{-2}\ s^{-1}$$

$$L_{H\beta} = 7.44 \times 10^{-6} \left[\frac{V_s}{100kms^{-1}}\right]^{2.41} \left[\frac{n}{cm^{-3}}\right] \quad erg\ cm^{-2}\ s^{-1}$$

If the photoionised precursor is included, the shock structure is about a factor of two brighter in the Balmer lines. Typically, only about 1% of the energy emerges as H_β photons. The decrease in relative emissivity in the Balmer lines at higher shock velocities is a result of the higher temperature in the photoabsorption and recombination zone (and the consequent decrease in the recombination rate). However, because of the increase in collisional excitation rates, the optical forbidden lines are emitted with higher efficiency, so the ratio of forbidden lines to recombination lines is a sensitive function of shock velocity. This is analogous to the measurement of Stoy temperatures in planetary nebulae.

Apart from the shock velocity, the most important factor which controls the nature of the emergent spectrum is the transverse magnetic field at high shock velocities, this serves to limit the compression through the shock. When the pressure of the transverse field dominates in the cool zone, we can set the pressure equal to the ram pressure of the shock:

$$P = \frac{B^2}{8\pi} = \mu m_H n V_s^2$$

where n is the pre-shock density, B is the pre-shock transverse magnetic field, and V_s is the shock velocity. Assuming that the magnetic field is frozen into the flow, the maximum compression factor through the shock is:

$$\frac{n_f}{n} = \frac{(8\pi\mu m_H)^{1/2} V_s}{(B/n^{1/2})}.$$

Since the number of ionising photons produced in the cooling zone is effectively independent of the initial magnetic field, the effective ionisation parameter in the photoionisation and recombination zone is therefore higher the higher the magnetic field. This effect is clearly seen in the structure of this zone as a function of the magnetic field, shown in Fig 2. The appropriate magnetic parameter is the ratio $B/n^{1/2}$. Likely values of this ratio can be estimated on the assumption that the magnetic field in the pre-shock ISM is saturated when the magnetic rigidity is high enough to damp out the turbulence.This condition is met when the Alvén velocity is of the same order of magnitude as the thermal sound speed. For a warm interstellar medium, having a sound speed ≈ 8 km s^{-1}, this implies:

$$\left[\frac{B}{\mu G}\right] = 3\left[\frac{n}{cm^{-3}}\right]^{1/2}.$$

We have therefore computed shock models in the range $0 < B/n^{1/2} < 4$. The shock velocities are in the range $150 < V_s < 500$ km s^{-1}. These parameters control the mapping of the models onto the line diagnostic planes formed by the Baldwin, Phillips and Terlevich (1981), Veilleux and Osterbrock (1987), or Osterbrock, Tran and Veilleux (1992) plots. In the photoionisation models, the parameters are the spectral index and the ionisation parameter. In the shock models the corresponding parameters are shock velocity (which controls the flatness of the UV field, see Fig 1), and the magnetic parameter, which controls the ionisation parameter experienced in the optically emitting photoionisation recombination region of the shock.

In the shock models, the spectrum is also largely determined by the presence or absence of a photoionised precursor. The gas ahead of the shock experiences a radiation field with a very high ionisation parameter. Values of $U \approx 10^{-2}$ in the precursor gas are typical for velocities typical of Seyfert NLRs ($V_s \approx 500$ km s^{-1}) (c.f. Table 1; where $U \approx Q/c$). The value of U which characterises the recombination zone is two orders of magnitude lower thanks to the compression of the gas through the shock. In the shock model, this distinction is the fundamental difference between objects producing Seyfert-like NLRs and objects which display LINER spectra, and reflects simply upon the column density of the pre-shock gas; Seyferts have at least a Strömgren column of gas in the vicinity of the shock, whereas LINERS are (or cooling flows) relatively gas-poor objects. This provides a natural unification of narrow line emission regions in galaxies, and explains why the two classes of object appear to group into two separate regions in diagnostic plots. In the context of this review, there is no space to show all the diagnostic plots. However, figure 3 presents one particularly fine (and easy to measure) example. The observational data has been drawn from a wide selection of sources already referred to in this paper.

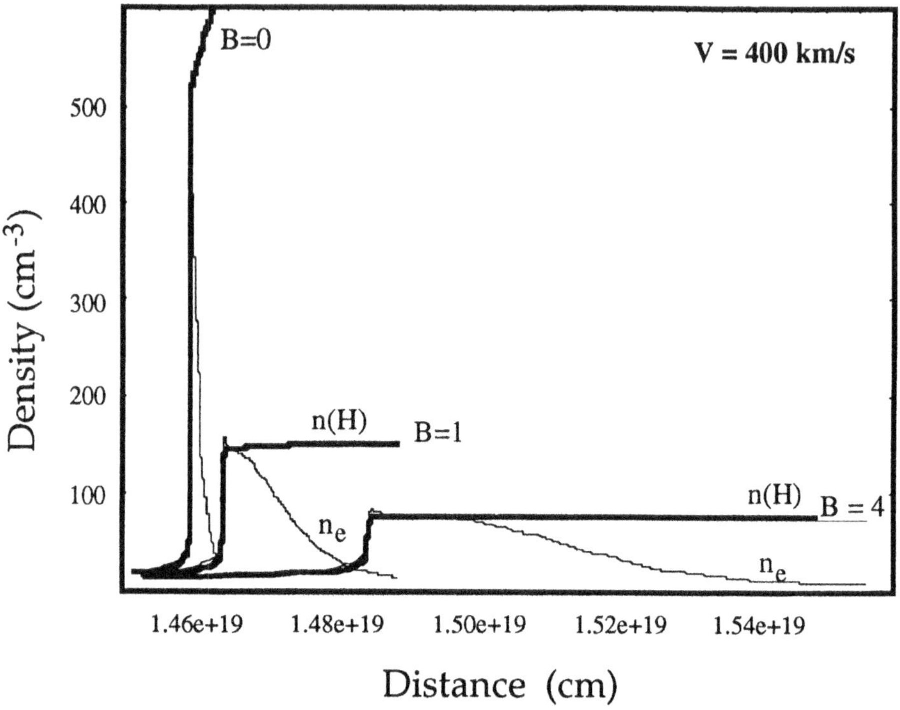

Fig. 2. The structure of the photoionisation/recombination region of an $n(H) = 1.0\ cm^{-3}$, $400\ km\ s^{-1}$ shock with three values of the pre-shock transverse magnetic field, 0, 1 and 4 μG. Note the radical differences in the structure.

A notorious problem in the analysis and interpretation of the spectra of Seyfert galaxies has been the so-called 'temperature problem' (Tadhunter et al. 1989). In brief, whenever the electron temperature could be measured from the [O III] $\lambda 4363/\lambda 5007$ line ratio, it was found to be high (15,000 K or higher). However, photoionisation models give much lower electron temperatures with normal abundances, and this problem gets worse as the ionisation parameter is decreased. Although the high [O III] ratio had been ascribed to high density in Seyfert galaxies, this conclusion is not supported by the observed ratio of the density sensitive [S II] lines. However, as pointed out by Sutherland, Bicknell and Dopita (1993), shocks provide a region of high temperature to increase the relative emissivity of the $\lambda 4363$ line. This high-temperature region is relatively more dominant in (LINER) shocks without precursor HII regions which are characterised by even higher electron temperatures. For the few objects for which the [O III] line has been observed, the photoionisation models predict a $\lambda 4363$ line intensity some 50 times weaker than seen. Shocks, on the other hand, yield predicted line ratios very similar

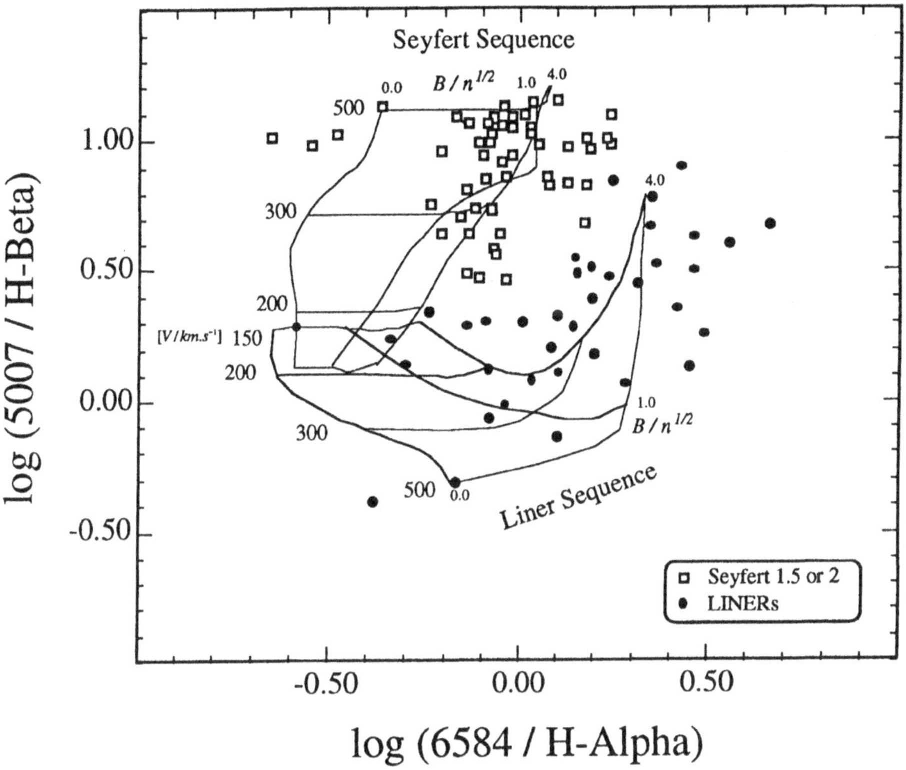

Fig. 3. Diagnostic plot for Seyferts and LINERs, with model grids superimposed. Note how both classes of object are fitted by fast shocks having $B/n^{1/2}$ in the range 2-4. However, the LINERs fit to a model without a photoionised precursor, and the Seyfert galaxies require a precursor.

to those observed. To achieve a definitive distinction between the two models from emission line diagnostics alone, we really need to make observations in the UV. Those resonance lines which are predominantly produced in the cooling zone of the shock are predicted to be much stronger than in a photoionisation model that would give a similar optical spectrum. This will be particularly marked in the case of LINERs. Also, the UV line ratios that are temperature sensitive are expected to give a much higher temperature than the [O III] line ratio. In this respect the HUT (Kriss et al. 1992) on NGC1068 provide an unambiguous diagnostic. In this object the CIII $\lambda(1907{+}9)/\lambda977$ intensity ratio gives an electron temperature in excess of 27000 K, and the NIII $\lambda1750/\lambda991$ ratio suggests temperatures in excess of 24000 K. In addition, the absolute intensity of the CIV line is much higher than would be predicted from a photoionised model, even accounting from the flux in the broad wings, which are probably derived from scattering from the hidden central object. In NGC1068, there is therefore little doubt that shocks are of paramount importance in understanding the emission line spectrum.

References

Antonucci, R.R.J., and Miller, J.S.: 1985, *Astrophys. J.* **297**, 621
Barthel, P.: 1989, *Astrophys. J.* **336**, 606
Baldwin, J.A., Phillips, M.M., and Terlevich, R.: 1981, *Publ. Astron. Soc. Pac.* **93**, 5
Baum, S.A., et al.: 1988, *Astrophys. J. Suppl.* **68**, 643
Baum, S.A., and Heckman, T.M.: 1989, *Astrophys. J.* **336**, 702
Baum, S.A., Heckman, T.M., and van Breugel, W.: 1992, *Astrophys. J.* **389**, 208
Bicknell, G.V.: 1990, *Astrophys. J.* **354**, 98
Bicknell, G.V.: 1993, 'The Physics of Active Galaxies' in G.V. Bicknell, M.A. Dopita and P.J. Quinn, ed(s)., *Publ. Astron. Soc. Pac. Conf. Ser. No. 54*, , 231
Binette, L., Dopita, M.A., and Tuohy, I.R.: 1985, *Astrophys. J.* **297**, 476
Binette, L, Parker, D., and Fosbury, R.A.E.: 1993, , preprint
Blair, W.P., Kirshner, R.P., and Winkler, P.F.: 1983, *Astrophys. J.* **272**, 84
Blair, W.P., Raymond, J.C., Danziger, J., and Matteucci, F.: 1989, *Astrophys. J.* **338**, 812
Cecil, G., and Bland, J.: 1989, 'IAU Symp No.134, Active Galactic Nuclei' in D.E. Osterbrock and J.S. Miller, ed(s)., *Kluwer*, Dordrecht, p345
Chevalier, R.A., and Kirshner, R.P.: 1979, *Astrophys. J.* **233**, 154
Cowie, L.L., Hu, E.M., Jenkins, E.B., and York, D.G.: 1983, *Astrophys. J.* **272**, 29
Crotts, A.P.S., Kunkel, W.E., and McCarthy, P.J.: 1989, *Astrophys. J.* **347**, L61
Daltabuit, E. and Cox, D.: 1972, *Astrophys. J.* **173**, L13
De Bruyn, A.G., and Wilson, A.S.: 1978, *Astron. Astrophys.* **64**, 433
Dopita, M.A.: 1987, *Aus. J. Phys.* **40**, 789
Dopita, M.A., Binette, L., D'Odorico, S. and Benvenuti, P.: 1984a, *Astrophys. J.* **276**, 653
Dopita, M.A., Binette, L., and Schwartz, R.D.: 1982, *Astrophys. J.* **261**, 183
Dopita, M.A., Binette, L., and Tuohy, I.R.: 1984b, *Astrophys. J.* **282**, 142
Dopita, M.A. and Tuohy, I.R.: 1984, *Astrophys. J.* **282**, 135
Dopita, M.A., Tuohy, I.R., and Mathewson, D.S.: 1981, *Astrophys. J.* **248**, L105
Ferland, G.J., and Netzer, H.: 1983, *Astrophys. J.* **264**, 105
Forbes, D.A., and Ward, M.J.: 1993, *Astrophys. J.* **416**, 150
Fosbury, R.A.E.: 1989, 'Extranuclear Activity in Galaxies' in E.J.A. Meurs and R.A.E. Fosbury, ed(s)., *ESO*, Garching, p169
Fosbury, R.A.E., Mebold, U., Goss, W.M. and Dopita, M.A.: 1978, *Mon. Not. R. astr. Soc.* **183**, 549
Fransson, C. et al.: 1989, *Astrophys. J.* **347**, L61
Goss, W.M. et al.: 1979, *Mon. Not. R. astr. Soc.* **188**, 357
Heckman, T.M.: 1980, *Astron. Astrophys.* **87**, 142
Heckman, T.M.: 1981, *Astrophys. J.* **250**, L59
Heckman, T.M., Baum, S.A., van Breugel, W.J.M., and McCarthy, P.: 1989, *Astrophys. J.* **338**, 48
Ho, L.C., Filippenko, A.V., and Sargent, W.L.: 1993, *Astrophys. J.* **417**, 63
Innes, D.E.: 1992, *Astron. Astrophys.* **256**, 660
Innes, D.E., Giddings, J.R., and Falle, S.A.E.G.: 1987a, *Mon. Not. R. astr. Soc.* **224**, 179
Innes, D.E., Giddings, J.R., and Falle, S.A.E.G.: 1987b, *Mon. Not. R. astr. Soc.* **226**, 67
Itoh, H.: 1981a, *Publ. Astron. Soc. Japan* **33**, 121
Itoh, H.: 1981b, *Publ. Astron. Soc. Japan* **33**, 521
Jacobsen, P. et al.: 1991, *Astrophys. J.* **369**, 63

Johnstone, R.M., Fabian, A.C., and Nulsen, P.E.J.: 1987, *Mon. Not. R. astr. Soc.* **224**, 74
Kamper, K.W., and van den Bergh, S.: 1976, *Astrophys. J. Suppl.* **32**, 351
Keele, W.C.: 1983a, *Astrophys. J. Suppl.* **52**, 229
Keele, W.C.: 1983b, *Astrophys. J.* **269**, 466
Khachikian, E. Ye., and Weedman, D.W.: 1974, *Astrophys. J.* **192**, 581
Kirshner, R.P. et al.: 1989, *Astrophys. J.* **342**, 260
Koski, A.T.: 1978, *Astrophys. J.* **223**, 56
Kriss, G.A., et al.: 1992, *Astrophys. J.* **394**, L37
Lasker, B.M.: 1978, *Astrophys. J.* **223**, 109
Long, K.S., Helfand, D.J., and Grabelsky, D.A.: 1981, *Astrophys. J.* **248**, 925
Lundqvist, P., and Fransson, C.: 1987, 'Proc ESO/EIPC Workshop, SN1987A and other Supernovae' in I.J. Danziger and K. Kjär, ed(s)., *ESO*, Garching, p607
Lundqvist, P., and Fransson, C.: 1991, *Astrophys. J.* **380**, 575
Luo, D, McCray, R., and Slavin, J.: 1994, *Astrophys. J.* , in press
McCarthy, P.J., van Breugel, W., Spinrad, H., and Djorgovski, S.: 1987, *Astrophys. J.* **321**, L29
Meisenheimer, K., Hippelein, H., and Neeser, M.: 1994, 'The Physics of Active Galaxies' in G.V. Bicknell, M.A. Dopita and P.J. Quinn, ed(s)., *Publ. Astron. Soc. Pac. Conf. Ser. No. 54*, , p397
Meurs, E.J.A.et al.: 1989, 'Extranuclear Activity in Galaxies' in E.J.A. Meurs and R.A.E. Fosbury, ed(s)., *ESO*, Garching, p249
Morganti, R. et. al.: 1991, *Mon. Not. R. astr. Soc.* **249**, 91
Morganti, R. et al.: 1992, *Mon. Not. R. astr. Soc.* **256**, 1P
Morganti, R., Fosbury, R.A.E., Oosterloo, T.A. and Tadhunter, C.N. : 1994, 'The Physics of Active Galaxies' in G.V. Bicknell, M.A. Dopita and P.J. Quinn, ed(s)., *Publ. Astron. Soc. Pac. Conf. Ser. No. 54*, , p267
Mulchaey, J.S., Tsvetanov, Z., Wilson, A.S., and Pérez-Fournon, I.: 1992, *Astrophys. J.* **394**, 91
Osterbrock, D.E.: 1971, 'Pontificiae Academiae Scientiarum, Scripta Varia' in D.J.K. O'Connell, ed(s)., , Amsterdam, No. 35, p151
Osterbrock, D.E., Tran, H.D., and Veilleux, S.: 1992, *Astrophys. J.* **389**, 196
Phillips, M.M. et. al.: 1986, *AJ* **91**, 1062
Plait, P.C., Lundqvist, P., Chevalier, R.A., and Kirshner, R.P.: 1994, *Astrophys. J.* , in press
Pogge, R.W.: 1988a, *Astrophys. J.* **328**, 519
Pogge, R.W.: 1988b, *Astrophys. J.* **332**, 702
Pogge, R.W.: 1989, *Astrophys. J.* **345**, 730
Prieto, A., di Serego Alighieri, S., and Fosburry, R.A.E.: 1989, 'Extranuclear Activity in Galaxies' in E.J.A. Meurs and R.A.E. Fosbury, ed(s)., *ESO*, Garching, p31
Raymond, J.C.: 1976, *PhD Thesis*, University of Wisconsin
Raymond, J.C., and Smith, B.W.: 1977, *Astrophys. J. Suppl.* **35**, 419
Schmutzler, T., and Tscharnuter, W.M.: 1993, *Astron. Astrophys.* **273**, 318
Stasinska, G.: 1984,, *Astron. Astrophys.* **135**, 341
Sutherland, R.S.: 1993, *PhD Thesis*, The Australian National University
Sutherland, R.S., Bicknell, G.V., and Dopita, M.A.: 1993, *Astrophys. J.* **414**, 510
Sutherland, R.S., and Dopita, M.A.: 1993, *Astrophys. J. Suppl.* **88**, 253
Sutherland, R.S., and Dopita, M.A.: 1995a, *Astrophys. J.* **439**, 365
Sutherland, R.S., and Dopita, M.A.: 1995b, *Astrophys. J.* **439**, 381
Tadhunter, C.N., Robinson, A., and Morganti, R.: 1989, 'Extranuclear Activity in Galaxies' in E.J.A. Meurs and R.A.E. Fosbury, ed(s)., *ESO*, Garching, p293
Tadhunter, C.N. and Tsvetanov, Z.I.: 1989, *Nature* **341**, 422
Tsvetanov, Z.I., Fosbury, R.A.E., and Tadhunter, C.N.: 1994, *Astrophys. J.* , in press
Unger, S.W., et al.: 1987, *Mon. Not. R. astr. Soc.* **228**, 671
Veilleux, S.: 1991a, *Astrophys. J. Suppl.* **75**, 357
Veilleux, S.: 1991b, *Astrophys. J. Suppl.* **75**, 383

Veilleux, S., and Osterbrock, D.E.: 1987, *Astrophys. J. Suppl.* **63**, 295
Veilleux, S., Tully, R.B., and Bland-Hawthorn, J.: 1993, *AJ* **105**, 1318
Viegas-Aldrovandi, S.M., and Contini, M.: 1989, *Astrophys. J.* **339**, 689
Wampler, E. et al.: 1990, *Astrophys. J.* **362**, L13
Whittle, M.: 1992, *Astrophys. J. Suppl.* **79**, 49
Whittle, M.: 1994, *Astrophys. J. Suppl.* **91**, 491
Wilson, A.S., and Tsvetanov, Z.I.: 1994, *AJ* **107**, 1227
Wilson, A.S., and Willis, A.G.: 1980, *Astrophys. J.* **240**, 429
Winkler, P.F., and Kirshner, R.P.: 1985, *Astrophys. J.* **299**, 981

DIAGNOSTICS OF SUPERNOVA REMNANT SHOCK WAVES

J.C. RAYMOND
Center for Astrophysics, 60 Garden Street
Cambridge, MA 02138, USA

Abstract. Fairly simple models can explain the emission from non-radiative shock waves in supernova remnants. This talk reviews some of the more robust diagnostics of shock parameters and some of the implications for the physics of collisionless shock waves in interstellar gas. The Hopkins Ultraviolet Telescope observed several non-radiative shocks during the ASTRO-2 mission, and the spectra have interesting implications for the physics of collisionless shocks.

Key words: Shock Waves – Supernova Remnants

1. Introduction

Astronomers who observe supernova remnants at optical wavelengths are most familiar with radiative shocks – those which convert the energy dissipated into visible and UV light as the shocked gas cools and recombines. Astronomers who observe in X-rays are accustomed to non-radiative shocks, because the radiative cooling times of the shocks which produce X-rays are generally much longer than the relevant dynamical times, such as the age of the supernova remnant. It is sometimes possible to observe non-radiative shocks at visible and UV wavelengths, though they are faint enough that this is fairly difficult. Those observations provide a set of unique diagnostics for the physical processes in high Mach number collisionless shocks. An excellent broad review of radiative and non-radiative shocks is given by Draine and McKee (1993), and a grid of radiative shock models is given in Hartigan, Raymond and Hartmann (1987). Raymond (1991) gives a more specialized discussion of non-radiative shock diagnostics. Here we review the basic mechanism for the production of optical and UV radiation in the ionization zone just behind a fast non-radiative shock, then consider the implications of some recent observations for electron-ion temperature equilibration and for the acceleration of cosmic rays. A detailed calculation of the hydrogen velocity distribution in a collisionless shock is given by Lim and Raga in this volume.

2. Collisionless Shock Physics

A shock wave converts a supersonic pre-shock flow (in the reference frame in which the shock is at rest) into a subsonic post-shock flow by thermalizing

Astrophysics and Space Science **233**: 231–237, 1995.

some of the bulk velocity of the particles. For a strong shock in a gas with adiabatic constant γ=5/3, the shock reduces the bulk velocity by a factor of 4, compressing the gas by the same factor to satisfy mass conservation. If $T_e = T_i$, the gas is heated to about 1.4×10^5 v_7^2, where v_7 is the shock velocity in units of 100 km s^{-1}, but the electron and ion temperatures may not be equal. A collisionless shock in a plasma thermalizes the velocities by way of electromagnetic fields and scattering from plasma turbulence, rather than by particle-particle scattering. The ions carry very nearly all of the mass and therefore nearly all the kinetic energy of the incoming flow, so it is quite likely that when the bulk velocity is thermalized, the ions will carry nearly all the thermal energy. On the other hand, it is plausible that the plasma turbulence which thermalizes the ion velocities might also transfer kinetic energy from ions to electrons. As it is the electron temperature which is inferred from X-ray spectral measurements, the relationship between T_e and the shock velocity is vital to interpreting X-ray observations in terms of the evolutionary state of the supernova remnant. Thus electron-ion temperature equilibration is a major area of uncertainty which we would like to address.

A second major issue is the question of cosmic ray acceleration. The cosmic ray community generally accepts the idea that the diffusive shock mechanism in supernova remnant shocks accelerates relativistic particles to energies up to 10^{14} or 10^{15} GeV. The main arguments for this idea are the agreement of the observed power law index with that expected from the diffusive shock mechanism and the difficulty in finding any power source other than supernovae which can match the energy loss of cosmic rays. Diffusive shock acceleration is reviewed by Blandford and Eichler (1987). The gap between the acceleration models and observations is enormous, however, and any constraints which can be obtained from optical observations would be extremely useful.

The following sections describe the atomic processes which produce observable emission, the existing observations, the inferences which can be made regarding electron-ion equilibration, and the implications of possible shock precursors for cosmic ray acceleration.

3. Atomic Processes

A collisionless shock thermalizes the electron and ion velocities by means of plasma turbulence or electromagnetic fields (e.g. Tidman and Krall 1971), but a neutral atom is unaware of the shock front. A hydrogen atom, for instance, suddenly finds itself immersed in hot plasma without warning. If T_e and T_i are above 10^5 K, the atoms will be quickly ionized in a time

$$t_i = \frac{1}{n_e q_i} \sim \frac{10^8}{n_e} \quad \text{seconds} \tag{1}$$

where q_i is the ionization rate coefficient and n_e is the electron density. While it awaits the impact which leads to ionization, the atom has some chance of undergoing an excitation. For most ions, the excitation rates of the resonance lines are several times the ionization rates, so that several photons are produced by each atom (or ion which is not fully stripped) passing through the shock. The number of photons can be greatly enhanced by the ratio of Boltzmann factors if kT_e is smaller than the ionization potential, but larger than the excitation potential of the line. The surface brightness of an emission line will be

$$I = \frac{1}{4\pi} n_i v_s \frac{q_{ex}}{q_i} \qquad \text{photons cm}^{-2}\ \text{s}^{-1}\ \text{sr}^{-1} \tag{2}$$

where n_i is the density of ion i (or lower ions of the same element) ahead of the shock, v_s is the shock velocity and q_{ex} is the excitation rate coefficient (Chevalier and Raymond 1978; Bychkov and Lebedev 1979). For strong shocks this amounts to about 0.2-0.3 Hα photons per neutral hydrogen atom passing though the shock.

The profile of the Hα line provides several important diagnostics. Some of the Hα photons will be produced by atoms which have not felt the effects of the shock, so they should have the kinetic distribution of the preshock gas, and the line profile should be less than 20 km s^{-1} wide. On the other hand, resonant charge transfer has a very large cross section. A charge transfer event leads to a stationary ion (which quickly joins the ion fluid through interaction with the magnetic field and through Coulomb collisions) and a neutral with a velocity typical of the post-shock ion velocity distribution. These neutrals will produce an Hα emission component whose width is that of the post-shock ion temperature, somewhat modified by the weighting of the charge transfer process $< v\sigma_x >$. The ratio of the intensities of the two components depends on the ratio of charge transfer rate to ionization rate, and it therefore depends on T_e. An additional term enters at high shock velocities, where change transfer into excited states becomes significant (Chevalier, Kirshner and Raymond 1980; hereafter CKR). A further complication is the absorption of Lyβ by neutral hydrogen atoms and the conversion of a Lyβ photon to Hα plus a two-photon pair (CKR). Figure 1 shows the Hα profile of a non-radiative filament in the Cygnus Loop, illustrating the two component nature of the profile.

The intensity of any other line relative to Hα is determined by the relative abundance of the element and the number of photons emitted by an average atom passing through the shock. Optical forbidden lines, such as [O III] produce about 1 photon per oxygen atom at low shock velocities (200 km s^{-1}) and a smaller number in faster shocks. Since the oxygen abundance is less than 1/1000 that of hydrogen, the [O III] lines are on the order of 1% the intensity of Hα in the Cygnus Loop non-radiative shocks, and they are

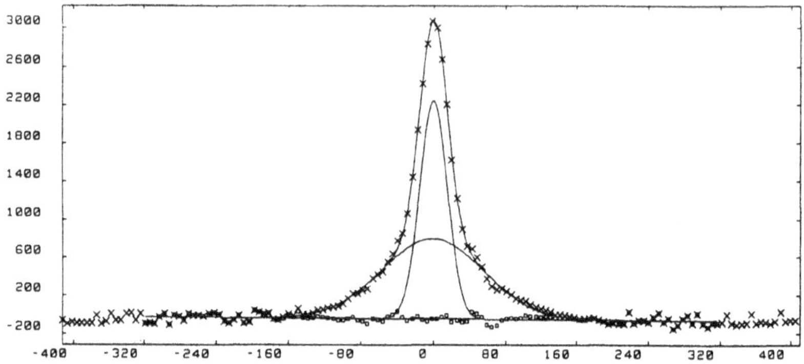

Fig. 1. Hα profile from Hester, Raymond and Blair (1994).

undetectably weak in other SNRs. Therefore, non-radiative shocks in partially neutral gas appear as pure Balmer line filaments in the optical. The ultraviolet He II λ1640 line is equivalent to Hα, so its intensity is of order 1/10 that of Hα (in photon units). The ultraviolet resonance lines of C IV, N V and O VI are produced rates of ~100 photons per atom, so these lines tend to be of order 1/10 Hα in photon units or comparable to Hα in energy units. The UV spectra are completely dominated by these lines, while spectra of radiative shocks are generally dominated by intercombination lines such as C III] and O IV].

4. Observations

Pure Balmer line filaments have been observed in the young SNRs Tycho (CKR), SN1006 (Long, Blair and van den Bergh 1988) and Kepler (Fesen et al. 1989), in the Cygnus Loop (Raymond et al 1983), in 4 LMC remnants (Tuohy et al 1982; Smith et al 1991), and two pulsar-driven bow shocks (1957+20; Aldcroft, Romani and Cordes 1992 and the Guitar nebula; Cordes, Romani and Lundgren 1993). Velocities inferred from broad component widths range from 150-250 km s^{-1} in the Cygnus Loop (Long et al 1993; Hester, Raymond and Blair 1994) to over 2000 km s^{-1} in SN1006, and the inferred shock velocities are comparable to the broad component widths.

It has also been possible to observe several non-radiative shocks in the UV. Raymond et al (1983) and Hester, Raymond and Blair (1994) present IUE spectra of filaments in the northeastern part of the Cygnus Loop, and

Long et al (1993) give the Hopkins Ultraviolet Telescope spectrum of one of these from the ASTRO-1 mission. These filaments were chosen for their brightness, and unfortunately they are both beginning to enter the radiative phase. The Hopkins Ultraviolet Telescope observed two other positions in the Cygnus Loop and one in SN1006 in March 1995. The data analysis is not yet complete, but lines of O VI, C IV and He II were detected.

5. Electron-ion Equilibration

Cargill and Papadopoulos (1988) used a hybrid plasma code to model a strong collisionless shock. They found that two types of plasma turbulence transferred some thermal energy from ions to electrons, but that the electron temperature was only about 1/5 that of the ions after the shock. Subsequently, Coulomb collisions would eventually bring the two temperatures into equilibrium, but only over a distance much greater than that over which the optical and UV lines are formed. There are two diagnostics for electron-ion equilbration. Smith (1991) and Smith, Laming and Raymond (1995) use the ratio of broad to narrow component intensities to infer the ratio of charge transfer rate to ionization rate, and from that they deduce that $T_e \ll T_i$, in keeping with the prediction of Cargill and Papadopoulos. The UV observations of filaments in the Cygnus Loop produced ambiguous results because of the contamination by cooling gas and the possibility that the shock velocity is rapidly changing. Recent ASTRO-2 observations with HUT suggest that $T_e \ll T_i$ in SN1006, but that pure Coulomb equilibration would give an O VI:C IV ratio larger than observed. When the analysis is complete, it should provide a very good test of the Cargill and Papadopoulos prediction (Raymond, Blair and Long 1995).

6. Cosmic Ray Precursors

High resolution line profiles reveal the width of the narrow component. While it was expected to be smaller than 20 km s^{-1}, in accordance with the temperature required for partially neutral gas, it has been measured to be 30-50 km s^{-1} in two LMC remnants (Smith, Raymond and Laming 1994), several positions in the Cygnus Loop (Hester, Raymond and Blair 1994; W. Blair, private communication) and Tycho (W. Blair, private communication). If this width is thermal, it implies a temperature of order 40,000 K, which would be hot enough that hydrogen would be fully ionized; no neutrals would pass through the shock and no Balmer filament would be seen. Smith et al (1994) and Hester, Raymond and Blair (1994) discuss many possible interpretations for this width, the most attractive (or least unattractive) of which was a cosmic ray precursor. A precursor in which cosmic rays are trapped between the shock discontinuity and upstream plasma turbulence is essential to diffusive shock acceleration models (though not necessarily all shock acceleration models; Jokipii 1987). Models for such a structure in partially neutral gas show that the precursor heats the gas to 90,000 K (Boulares and Cox 1988). If the precursor is sufficiently thin, there is not enough time for the gas to be ionized before it encounters the shock, and there is no difficulty with the Hα intensity. It was recently realized, however, that the echelle profiles from the centers of the LMC remnants 0519-69.0 and 0509-67.9 ought to show the bulk velocities of the precursors on the approaching and/or receding sides of the SNRs (Ellison et al. 1995), and that this ought to be far larger than the upper limits of Smith, Raymond and Laming (1994). This may mean that some other means must be found to explain the anomalous Hα narrow component widths. It also may mean that shocks in partially neutral gas do not accelerate cosmic rays, at least through the diffusive shock mechanism. This may present a problem for the cosmic ray energy budget, in that most SNRs probably expend much of their kinetic energy moving into partially neutral gas. If they do not accelerate cosmic rays, there may not be enough power available for cosmic ray generation.

7. Surprise Ending

The HUT spectrum of SN1006 obtained during the ASTRO-2 mission has 3Å resolution, and it shows C IV and He II line widths exceeding 2000 km s^{-1}. (The width of the O VI line is difficult to measure, because it is close to a strong airglow line.) These widths are close to the Hα broad component width reported by Smith et al (1991). If the kinetic temperatures of protons, He ions and C ions were equal, the He II and C IV line widths would be around 1200 km s^{-1} and 700 km s^{-1}, respectively. Thus neither plasma turbulence nor Coulomb collisions redistributes the energy among ions of differing mass between the shock front and the zone where these lines are emitted (Raymond, Blair and Long 1995).

References

Aldcroft, T.L., Romani, R.W., and Cordes, J.M.: 1992, *Astrophys. J.* **400**, 638
Blandford, R., and Eichler, D.: 1987, *Phys. Reports* **154**, 1
Boulares, A., and Cox, D.P.: 1988, *Astrophys. J.* **333**, 198
Bychkov, K.V., and Lebedev, V.S.: 1979, *Astron. Astrophys.* **80**, 167
Cargill, P.J., and Papadopoulos, K.: 1988, *Astrophys. J.* **329**, L27
Chevalier, R.A., and Raymond, J.C.: 1978, *Astrophys. J.L* **225**, L27
Chevalier, R.A., Kirshner, R.P., and Raymond, J.C.: 1980, *Astrophys. J.* **235**, 186
Cordes, J.M., Romani, R.W., and Lundgren, S.C.: 1993, *Nature* **362**, 133

Draine, B.D., and McKee, C.F.: 1993, *Ann. Rev. Astron. Astrophys.* **31**, 373
Ellison, D.C., et al.: 1994, *PASP*, submitted
Fesen, R.A., Becker, R.H., Blair, W.P., and Long, K.S.: 1989, *Astrophys. J.* **338**, L13
Hester, J.J., Raymond, J.C., and Blair, W.P.: 1994, *Astrophys. J.* **420**, 721
Hester, J.J., Raymond, J.C., and Danielson, G.E.: 1986, *Astrophys. J.L* **303**, L17
Hartigan, P., Raymond, J., and Hartmann, L.: 1987, *Astrophys. J.* **316**, 323
Jokipii, J.R.: 1987, *Astrophys. J.* **313**, 842
Long, K.S., Blair, W.P., and van den Bergh, S.: 1988, *Astrophys. J.* **333**, 749
Long, K.S., Blair, W.P., Vancura, O., Bowers, C.W., Davidsen, A.F., and Raymond, J.C.: 1993, *Astrophys. J.* **400**, 214
Raymond, J.C.: 1991, *PASP* **103**, 781
Raymond, J.C., Blair, W.P., Fesen, R.A., and Gull, T.R.: 1983, *Astrophys. J.* **275**, 636
Raymond, J.C., Blair, W.P., and Long, K.S.: 1995, *BAAS*, in press
Smith, R.C.: 1991, Thesis, Harvard University
Smith, R.C., Kirshner, R.P., Blair, W.P., and Winkler, P.F.: 1991, *Astrophys. J.* **375**, 652
Smith, R.C., Laming, J.M., and Raymond, J.C.: 1995, in preparation
Smith, R.C., Raymond, J.C., and Laming, J.M.: 1994, *Astrophys. J.* **420**, 286
Tidman, D.A., and Krall, N.A.: 1971, *Shock Waves in Collisionless Plasmas*, Interscience, New York
Tuohy, I.R., Dopita, M.A., Mathewson, D.S., Long, K.S., and Helfand, D.J.: 1982, *Astrophys. J.* **261**, 473

SHOCK INSTABILITIES

S.A.E.G. FALLE
Department of Applied Mathematics,
University of Leeds, Leeds LS2 9JT, UK

Abstract. It is well known that adiabatic shocks in ordinary gases are stable to both tranverse and longitudinal perturbations, but this need not be true if there are significant thermal effects due to chemical reactions or cooling processes. For example, detonation waves in gases are observed to form cellular structures if the chemical reaction is sufficiently temperature sensitive and a similar instability occurs in radiative shocks in the ISM if their speed exceeds 150 km s^{-1}. This means that interstellar shocks will be subject to this radiative instability in many cases. The temperature sensitivity of the nuclear reactions in Type I supernovae is also such that we would expect detonation waves in these objects to have a cellular structure.

Key words: Shock Stability – Detonation Waves – Radiative Shocks

1. Introduction

Although adiabatic shock waves in polytropic gases are stable to both longitudinal and transverse perturbations, instabilities can occur in detonation waves which are driven by the heat released by a chemical reaction (e.g. Fickett & Davis 1979). The stability criterion is rather complicated, but instability seems to be unavoidable if the chemical reactions are sufficiently temperature sensitive. Since the nuclear reactions that are believed to power Type I supernova explosions are extremely sensitive to temperature, nuclear detonation waves are almost certainly subject to this kind of instability.

These instabilities cause initially plane detonations to develop a cellular structure whose length can be an order of magnitude larger than the thickness of the reaction zone. Although this has little effect on the speed of the wave, it does influence the conditions under which such detonations can propagate. The final composition may also be affected if some of the reactions are reversible.

Several authors (e.g. Falle 1975, 1981; Langer, Chanmugam & Shaviv 1981, 1982; Chevalier & Imamura 1982; Imamura, Wolff & Durisen 1984; Bertschinger 1986) have shown that radiative shocks are also unstable if the cooling rate does not increase sufficiently rapidly with temperature. Although the instability was first discovered for simple model cooling laws, Innes, Falle & Giddings (1987a,b,c) found that it also occurs in more realistic calculations and that it can have a profound effect upon the emission line ratios. This is unfortunate since these ratios are often used to determine the speed of radiative shocks.

Astrophysics and Space Science **233**: 239–249, 1995.

The cellular structure in detonation waves is well known from experiments and has been reproduced in numerical calculations (e.g. Bourlioux & Majda 1992), but the only multidimensional work on radiative shocks is a linear stability analysis by Bertschinger (1986). Since these instabilities seem to be related, it makes sense to look for cellular structure in radiative shocks. The resemblance between schlieren photagraphs of unstable detonations and the filamentary structures in old supernova remnants suggests that radiative shocks do indeed contain such features.

2. Detonations

The simplest model of a detonation wave has a single fuel which we can describe by a mass fraction c, where $c = 1$ is unburnt and $c = 0$ is burnt. c satisfies the equation

$$\frac{\partial}{\partial t}(\rho c) + \nabla.(\rho \mathbf{v} c) = -W,$$

where $W(\rho, c, T)$ is the reaction rate per unit volume. If Q is the heat released per unit mass, then the energy equation becomes

$$\frac{\partial e}{\partial t} + \nabla.[\mathbf{v}(e+p)] = QW$$

where e is the total energy per unit mass. For a binary reaction we might have

$$W = K\rho^2(1-c)^2 \exp(-T_{\mathrm{A}}/T)$$

where K is a rate constant and T_{A} is an activation temperature.

A detonation wave consists of a shock which heats the gas to a temperature at which the reaction can begin and the resulting heat release sustains the shock. If the reaction zone is thin compared to any other length scales, then we can model this as a discontinuity. Suppose that the front moves with speed D. Then in the frame of the detonation, the conservation laws demand that

$$\begin{aligned} \rho_{\mathrm{d}} v_{\mathrm{d}} &= \rho_{\mathrm{u}} D, \\ p_{\mathrm{d}} + \rho_{\mathrm{d}} v_{\mathrm{d}}^2 &= p_{\mathrm{u}} + \rho_{\mathrm{u}} D^2, \\ I_{\mathrm{d}} + v_{\mathrm{d}}^2 &= I_{\mathrm{u}} + D^2 + Q, \end{aligned}$$

where the suffices u, d refer to the upstream and downstream side of the wave and I is the internal energy per unit mass.

If there are no other sources of energy and the reactions are entirely exothermic, then the downstream flow must be sonic in the frame of the

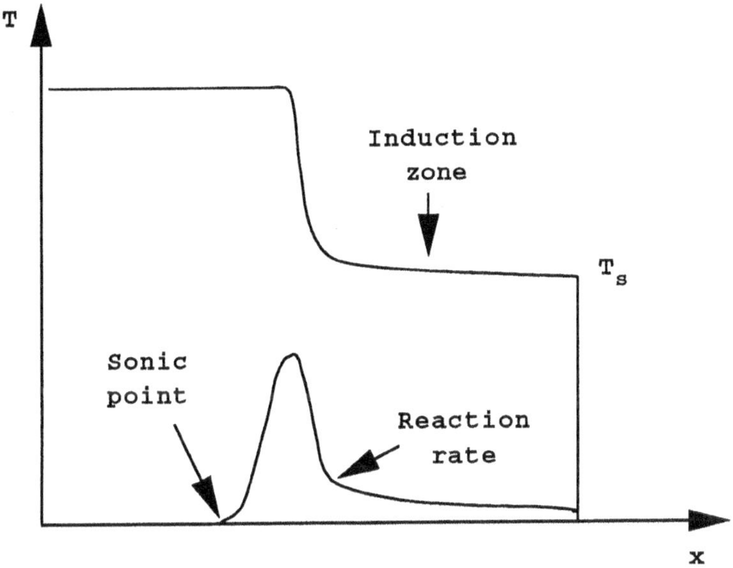

Fig. 1. Structure of a steady detonation

wave (Chapman-Jouquet detonation). This, together with the conservation conditions determines the speed of the detonation. For a strong shock in a perfect gas we get

$$D = [2(\gamma^2 - 1)Q]^{1/2}.$$

Note that this is entirely determined by the properties of the upstream gas.

Figure 1 shows the steady structure within such a detonation. If the activation temperature, T_A, is much greater than the post-shock temperature, T_s, then reaction rate immediately behind the shock will be small and there will be a long induction zone before the reaction really takes off. It is this kind of structure which is prone to instability. If the wave propagates within a tube whose width is not too much larger than the thickness of the reaction zone, then this instability generates transverse waves as shown in figure 2. What happens is that the main shock does not trigger the reaction, but that a series of explosions occur, which generate transverse shocks. These shocks collide and create the conditions for an explosive reaction, which in turn regenerates the transverse waves. We shall see that the same sort of thing happens with radiative shocks.

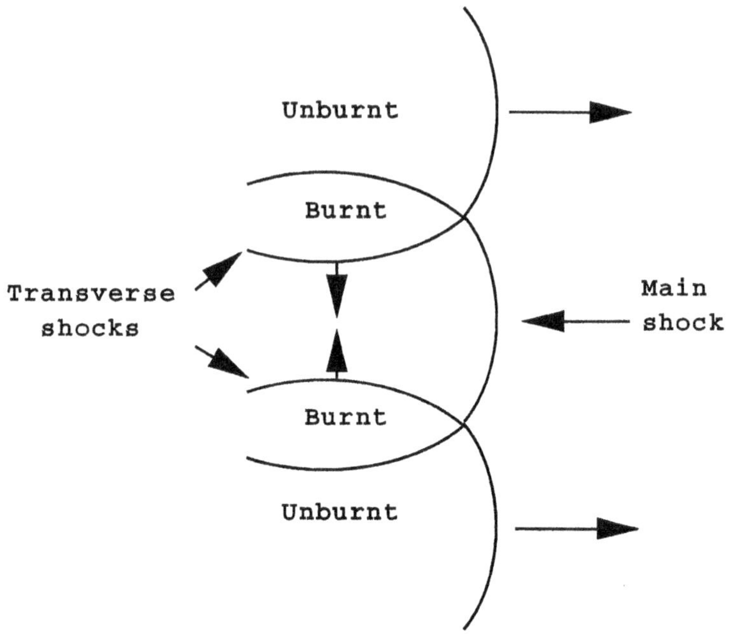

Fig. 2. Transverse waves in a cellular detonation

3. Radiative Shocks

The energy equation is now

$$\frac{\partial e}{\partial t} + \nabla.[\mathbf{v}(e+p)] = -L, \tag{1}$$

where $L = L(\rho, \mathbf{c}, T)$ is the radiative loss rate per unit volume. Here $\mathbf{c} = (c_1, \ldots c_n)$ and c_i is the ionisation fraction of the ith ionic species. These are governed by the equations

$$\frac{\partial}{\partial t}(\rho c_i) + \nabla.(\rho \mathbf{v} c_i) = W_i, \tag{2}$$

where $W_i = W_i(\rho, \mathbf{c}, T)$ is the net rate at which the ith ionic species is created.

The trouble is that there are a large number of ions and they are coupled by the radiation field, which means that their abundance cannot be determined from such purely local equations. Although it is possible to take all this into account in one dimensional calculations (e.g. Innes, Giddings & Falle 1987a,b,c), this is, as yet, impossible in multi-dimensions. One can,

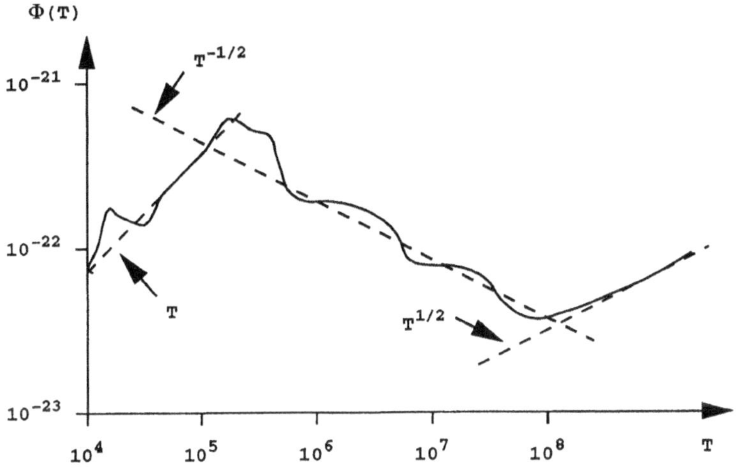

Fig. 3. Cooling law for a gas cooling from a high temperature

however, assume that the ionisation is in equilibrium, which is a reasonable approximation except when the gas is cooling rapidly in the temperature range $10^4 \leq T \leq 10^6$. In that case we have

$$\mathbf{c} = \mathbf{c}(\rho, T),$$

which means that we can write

$$L = \frac{\rho^2}{m^2}\Phi(T). \tag{3}$$

Here m is the mean particle mass. The function $\Phi(T)$ is sketched in figure 3 for a gas cooling from a high temperature. Although this function is quite messy, it is quite well approximated by the three power laws shown.

Consider a one dimensional flow, in which a shock is driven into stationary gas by a piston moving with speed $V_{\rm p}$. If the gas cools to a temperature which is much lower than that behind the shock, then the swept up gas accumulates in a very thin dense region near the piston. The speed of the shock must therefore be approximately equal to $V_{\rm p}$ and the flow is quasi-steady. The continuity and momentum equations integrate immmediately to give

$$\rho v = -\rho_{\rm u} V_{\rm p}, p + \rho v^2 = -\rho_{\rm u} V_{\rm p}^2, \tag{4}$$

where $\rho_{\rm u}$ is the upstream density and we have assumed that the shock is strong.

If these relations are used in the one dimensional, steady versions of the energy and ionisation balance equations (1) and (2), then it is a relatively simple matter to calculate the structure of such a flow even with very detailed physics (e.q. Raymond 1979; Shull & McKee 1979; Innes, Giddings & Falle 1987a). The only complication is the non-local effects due to the radiation field, but this can be dealt with by an iterative process. The nice thing about this is that there a number of emission line ratios that depend only on the shock velocity. Since these ratios can readily be measured, this is a very effective way of determining the velocity of interstellar shocks.

If we ignore all the complexities of ionisation state and suppose that the gas has a simple equation of state and that L is given by equation (3), then the energy equation becomes

$$\frac{d}{dx}(C_{\rm p}T + v^2) = \frac{\rho^2\Phi(T)}{m^2\rho_{\rm u}V_{\rm p}},$$

where $C_{\rm p}$ is the specific heat at constant pressure. Near the piston the velocity, v, is small, the pressure, $p = const. = \rho_{\rm u}V_{\rm p}^2$ and we can approximate Φ by a power law, $\Phi(T) \propto T^{\alpha}$. Equation (4) then reduces to

$$\frac{dT}{dx} \propto T^{\alpha-2}.$$

Clearly if $\alpha < 2$, as it is for small T, then the temperature gradient becomes very large near the piston. For shock velocities greater than 120 km s^{-1}, the post-shock temperature is greater than that at the peak of the cooling curve and the cooling distance behind the shock is large. The temperature profile is then as shown in figure 4, which is an inverted version of that in a detonation wave with a large induction region. We might therefore expect a similar kind of instability to occur in this case.

Such instabilities have indeed been found in a number of one dimensional numerical calculations with model cooling laws (e.g. Falle 1975, 1981; Langer, Chanmugam & Shaviv 1981, 1982; Imamura, Wolff & Durisen 1984). Chevalier & Imamura (1982) showed that if $\Phi(T) \propto T^{\alpha}$, then a one dimensional shock is linearly unstable if $\alpha \leq 0.8$, while Bertschinger (1986) found that in two dimensions it is unstable if $\alpha \leq 1$. Innes, Giddings & Falle (1987b) showed that when all the important physical processes are taken into account, then shocks are unstable if their velocity is greater than 150 km s^{-1}. Unfortunately this leads to wild fluctuations in the very line ratios that are usually used to determine shock velocity. Even worse, there are times during the evolution of a fast unstable shock when it can look like a much slower stable one. However, all is not lost. Innes (1992) has shown that there are certain features, such as multiple line profiles, which can be used to detect unstable shocks.

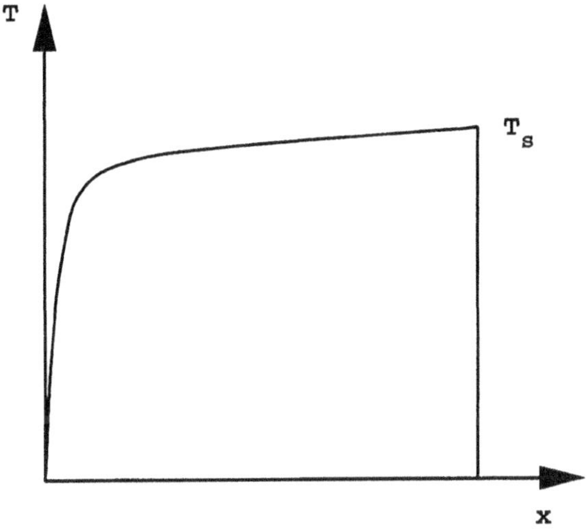

Fig. 4. Structure of a steady radiative shock

4. A Model Radiative Shock Calculation

In order to study the two dimensional behaviour of the instability, we use the following model cooling law

$$L = A\rho^2\Phi(T),$$

with

$$\begin{aligned}\Phi(T) &= \mid T-1 \mid (T-1)^3 \text{ for } T \leq 5, \\ &= BT^{-1/2} \text{ for } T \leq 5.\end{aligned}$$

where $T = p/\rho$ and A, B are constants. A determines the width of the cooling region and B is chosen to make Φ continuous at $T = 5$. Although this is not a particularly good approximation to the function shown in figure 3, it does capture its flavour, and the fact that it is stable for $T \leq 5$ makes it rather more convenient for numerical computation.

Even for this model cooling law, the temperature gradient near the piston becomes so large that a two dimensional calculation on a uniform grid would be very expensive. We have therefore used the adaptive grid code, μCobra, to give us the required resolution (Falle & Giddings 1993). μCobra uses a hierarchy of grids G^0, $G^1 \ldots G^{\rm N}$ such that the mesh spacing on $G^{\rm n}$ is $h/2^{\rm n}$, where h is the spacing on the coarsest grid, G^0. Grids G^0 and G^1 cover the whole domain, but the finer grids need only exist in regions which require high resolution. The grid hierarchy is used to generate an estimate of the truncation error by comparing solutions on grids with different mesh spacings and the grid refines if this error exceeds a given tolerance. Refinement also occurs in time so that if the time step on G^0 is Δt, then it is $\Delta t/2^n$ on G^n. The code is extremely general and works equally well in one, two and three dimensions.

We consider the two dimensional domain defined by $0 \leq x \leq 10, 0 \leq y \leq 5$ with symmetry boundaries at $x = 0, y = 0, y = 5$ and a flow boundary at $x = 10$. Initially the entire domain is filled with gas with $\rho = 1$, $p = 10^{-3}$, $v = -10$. If there were no perturbations then the flow would remain one dimensional and the symmetry boundary at $x = 0$ acts like a piston moving with speed 10 relative to the upstream gas. Initially the flow is adiabatic and the post-shock temperature is 33.33 if $\gamma = 5/3$, which is well into the region of unstable cooling. Even when the shock becomes radiative, the post-shock temperature in the steady solution would be 18.75, which should still be unstable.

Figure 5 shows how the shock velocity oscillates in a one dimensional calculation. The oscillations do eventually become roughly periodic, but this takes a long time and there is little to be gained by waiting for this before imposing a two dimensional perturbation. We therefore perturb the flow upstream of the shock at $t = 2$ by increasing the density by 50% in the rectangle $6 \leq x \leq 7, 0 \leq y \leq 2.5$. This breaks the symmetry and eventually something interesting happens.

Figure 6 shows the pressure at $t = 10$, at which time the primary shock velocity is at a minimum. A secondary shock has formed near the piston and is propagating towards the primary shock. For $y \leq 2.24$, this shock is nearly plane and parallel to the piston, but at $y = 2.24$ it splits into two at a triple point. This triple point moves in the direction of increasing y until it reaches the upper boundary, at which point it reflects. This is very similar to the behaviour of the transverse waves in a cellular detonations.

This has some interesting implications for the filamentary structure in old supernova remnants such as the Cygnus loop. Most of the time there is very little radiative cooling near the primary shock, which would therefore be invisible. However, the cooling rate behind the secondary shock is large when it first forms and it is this structure which should be identified with the optical filaments. The secondary shock structure in figure 6 would produce an apparent splitting of the filaments, something which is seen all over the place in old supernova remnants. The $H\alpha$ image of the Cygnus Loop in Hester & Raymond (1988) shows a particularly fine example of this.

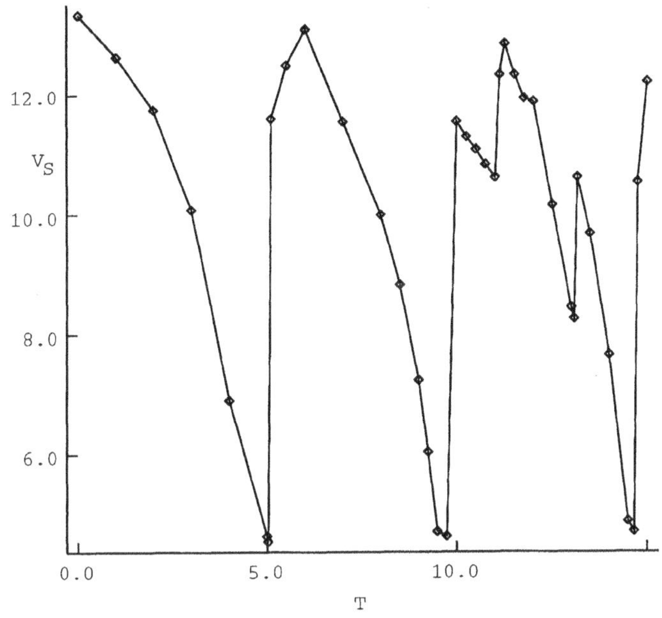

Fig. 5. Shock velocity in a one dimensional unstable radiative shock

5. Conclusions

Although the above arguments and numerical results are obviously not conclusive, they do suggest that the instabilities in detonation waves and radiative shocks are related in the sense that they tend to generate the same kind of structure. In the case of radiative shocks, there is some observational evidence for this phenomenon in the filamentary structures seen in old supernova remnants and it ought to be possible to confirm this by a careful analysis of the velocity field associated with split filaments. This would be very useful since it could provide evidence for the instability that is independent of line ratio measurements.

On the other hand, any cellular structure in Type I supernovae would be impossible to detect directly. However, our experience with terrestrial detonations suggest that it should occur in Type I supernova explosions, provided these involve detonations waves (Khokhlov 1989). As we have already point-

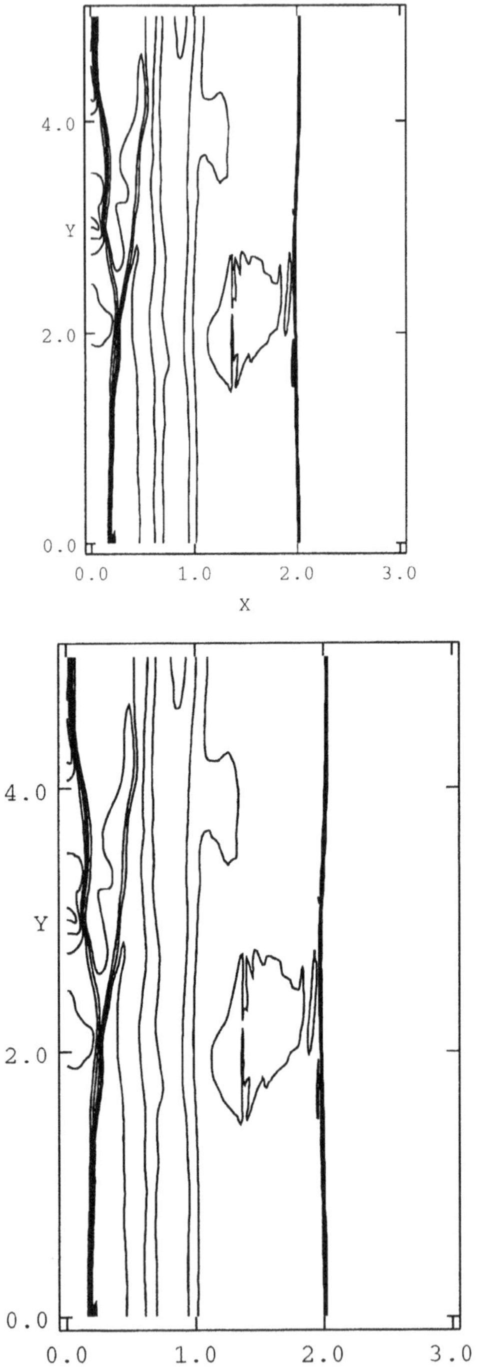

Fig. 6. Linear pressure contours in an unstable two dimensional radiative shock

ed out, this does not affect the speed of detonations waves, but the nuclear reactions are, in part, reversible and we might therefore expect the instability to affect the production of intermediate elements. It is also relevant to the question of whether detonation waves can form since it is known from terrestrial experiments that cellular detonations can propagate in conditions where a plane steady detonation would simply die out.

References

Bertschinger, E.: 1986, *Astrophys.J.* **304**, 154

Bourlioux, A. & Majda, A.J.: 1992, *Combust.Flame* **90**, 211

Chevalier, R.A. & Imamura, J.N.: 1982, *Astrophys.J.* **261**, 543

Falle, S.A.E.G.: 1975, *Mon. Not. R. astr. Soc.* **172**, 55

Falle, S.A.E.G.: 1981, *Mon. Not. R. astr. Soc.* **195**, 1011

Falle, S.A.E.G., Giddings, J.R.: 1981, in M.J. Baines & K.W. Morton, ed(s)., *Numerical Methods for Fluid Dynamics 4*, Clarendon Press: Reading, 337

Fickett, W. & Davis, W.C.: 1979, *Detonation*, University of California Press: Berkeley

Hester, J.J. & Raymond, J.C.: 1988, in R.S. Roger & T.L. Landecker, ed(s)., *IAU Colloquium 101*, CUP: Cambridge, 415

Imamura, J.N., Wolff, M.T. & Durisen, R.H.: 1984, *Astrophys. J.* **276**, 667

Innes, D.E.: 1992, in J.E. Dyson & E.B. Carling, ed(s)., *Kinematics & Dynamics of Diffuse Astrophysical Media*, Kluwer: Dordrecht, 311

Innes, D.E., Giddings, J.R. & Falle, S.A.E.G.: 1987a, *Mon. Not. R. astr. Soc.* **224**, 179

Innes, D.E., Giddings, J.R. & Falle, S.A.E.G.: 1987b, *Mon. Not. R. astr. Soc.* **226**, 67

Innes, D.E., Giddings, J.R. & Falle, S.A.E.G.: 1987c, *Mon. Not. R. astr. Soc.* **227**, 1021

Khokhlov, A.: 1989, *Mon. Not. R. astr. Soc.* **239**, 785

Langer, S.H., Chanmugam, G. & Shaviv, G.: 1981, *Astrophys.J.* **245**, L23

Langer, S.H., Chanmugam, G. & Shaviv, G.: 1982, *Astrophys.J.* **258**, 289

Raymond, J.C.: 1979, *Astrophys. J. Suppl.* **39**, 1

Shull, J.M. & McKee, C.F.: 1979, *Astrophys. J.* **227**, 131

PARTICLE ACCELERATION IN SHOCKS

L. O'C. DRURY
Dublin Institute for Advanced Studies, School of Cosmic Physics
5 Merrion Square, Dublin 2, Ireland

Abstract. The mechanism of particle acceleration by shocks is reviewed, and the resulting modifications of the shock structure contrasted with those which occur in molecular shocks with magnetic precursors. The danger associated with the use of steady models is emphasised and the applicability of the theory to cosmic ray acceleration in supernova remnants briefly discussed.

Key words: Particle Acceleration – Cosmic Rays – Shock Structure

1. Introduction

A reviewer may occasionally be forgiven for going back to basics, and starting with some simple questions. I propose to take advantage of this freedom and begin by asking "What are shocks?" and "Why do we (or, more importantly, Nature) need them?". To an applied mathematician shocks are simply generic features, associated with a local loss of smoothness, which occur when characteristics intersect in the solution of nonlinear hyperbolic systems. A more physical way of saying essentially the same thing is that shocks are associated with wave pile-up.

If, as is the case in gas dynamics and very generally in other continuum systems, small disturbances propagate as waves with definite velocities (sound waves in the case of gas dynamics, magnetosonic and Alfvén modes in the case of ideal MHD) and if these characteristic velocities vary with position and the state of the system, as they generally will, it is possible for waves to pile-up at one point. Specifically, let us imagine a flow from left to right which is initially faster than the local small disturbance propagation speed, but drops to a slower speed. Then small disturbances far downstream on the righthand side of the transition can propagate back against the flow until they reach the point where their propagation speed is exactly balanced by the flow speed (in gas dynamics this is the sonic point). Similarly waves from the upstream region will be carried back to this point where they will stagnate. Clearly there is a break-down of the physical model at this point. Microphysical processes, which have been ignored in the macroscopic continuum description, become important because structures are created on small length scales where these processes are dominant. These structures must be dissipative and entropy generating because all the information associated with leftward travelling waves flows into this one point.

Astrophysics and Space Science **233**: 251–260, 1995.

There are interesting analogies with the phenomena of turbulence and magnetic reconnection. Inviscid fluid flow, in general, becomes turbulent and generates vortical structures on smaller and smaller scales until dissipative structures are created on very small length-scales where the neglected viscous effects become important. Similarly in ideal MHD theory the magnetic field lines can be twisted and distorted to the point where small-scale resistive effects become important and reconnection sets in. However in both magnetic reconnection and turbulence the small-scale dissipative structures are geometrically complicated and possibly fractal. By contrast shocks transform ordered large-scale motion into small-scale random motion in relatively simple one dimensional structures.

Thus shocks are important, not just because they occur so generally, but because they are the one example of a self-organising dissipative structure which is relatively easy to model. They are also important for their effects. In astrophysics shocks heat gas (and thereby cause the emission of thermal radiation); drive interstellar chemistry; destroy dust grains; focus, deflect and dissipate flows; and finally, the subject of this article, accelerate particles.

2. Collisionless Plasma Shocks

The shocks which are important for particle acceleration are collisionless plasma shocks. Collisionless means that two-body collisions can be neglected with respect to interactions between the charged particles constituting the plasma and collective modes of the plasma. This is nearly always the case under astrophysical conditions. Plasma means of course that the medium in which the shock is propagating is ionized and may contain a magnetic field.

For simplicity, let us now consider a shock which is propagating parallel to the (mean) magnetic field. There have been many recent hybrid computer simulations of such shocks (hybrid in the sense that the electrons are treated as a fluid, but the ions as individual particles) (e.g. Quest, 1988; Scholer, 1990; Giacalone et al, 1992). It is instructive to look at the ion velocity phase space structure as displayed in, for example, fig. 15 of Quest (1988). Coming in from the left we have the cold upstream ion distribution with a large positive bulk velocity. In the shock front these particles are scattered into a downstream distribution which is essentially isotropic about the downstream velocity which is roughly one quarter of the upstream velocity (this follows from mass, momentum and energy conservation; as long as the distributions are isotropic, though not necessarily thermal, in the local comoving frame the standard Rankine-Hugoniot relations follow). It follows that there is inevitably a population of ions with negative velocity moving back towards the shock front, and some few of these may penetrate through the front into the upstream region as backstreaming ions. In fact this backstreaming population appears to be essential for the existence of the shock. It is the instabilities associated with this ion distribution which generate the magnetic field fluctuations which then provide the scattering needed for the dissipation in the shock front.

Some of the backstreaming ions will be scattered by the upstream magnetic fluctuations and turned back towards the shock. Because this is scattering by a magnetic field structure the particle conserves its energy in the frame of the scatterer; to put it another way, magnetic fields change the direction, but not the magnitude, of a particle's momentum. Now clearly the backstreaming ions have a substantially higher energy in the upstream bulk plasma frame than the incoming cold ions, and once turned around and heading back towards the shock an even higher energy in the downstream frame. If we now scatter some of these again from the downstream region back into the upstream region, we clearly have the possibility of a Fermi type acceleration where the particles gain energy each time they complete a cycle of shock crossings.

A question which is often asked, and which can be very confusing, is where exactly the acceleration occurs. The key to understanding the process is to note that energy is not a scalar quantity (even in Newtonian physics) but depends on the choice of a reference frame. Each process of the cycle can be regarded as having no associated acceleration in some frame. The upstream scattering is at constant energy in the upstream frame; the shock crossings are at constant energy in the shock frame; and the downstream scattering is at constant energy in the downstream frame. However there is no single global frame in which there is no acceleration. All that one can really say is that a particle which has completed a cycle of shock crossings and scatterings will have increased its energy if the initial and final energies are measured in the same local frame; to ask where exactly the change occured is analogous to asking where the twist in a Möbius band is located.

However it is convenient mathematically to make a specific choice. If we agree to measure particle energies in a frame moving with the mean velocity of the scattering centres (this will differ from the bulk plasma speed by an amount at most of order the local Alfvén speed; for simplicity I will ignore this effect here; for a recent discussion of the importance of Alfvénic effects see Jones, 1993) then *all* upstream and downstream scattering events are energetically neutral (strictly there may be a small amount of second order Fermi acceleration if the scattering centres are not all moving at exactly the same velocity; again I ignore this effect which is discussed by Ostrowski and Schlickeiser, 1993), however every time a particle crosses the shock front we shift reference frames in such a way that the energy increases. Once the particle velocity v has become much larger than the bulk plasma speed U there is a simple analytic theory (Krymsky, 1977; Axford et al, 1977; Bell 1978;

Blandford and Ostriker, 1978). In this limit, $v \gg U$, the particle distributions are close to isotropy and the particle transport can be described by simple diffusion. Hence the theory is often called diffusive shock acceleration.

It turns out to be more convenient mathematically to work with the modulus of the particle's momentum, $p = |\mathbf{p}|$, than the kinetic energy. If the particle crosses the shock from the upstream region, where we measure p in a frame moving towards the front at speed U_1, to the downstream region where we measure in a frame moving at velocity U_2 away from the shock, it is easy to calculate that to lowest order in U/v the change in p is

$$\Delta p = p\frac{U_1 - U_2}{v}\cos\theta$$

where θ is the angle between the shock normal and the trajectory of the particle. If the particles are coming from an upstream distribution which is isotropic to lowest order in U/v, the average change is

$$\langle\Delta p\rangle = p\frac{U_1 - U_2}{v}\frac{\int_0^{\pi/2}\cos^2\theta\sin\theta d\theta}{\int_0^{\pi/2}\cos\theta\sin\theta d\theta} = \frac{2}{3}p\frac{U_1 - U_2}{v}.$$

All particles entering the upstream region (at least in simple plane geometry, and with scattering at some level throughout the entire region) will be swept back to the shock, although the return time distribution is very skewed. Most particles return quite quickly, but there is a very long slowly-falling tail of particles which take a long time to return. However some of the particles entering the downstream region will be carried off by the flow and never return to the shock. The probability of escaping from the shock in this way is easily calculated. We note that a uniform number density of energetic particles, n, throughout the entire downstream region is the only steady solution allowed in the diffusion approximation. This implies a flux of particles escaping to the right of nU_2. However the flux of particles into the downstream region across the shock is $\frac{1}{2}n\int_0^{\pi/2} v\cos\theta\sin\theta d\theta = nv/4$ and thus the escape probability is simply $4U_2/v$.

Combining these two results we easily see that the steady state solution for the accelerated particle distribution function at the shock is a power-law in momentum,

$$f(p) \propto p^{-3U_1/(U_1-U_2)}.$$

There are two important things to note about this result. First, this is the only known acceleration process which *naturally* predicts a power-law. In all Fermi type processes the spectral slope is determined by the ratio of the acceleration time-scale to the escape time-scale. However these are normally determined by different physical processes and it is not obvious why the ratio should be constant over many decades. Here it is determined simply by the compression of the shock. The second point is that for the compression expected in a strong shock, $U_1 \approx 4U_2$, the resultant spectrum, $f(p) \propto p^{-4}$, is close to what is seen in the Galactic cosmic rays and inferred in many non-thermal radio sources.

An interesting way to think of the p^{-4} distribution function predicted by this simple test-particle theory is to look at the energy per logarithmic interval, $\mathcal{E}$, in the spectrum,

$$\int \mathcal{E} d\ln(p) = \int 4\pi p^2 f(p) T(p) dp,$$

where T is the kinetic energy of the particle. In the non-relativistic region, $p \ll m_0 c$ where m_0 is the particle rest-mass, we have

$$\mathcal{E} = 4\pi p^2 \frac{p^2}{2m_0} f(p) p \propto p$$

and in the relativistic region, $p \gg m_0 c$,

$$\mathcal{E} = 4\pi p^2 c p f(p) p = \text{const.}$$

Thus in the non-relativistic region the energy density, or equivalently the pressure, associated with the accelerated particles at a given momentum increases linearly with momentum, and then flattens off to become constant in the relativistic region. This has two important consequences.

The hybrid collisionless shock simulations suggest that the initial population of backstreaming ions has to have enough energy to disturb the upstream magnetic field sufficiently to create the shock in the first place. If there is enough energy to produce a self-generated scattering wave field at a few times thermal energy, there will then certainly be more than enough at higher energies. As first pointed out by Bell (1978), this is an important aspect of shock acceleration. The scattering near the shock which is responsible for the acceleration can be generated by the accelerated particles themselves and be very intense. In fact the nature of this strongly nonlinear magnetic turbulence is an interesting unsolved problem in plasma physics; the usual assumption made is that the particle scattering mean free path drops to something close to the gyroradius (Bohm diffusion approximation) and the excess wave energy is dissipated locally as heat.

Secondly, it is clear that the total amount of energy in the accelerated particles can easily be dynamically significant. For applications this is important because it means that the process can be efficient, in the sense of putting a significant part of the mechanical energy dissipated in the shock into particle acceleration. However it also means that the test-particle assumption which forms the basis of the elementary theory is invalid. What is needed is a nonlinear theory of shock acceleration. All that is needed is to solve the coupled system comprising the usual hydrodynamical (or MHD) equations with an additional pressure P_C in the momentum equation, *ie*

$$\frac{\partial(\rho \mathbf{U})}{t} + \nabla(\rho \mathbf{U}\mathbf{U}) + \nabla(P_G + P_C) = 0,$$

the diffusive transport equation for the accelerated particle distribution function $f(p, x, t)$,

$$\frac{\partial f}{\partial t} + \mathrm{U} \cdot \nabla f = \nabla(\kappa(p, x, t)\nabla f) + \frac{1}{3}\nabla \cdot \mathrm{U}p\frac{\partial f}{\partial p} + Q,$$

some model for the evolution of the diffusion coefficient $\kappa(p, x, t)$ as a function of particle momentum p, position x and time t, and a prescription for the "injection" Q of particles from the thermal population into the non-thermal accelerated population. By analogy with "radiation hydrodynamics" one could call this "energetic particle hydrodynamics".

Clearly this is not a very tractable problem. Separate parts of it have been attacked with varying degrees of success using a variety of techniques. There exist a small number of analytic results. These suffer from the disadvantage of being very special and non-generic, though they are invaluable as test cases for the various numerical methods. A good recent review, with many references to the earlier literature, is that by Jones and Ellison (1991).

One method of proceeding, associated mainly with Don Ellison, is to use a Monte-Carlo approach to directly follow the particle scattering. This has the advantage that there is no injection problem; the thermal and non-thermal particles are treated in exactly the same way. However the calculations published to date use a simple large angle scattering model at all energies. This is undoubtedly a gross over-simplification of the plasma physics of the collisionless shock, but empirically it does seem to reproduce reasonably well both experimental observations of the thermal to superthermal transition at the Earth's bow shock and the results of detailed hybrid simulations. Computationally the Monte-Carlo approach is quite slow and, until recently (D Ellison, personal communication), it has only been possible to consider steady systems (requiring the imposition of free escape boundaries and upper momentum cut-offs).

Making *ad hoc* assumptions for κ and Q it is possible to attempt a direct numerical integration of the full system. This has the advantage that it can handle time-dependent problems, but it is still quite slow. A dangerous, but useful, simplification is the two-fluid approximation where the accelerated particles are treated as a second fluid with pressure and energy density, but negligible inertia. This is fast enough to allow reasonable exploration of parameter space, however results do need to be checked against full spectral or Monte-Carlo calculations. The problem is that the two fluid equations are not complete in themselves, and two closure parameters have to be estimated in an *ad hoc* fashion (in addition to the injection and the diffusion). In certain cases the results can be very sensitive to the exact closure parameters used.

Ideally, of course, all that is necessary is to run a very large plasma simulation, but this is completely impractical. For the moment the best we can do is use all the methods at our disposal while recognising the relative strengths and weaknesses of each.

3. Shocks Modified by Particle Acceleration

In general, for the reasons outlined above, we expect that collisionless plasma shocks under astrophysical conditions will often be strongly modified by the reaction of the accelerated particles. There is a close analogy to the theory of molecular shocks which is discussed extensively elsewhere in this volume.

In both cases we have the bulk flow component, the thermal plasma or the background neutrals, carrying most of the mass, being modified by coupling to a second component which has negligible mass but a dynamically significant pressure. In the case of shocks modified by particle acceleration the accelerated particles, which we may conveniently call the cosmic rays, provide the modifying agent. In the case of molecular shocks it is the magnetic field plus ions, which also have negligible mass, that can exert a significant pressure. In the cosmic ray case the accelerated particles "diffuse" through the background plasma as a result of wave-particle interaction. In the molecular case, the magnetic field with the attached ions drifts, by ambipolar diffusion, through the neutral component.

There are, however, some important differences. In the cosmic ray case the propagation speed of the cosmic rays is essentially infinite (the diffusion equation is elliptic). In the molecular case the maximum propagation speed of disturbances in the magnetic field and ionized component is the ion Alfvén speed. The scattering in the cosmic ray case is through collective effects, which allows amplification of small disturbances in the background density under certain circumstances (Drury and Falle, 1986), whereas in the molecular case the ambipolar diffusion is essentially a two-body scattering process. The cosmic ray pressure is essentially isotropic, whereas the magnetic pressure is highly anisotropic. Finally the cosmic rays have, in general, a "softer" equation of state than the background plasma (for a fully relativistic gas $\gamma = 4/3$ whereas the background typically has an adiabatic exponent of $\gamma = 5/3$) whereas in the molecular case the magnetic field, in general, has a "harder" equation of state than the background.

In view of the strong similarities it is interesting to ask how the particle pressure can be strongly amplified in a cosmic ray dominated shock, but not the magnetic field in a molecular shock. The key equation, for a cosmic ray dominated shock, is

$$\frac{\partial E_C}{\partial t} + \frac{\partial U E_C}{\partial x} = \frac{\partial}{\partial x}\left(\kappa \frac{\partial E_C}{\partial x}\right) - P_C \frac{\partial U}{\partial x}$$

where E_C is the energy density of the accelerated particles, P_C the associated pressure, κ the effective diffusion coefficient and U the velocity of the background plasma. The left hand side of this equation is the usual rate of change of the density plus divergence of the advected flux of cosmic ray energy. If there were no terms on the right hand side (RHS) this would be a simple conservation equation. The first term on the RHS is the diffusion term, which moves energy from high points to low points. The second term, which is the crucial one, is the "PdV" work done by the flow compression on the cosmic rays. In a shock-like transition, where $\partial U/\partial x$ is large and negative, this is a strong source of particle energy, some of which can then diffuse back into the upstream region and be amplified further.

In the case of molecular shocks, if we consider only the simple case of a perpendicular magnetic field, the field B satisfies the simple equation

$$\frac{\partial B}{\partial t} + \frac{\partial U B}{\partial x} = \frac{\partial}{\partial x}\left(\frac{\kappa \partial B}{\partial x}\right)$$

where κ is here the ambipolar diffusion coefficient. In terms of the magnetic energy density E_M and pressure P_M we can rewrite this as

$$\frac{\partial E_M}{\partial t} + \frac{\partial U E_M}{\partial x} = \frac{\partial}{\partial x}\left(\kappa \frac{\partial E_M}{\partial x}\right) - \frac{\kappa}{\mu_0}\left(\frac{\partial B}{\partial x}\right)^2 - P_M \frac{\partial U}{\partial x}.$$

Formally this is now exactly the same as the cosmic ray equation except for the extra negative term on the RHS. This term is easily seen to be the energy lost by friction between the ions and the neutrals. Thus one answer to why the magnetic field does not get diffusively amplified to very high levels is because the two-body collisions responsible for the ambipolar diffusion are dissipative and heat the background neutrals.

What does a shock modified by particle acceleration look like? In general terms this is simple. The accelerated particles diffusing ahead of the shock produce an adverse pressure gradient which decelerates and compresses the flow of plasma into the shock. This extended upstream region is generally called the cosmic ray precursor. The collisionless plasma shock itself is either completely wiped out by the precursor, which then consists of an extended smooth transition from upstream to downstream values, or survives as a somewhat weakened subshock between the precursor and the more or less uniform downstream region. The modifications are all in the upstream region (in cases where a subshock survives) because the additional pressure associated with the accelerated particles increases the propagation speed of long wavelength disturbances. The overall shock structure also becomes more compressive, reflecting in part the softer equation of state of the accelerated particles (if these are partially relativistic, as in the case of the cosmic rays) but also possible energy loss through escape of particles from the system and time dependent effects.

It is possible to produce a rather elegant theory for steady modified cosmic ray shocks, even including effects of wave excitation and dissipation, leading to generalised Rankine-Hugoniot relations (Drury, 1983, and references therein). However it must be emphasised that these results, although mathematically correct, probably have very little to do with real shocks. With realistic diffusion coefficients the shock structures are almost always time-dependent. In addition the presursor regions, at least in strongly modified shocks, are expected to be unstable; small density disturbances swept in from upstream can be strongly amplified to the point where an irregular series of weak shocks forms. It is important to realise that steady models of cosmic ray modified shocks, although easily calculated, can be very seriously misleading (Drury et al, 1995; Ko, 1995)

4. Application to Particle Acceleration in Supernova Remnants

One of the main motivations for the development of the theory of diffusive shock acceleration, at least the non-linear aspects, was the hope that this mechanism, applied to the strong interstellar shocks associated with supernova remnants, could at least partially explain the problem of the origin of the Galactic cosmic rays. There has been much work in this direction (e.g. Berezhko et al, 1994; Kang and Jones, 1991; Dorfi, 1990, 1991; Drury et al, 1989). The general result appears to be a partial success. There are still many *ad hoc* assumptions in all the models; however, it appears possible to get the correct total power, and roughly the right spectrum and composition. However this mechanism certainly does not solve the cosmic ray origin problem. Even with optimistic assumptions, acceleration in supernova remnant shocks is unable to accelerate protons to more than about 10^{14} eV and the observed cosmic ray spectrum continues for some six decades to at least 10^{20} eV! In the next few years it may be possible to test whether supernova remnants really are the main sources of Galactic cosmic rays below 10^{14} eV by looking for gamma-ray emission from favourably located remnants (Drury et al, 1994; Naito and Takahara, 1994).

Acknowledgements

The hospitality of the University of Manchester Institute of Science and Technology and the local organisers is gratefully acknowledged. The preparation of this review was partially supported by the Commission of the European Communities under contract ERBCHRXCT940604.

References

Aharonian, F., Drury, L. O'C. and Völk, H. J.: 1994, *Astron. Astrophys.* **285**, 645
Axford, W. I., Leer, E. and Skadron, G.: 1977, *Proc 15th Int. Cosmic Ray Conference (Plovdiv)* **11**, 132
Bell, A.: 1978, *Mon. Not. R. astr. Soc.* **182**, 147
Berezhko, E. G., Yelshin, V. K. and Ksenofontov, L. T.: 1994, *Astroparticle Phys.* **2**, 215
Blandford, R. D. and Ostriker, G.: 1978, *Astrophys. J.* **221**, L29
Dorfi, E. A.: 1990, *Astron. Astrophys.* **234**, 419
Dorfi, E. A.: 1991, *Astron. Astrophys.* **251**, 597
Drury, L. O'C.: 1983, *Sp. Sci. Rev.* **36**, 57
Drury, L. O'C., Aharonian, F. and Völk, H.J.: 1994, *Astron. Astrophys.* **267**, 959
Drury, L. O'C. and Falle, S. A. E. G.: 1986, *Mon. Not. R. astr. Soc.* **223**, 353
Drury, L. O'C., Markiewicz, W. and Völk, H. J.: 1989, *Astron. Astrophys.* **225**, 179
Drury, L. O'C., Berezhko, E. G., Völk, H. J.: 1995, *Astron. Astrophys.* in press,
Giacalone, J., Burgess, D., Schwartz, S. J. and Ellison, D, C,: 1992, *Geophys. Res. Lett.* **19**, 433
Jones, T. W.: 1993, *Astrophys. J.* **413**, 619
Jones, F. C. and Ellison, D. C.: 1991, *Space Sci. Rev.* **58**, 259
Kang, H. and Jones, T.: 1991, *Mon. Not. R. astr. Soc.* **249**, 439
Ko, M.: 1995, *Mon. Not. R. astr. Soc.* in press,
Krymsky, G. F.: 1977, *Dokl. Acad. Nauk SSSR* **234**, 1306
Ostrowski, M. and Schlickeiser, R.: 1993, *Astron. Astrophys.* **268**, 812
Naito, T. and Takahara, F.: 1994, *J. Phys. G* **20**, 477
Quest, K. B.: 1988, *J. G. R.* **93**, 9649
Scholer, M.: 1990, *Geophys. Res. Lett.* **17**, 1821

COMPOSITE MODELS (SHOCK AND PHOTOIONIZATION) FOR THE INTERPRETATION OF AGN SPECTRA

MARCELLA CONTINI
School of Physics and Astronomy, Tel-Aviv University, Ramat Aviv, 69978 Tel-Aviv, Israel

Abstract. AGN spectra are investigated by comparing model calculations with observations. Line intensity ratios show that the emitting gas is heated by shocks. Therefore, composite models which consistently account for both shocks, accompanying the radial motion of the gaseous clouds, and a photoionizing radiation flux are used. Liners are ambiguous objects between AGN and Starburst galaxies, therefore, power-law or black body radiations are considered in the models, depending on the characteristics of the line spectra. The SUMA code is used.

The results show that shocks definitively play an important role, even if a photoionizing flux is generally present. Particularly, the heterogeneous nature of the objects, which are classified as Liners, is confirmed.

Key words: - AGN – Liners – Radiation Mechanisms – Shock Waves

1. Introduction

It is generally considered that power-law radiation from the active center of Active Galactic Nuclei (AGN) is the main ionizing mechanism of gas in the clouds of the narrow line emission region (NLR). However, the spectra show that some line ratios are very similar to those of supernova remnants (SNR). The gas in the SNR clouds is heated and ionized by shocks. Therefore, both shocks and photoionization are active in AGN. Composite models which consistently account for shocks accompanying the radial motion of the clouds and for radiation from the active center (AC) are calculated by SUMA (Viegas & Contini 1994 and references therein). They are successful in resolving some problems which rise by using pure photoionization models : the fitting of the [O III] 5000/[O III] 4363 line ratio, of neutral line ratios to Hβ ([N I] 5200, [O I] 6300) and of ratios of lines from very high ionization levels (e. g. [Fe X] 6374). These problems are particularly critical in Liners (Low Ionization Narrow Emission-line Regions) spectra. Liners are included among AGN because in some of them a broad component of the Balmer lines has been observed, even if very weak. Compared with other AGN, they show a lower luminosity and the spectral line ratios are quite different than those emitted from Seyfert 2 galaxies or from the NLR of other AGN. Moreover, some objects show spectra which are very similar to those emitted from Starburst Galaxies (SB).

We investigate Liners by comparing calculated spectra with observations of individual objects. The sample is provided by Ho et al. (1993). The mod-

Astrophysics and Space Science **233**: 261–266, 1995.

els are characterized either by a power-law radiation, as for AGN, or by a black body radiation, as for SB. We consider the spectrum emitted by gas in a cloud, neglecting the contribution of regions in different physical conditions.

2. Composite Models

Composite models are fully described in Viegas & Contini (1994 and references therein). The shock front forms on the outer edge of the cloud in a outward motion and radiation from the AC hits the opposite (inner) edge. In the case of SB galaxies radiation from the stars hits either the inner or the outer edge of the cloud, depending on the location of the stars.

The gas entering the shock front is heated to temperatures which depend on the shock velocity, V_s, and is ionized at high levels when $V_s > 100\,km\,s^{-1}$. The physical conditions in the gas are calculated throughout a set of plane parallel slabs whose number and geometrical thickness depend on the shock parameters, V_s, B_0 (the magnetic field), n_0 (the preshock density), and on the geometrical thickness of the cloud. In each slab the density is calculated by the compression equation. Consequently, the density is not homogeneous through the cloud. The high temperature of the emitting gas in the immediate postshock region and the distribution of the density through the cloud are characteristic of the shock. In composite models the fractional abundances of the ions are calculated by ionization equilibrium, considering ionization rates due to the primary flux (from the AC or from the stars), collisional ionization rates, and ionization rates of diffuse radiation. The temperature is calculated by the energy equation near the shock front, and by thermal balancing in the slabs in which radiation ionization rates dominate. The radiation flux is determined by the spectral index, α, and by the flux intensity at 1 Rydberg, F, for a power-law, and by the star temperature, T_s and the ionization parameter, U, for a black body. Line intensities are calculated in each slab and integrated throughout the cloud.

The distribution of the electron temperature, of the electron density (refer to the scale on the right of the diagrams) and of the fractional abundances of oxygen ions as a function of distance from the shock front are shown in fig. 1. The shock front is on the left of each diagram. In the top diagram a shock dominated model is shown. Due to compression, the cooling rate increases rapidly and the temperature drops to $\leq 10^4$ K at a certain distance from the shock front. The gas recombines and a large zone of low ionization and neutral gas is present. So, strong high ionization lines and strong low ionization lines are expected, as well as X-ray emission.

The central diagram shows the case of a composite model in which the flux reaches the shock front edge. Notice that radiation maintains the gas

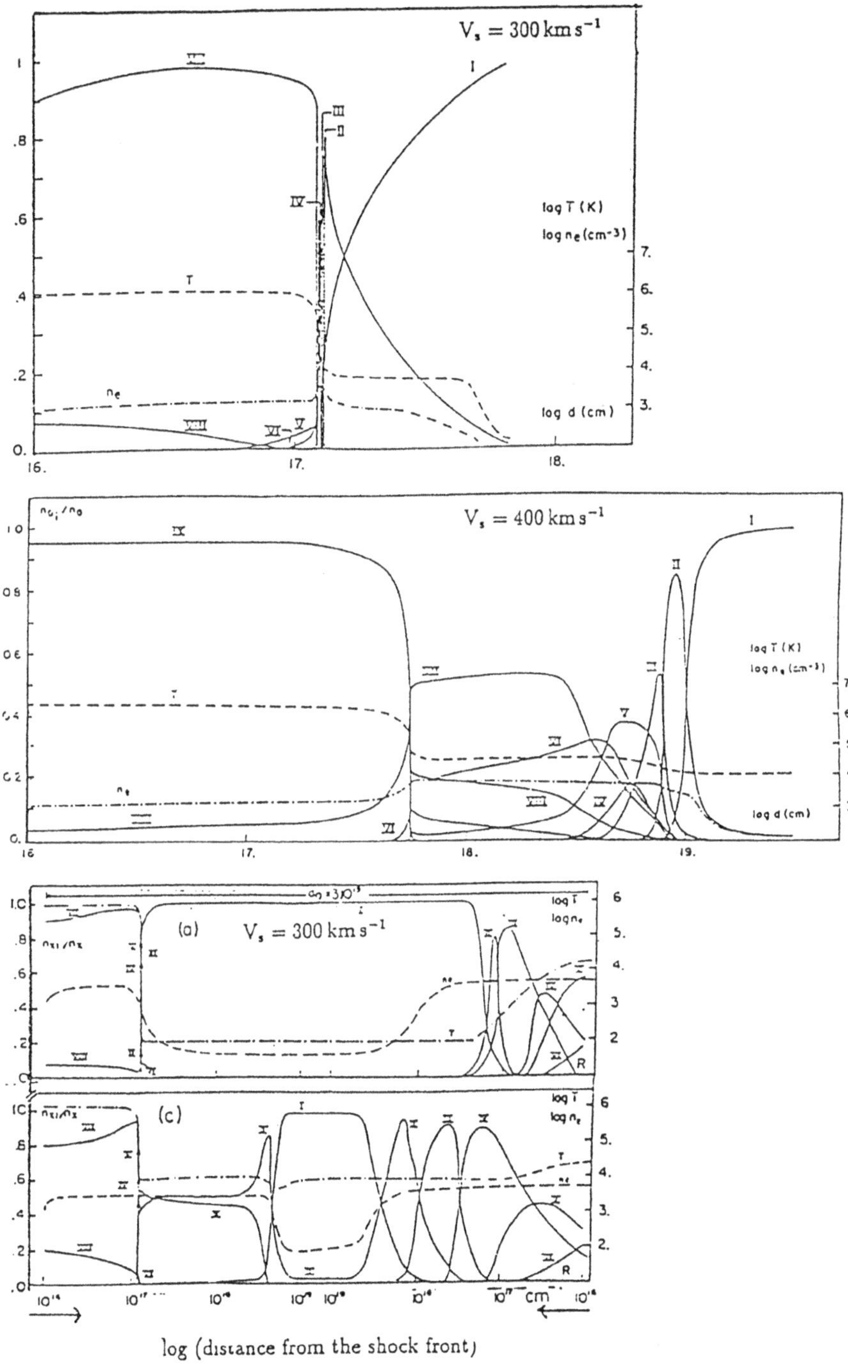

Fig. 1. The distribution of the electron temperature, of the electron density and of the fractional abundance of oxygen ions as a function of the distance from the shock front (on the left).

at a temperature $> 10^4$ K, so that intermediate lines are as strong as those observed in Seyfert 2 galaxies.

In the bottom diagram the shock and the radiation act on opposite edges. Notice that some lines can be emitted from both sides of the cloud. The intensity of neutral lines depends on the geometrical thickness of the cold region, which is reduced when the high energy cutoff of radiation is high or/and when the geometrical thickness of the cloud is small.

3. The Interpretation of the Spectra

The spectra observed by Ho et al. (1993) are rich in number of lines and ionization levels of many different elements. Therefore, the interpretation of the spectra is quite complicated and should be done in two steps. First, by using diagnostic diagrams of the most indicative ratios between lines of the same elements we neglect the problems connected with gas composition. Second, by fitting all observed line ratios to Hβ, the relative abundances of the elements to H are determined and a complete picture of each Liner can be obtained.

Sets of diagnostic diagrams are presented for [O II] 3727+ / [O I] 6300+ versus [O III] 5007+ / [O II] 3727+ (the + indicates that the doublet is considered), [S III] 9532+ / [S II] 6716+ versus [N II] 6548+ / [N I] 5200+, He II 4686 / He I 5876 versus [O III] 5007+/ [O II] 3727+, and for [Ne V] 3426+ / [Ne III] 3869+ versus [O III] 5007+ /[O II] 3727+ . The diagrams of each set represent different cases of shock velocities, densities and magnetic fields (the complete series of diagnostic diagrams is given in Contini, 1995). In each diagram model results are presented for different F ($\alpha = 1.5$) in the power-law radiation and for different U and T_s in the black body. Observed data refer to individual objects.

Comparing the data with calculation results, the set of physical parameters which provides the best fit in each diagram is deduced. However, the results must be consistent in all line ratios in order to represent actually the emitting gas. For some objects the results are ambiguous. An example is shown in fig. 2 for NGC 1275. $V_s = 150\,\mathrm{km\,s^{-1}}$, and $n_0 = 100\,\mathrm{cm^{-3}}$ are indicated by the diagrams. However, a power-law with $F \sim 10^9 - 10^{10}$ and/or a black body with $U = 10^{-2}$ and $T_s = 5 \times 10^4$ K are both roughly fitting.

Finally, comparing the complete spectrum for each Liner in the next step of the interpretation process most objects are definitively identified (cf. Contini 1995, Table 3), while some of them cannot be fitted by a unique model.

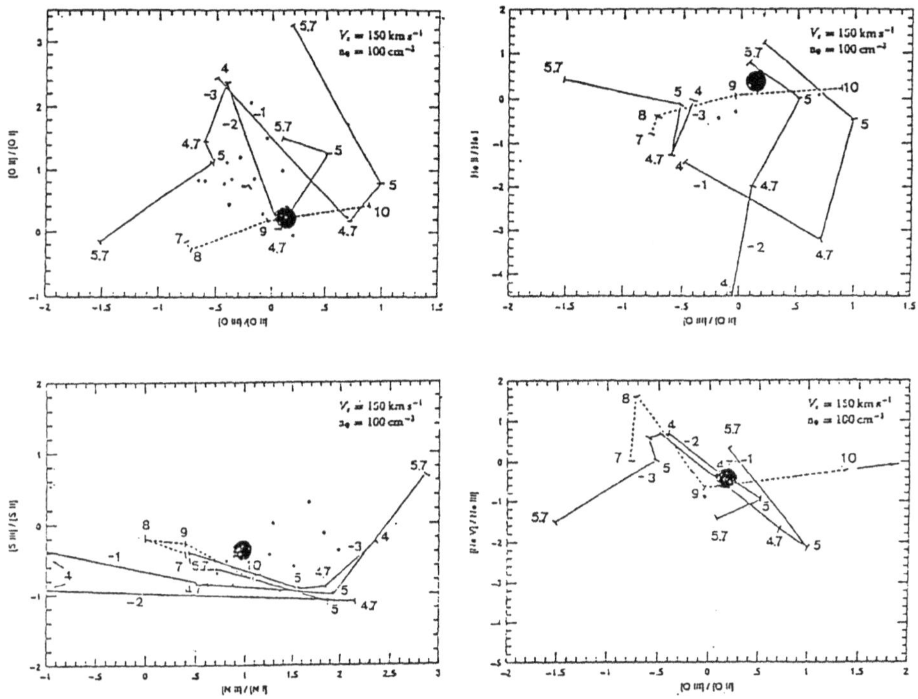

Fig. 2. Diagnostic diagrams for NGC 1275. (The observed data are indicated by a circle.)

4. Results.

The results for the complete sample of Liners is given by Contini (1995, Tables 2) and are summarized in fig. 3. Notice that some Liners in the observational sample have been previously classified as different objects : SB, H II galaxies, Seyfert 2, etc. We have found that out of 10 objects defined as Liners by Ho et al. (1993, Table 1), 3 are definitively AGN with low V_s, n_0, and F; 7 are SB galaxies with high T_s and low V_s. H II regions are explained by black body models with low T_s and low U. The spectra of the 3 Seyfert 2 galaxies are fitted by power-law models with relatively high V_s and n_0. The other 3 objects are ambiguous. This ambiguity is not surprising, because AGN and SB phenomena can easily coexist in a galaxy.

In conclusion, in order to better understand the spectra of Liners observations in different regions of a single galaxy must be studied in future work.

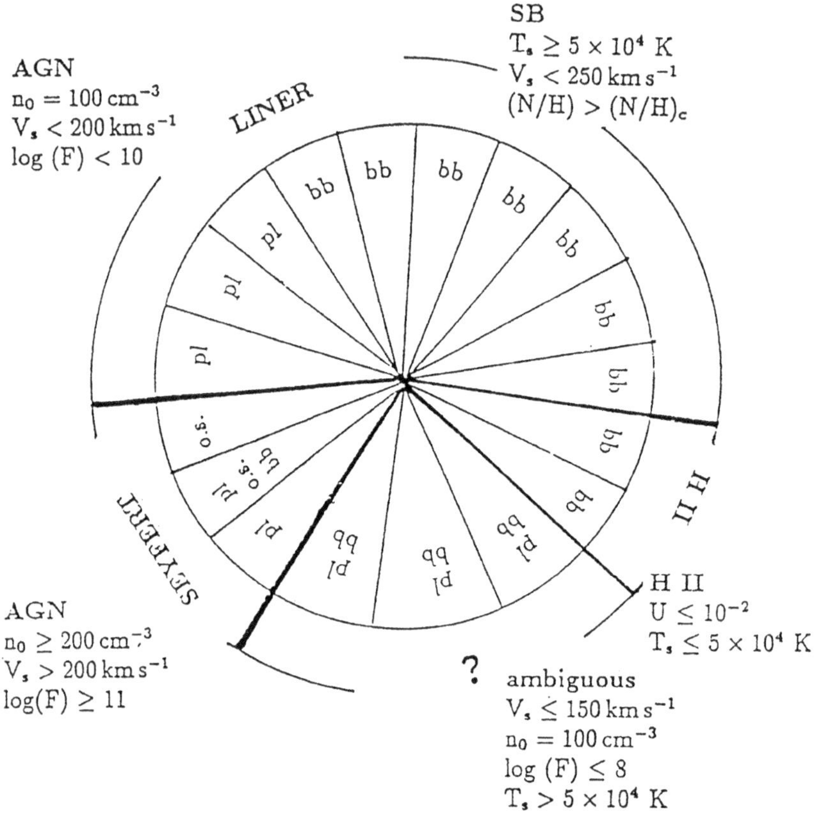

Fig. 3. Summary of results

References

Contini, M. 1995, submitted

Ho, L. C., Filippenko, A. V., & Sargent, W. L. W. 1993 ApJ 417, 63.

Viegas, M. S. & Contini, M. 1994 ApJ 428, 113.

MHD SIMULATIONS OF RAYLEIGH-TAYLOR INSTABILITY IN YOUNG SUPERNOVA REMNANTS

BYUNG-IL JUN and MICHAEL L. NORMAN
Laboratory for Computational Astrophysics
National Center for Supercomputing Applications
Department of Astronomy
University of Illinois at Urbana-Champaign

Abstract. Radio observations shows that young supernova remnants such as Tycho and Cas A generally exhibit a circular clumpy shell. This shell shows a radial magnetic field whose equipartition strength is 2 to 3 orders of magnitude higher than the interstellar field. A simple compression of the ambient field by the shock can explain neither of these observations. We show that the Rayleigh-Taylor instability which occurs at the ejecta/ISM interface can explain these observations. We have done MHD simulations of the instability in the shell of Type-I supernova remnants for the first time by utilizing moving grid technique. Our simulation shows that Rayleigh-Taylor and Kelvin-Helmholtz instabilities amplify ambient magnetic fields locally and produce the clumpy radio shell. Strong magnetic field lines draped around the Rayleigh-Taylor fingers produce the radial B-vector polarization, whereas thermal bremsstrahlung from the dense fingers themselves produce the clumpy X-ray emission.

Key words: Supernova Remnants – Instability – MHD Simulation

1. Introduction

Supernova remnants (hereafter SNRs) in the pre-Sedov stage are characterized by a two-shock structure in the shell. While the forward shock is moving outward, the reverse shock is formed inside of the contact discontinuity and moving inward with respect to the ejected material. Young SNRs generally show a circular, clumpy shell near the outer boundary of the radio and X-ray observations. The location of this clumpy shell corresponds to the intershock region. The magnetic fields in the radio shell have been inferred to be in the range of $10^{-4} \sim 10^{-3}$ G, in contrast to $\sim 10^{-6}$ G in ambient medium. Detailed radio polarization studies reveal that the magnetic field in the shell exhibits a cellular pattern with a net radial orientation (Downs and Thompson 1972; Milne 1987; Dickel et al. 1991).

The contact discontinuity separating the ejected stellar envelope and the ISM has been known to be Rayleigh-Taylor (hereafter R-T) unstable due to the strong deceleration felt by the ejecta as it sweeps up ambient gas (Gull 1973). Chevalier et al. (1992) have studied the instabilities in two dimensions by modelling the interaction region in the self-similar regime and without the inclusion of a magnetic field. In a previous work, we have explored the

Astrophysics and Space Science **233**: 267–272, 1995.

classical R-T instability in incompressible magnetic fluids by two- and three-dimensional MHD simulations (Jun, Norman, and Stone 1995). Although highly idealized, these simulations support Gull's suggestion that the radial magnetic field in the main shell of SNR are produced as the R-T fingers stretch existing field lines. We present the first study of the evolution of the R-T instability in young remnants of Type-I supernova by two-dimensional MHD simulations. Since we simulate the entire remnant, our calculation is not restricted to the self-similar stage; thus we are able to study the instabilities for the evolving remnant.

2. Methodology

We solve the ideal compressible MHD equations in a moving Eulerian grid using the ZEUS-3D code originally written by David Clarke and further developed by the Laboratory for Computational Astrophysics of the University of Illinois at Urbana-Champaign. Shock waves are treated using von-Neumann Richtmyer artificial viscosity. Fluid is advected through the mesh using the upwind, monotonic interpolation scheme of van Leer (second order) or piecewise parabolic (third order) advection. Magnetic fields are transported using the Constrained Transport method (Evans and Hawley 1988) modified with the Method of Characteristics (Stone and Norman 1992b) to upwind the treatment of Alfven waves. The reader is referred to Stone and Norman (1992a,b) for the detailed description of the algorithms.

We use the moving grid method to maintain high resolution in the region between the forward shock and the reverse shock. Basically, the velocity of the forward shock is measured numerically every time step and it is assigned to the grid velocity at the shock front. Then, grid velocities are computed for every other grid point within the shell assuming homologous expansion. The zones are uniform in the intershock region and ratioed with the increasing zone width toward the origin inside the reverse shock since we do not need high resolution inside the reverse shock. This method allows us to keep the high resolution in the intershock region and to avoid a very small timestep near the origin. We used a wedge of $\pi/4$ in spherical polar geometry, periodic boundary conditions on the angular boundaries and reflecting boundary condition at the origin. The intershock region is resolved with 200×400 grids and the region inside the reverse shock is assigned with 100×400 grids.

The outer 3/7 of the star of mass 1.4 $M_\odot$ is initialized with a power law density profile $\rho \sim r^{-7}$ and the inner 4/7 of the star with a constant density to mimic Type-Ia supernova remnants (Colgate and McKee 1969). An explosion of energy of 10^{51} ergs is deposited as kinetic energy where the velocity is assumed to be linearly proportional to the radius. The background density is chosen to be uniform at 1.67×10^{-24} g cm^{-3}. The temperature of

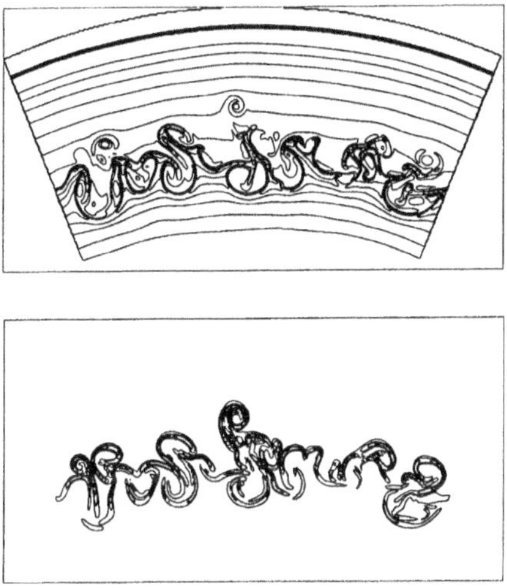

Fig. 1. Contour plots of the density (top) and the magnitude of magnetic fields (bottom) in the intershock region at 500 years.

the entire region is taken as 10^4 K. The background magnetic field is chosen to be 3.5μ G and assumed to be tangential to the shock front. The density is initially perturbed everywhere with a random noise of 2.0% amplitude.

3. Results

The flow is R-T unstable in regions where gradients of pressure and density are of the opposite sign (Jones et al. 1981). The region near the contact discontinuity satisfies this condition, and therefore is unstable. At early times, short wavelength modes appear first because the growth rate is inversely proportional to the square root of the wavelength. Large fingers become dominant at later stages because larger fingers feel less drag and reach higher terminal velocity than smaller fingers. As a finger grows, it streams ahead of the shocked ambient gas, and thus the boundary of the finger becomes sheared. As a consequence, the Kelvin-Helmholtz (hereafter K-H) shear wave instability develops so that the mushroom caps, which are regions of high vorticity, are generated at the tips of the fingers. Because of flux freezing, the magnetic fields are amplified by shearing and stretching at the finger boundaries. The magnetic field is also amplified by the compression near the tip of the finger due to ram pressure. At about 400 years, the maximum field strength reaches 2.1×10^{-4} G. Figure 1 shows the contour plots of density (top plot) and the magnitude of magnetic fields (bottom plot) in the intershock region of SNR at 500 years. Our results show well developed mushroom caps at the tips

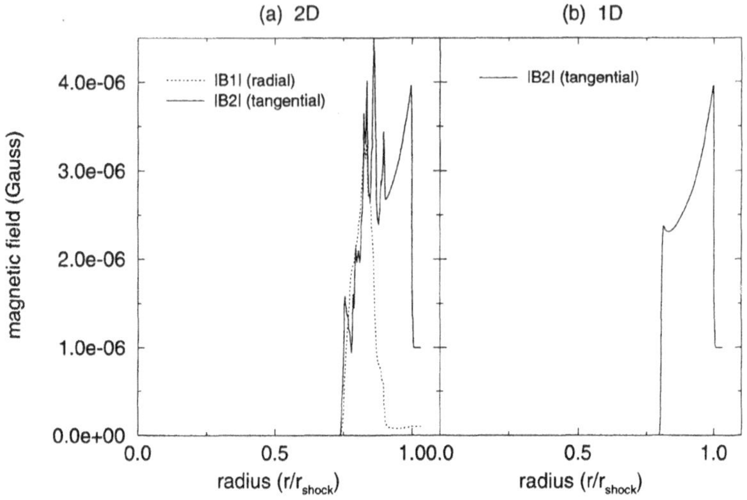

Fig. 2. Angular average of the magnitude of magnetic field at 500 years. (a) The angular average of the magnitude of a magnetic field from 2D. (b) The tangential component of a magnetic field from 1D simulation.

of fingers due to the K-H instability. The strongest magnetic fields are seen along the sides and tips of the fingers rather than at their base where bubbles of shocked ambient medium collect. This is because in the fluid frame the fingers have a higher speed than the bubbles. In our simulation, the remnant stays in Chevalier's self-similar stage until about 250 years. As the remnant evolves beyond the self-similar stage, the intershock region broadens, the instability weakens, and the magnetic energy density decreases. The evolution of the instability is sensitive to the deceleration of the ejecta and the evolutionary stage of the remnant (Jun and Norman 1995a).

Figure 2 shows angular averages of the magnetic field at 500 years and a comparison to the result of 1D simulation. The 2D result shows the magnetic fields peak near the contact discontinuity due to the amplification by the instabilities. In Fig. 2a, we can see the generation of a radial component of the magnetic fields quite clearly. Generally, the radial component of magnetic energy density is comparable to the circumferential (tangential) component. At 500 years, the radial field becomes the dominant component with an energy density of about 1.4 times the circumferential component when we consider only the contribution from the instability. It seems that the relative magnetic energy density of the radial component increases as the mixing region becomes wider allowing the formation of longer fingers. Our simulation is restricted to two-dimensional space. As seen by our previous simulations of classical R-T instability (Jun et al. 1995), the radial component of **B** is enhanced in 3D relative to the 2D results. This is because the mixing layer is

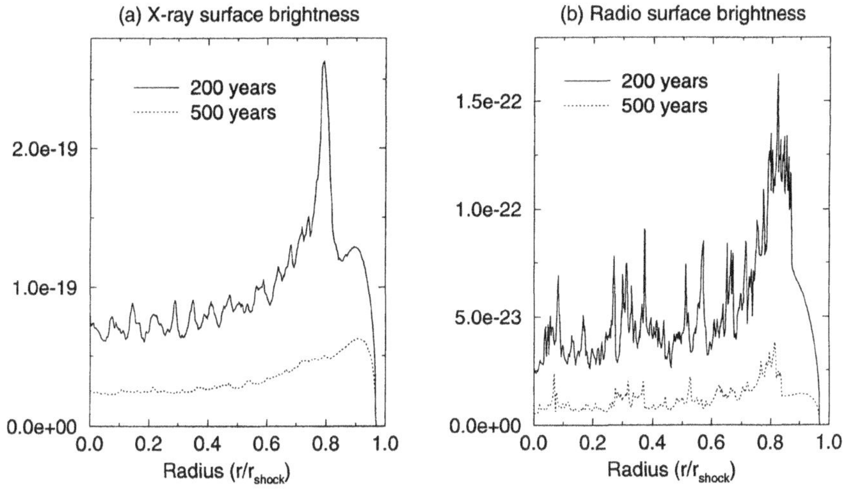

Fig. 3. Surface brightness profiles of radio and X-ray. The radius is normalized by the radius of a forward shock. The units of surface brightness are arbitrary.

thicker in 3D than in 2D. Currently a 3D simulation of the Rayleigh-Taylor instability in SNR is in progress (Jun and Norman 1995b) to confirm and extend this result.

Figure 3 shows the X-ray and radio surface brightness computed by integrating the emissivity along rays. X-ray emissivity (thermal bremsstrahlung) is given as

$$\epsilon_{ff} = 1.426 \times 10^{-27} \frac{\rho^2 T^{1/2}}{m_H^2} \tag{1}$$

while the nonthermal radio emissivity is taken to be

$$i(\nu) \propto p^{2\alpha} (B sin\psi)^{\alpha+1} \nu^{-\alpha} \tag{2}$$

where ρ is the density, T is the temperature, m_H is the mass of hydrogen, p is the gas pressure, B is the magnetic field strength, ψ is the angle between the local B field and the line-of-sight, ν is the frequency of radiation, and α is the spectral index and taken as 0.6. The radius is normalized by the radius of the outer shock. The bright shell exists inside of the outer shock in both radio and X-ray. The surface brightness at the outer shock is very weak due to the limb darkening effect. At 200 years, the remnant is within the self-similar stage and shows the strong X-ray emission near the reverse shock and the strong radio emission near the contact discontinuity (mixing region). Radio and X-ray emissions resemble each other closely, which may explain the coincidence of the inner radio shell and X-ray emission in Tycho's

SNR observed by Dickel et al. (1991). At about 500 years, the X-ray surface brightness peaks near the outer shock while the radio surface brightness is stronger in the inner region. The decrease of X-ray surface brightness in the inner region at this time is due to the density decrease at the reverse shock since the remnant is out of the self-similar stage. Thus, we find that a comparison of the radio and X-ray profiles is an indicator of the evolutionary stage of the remnant. Further observational features from our simulations are discussed in detail in Jun and Norman (1995a).

4. Conclusions

Our simulation shows that the R-T and K-H instabilities amplify the existing magnetic fields locally by as much as a factor of 60 due to the stretching and the compression of active fingers. As a consequence, our model produces the thick clumpy radio shell. We find that the evolutions of the surface brightness distributions in the radio and X-ray shell are sensitive to the evolutionary stage of the remnant. Our simulation supports Gull's picture that the dominant radial magnetic field in young SNR can be generated by the R-T instability.

Acknowledgements

The numerical simulations are performed on Cray C90 system in Pittsburgh Supercomputing Center.

References

Chevalier, R.A., Blondin, J.M. & Emmering, R.T: 1992, *Astrophys. J.* **392**, 118
Colgate, S.A. & McKee, C.: 1969, *Astrophys. J.* **157**, 623
Downs, G. S. & Thompson, A. R: 1972, *Astron. J.* **77**, 120
Dickel, J. R., van Breugel, W. J. M. & Strom, R. G: 1991, *Astron. J.* **101**, 2151
Evans, C.R. & Hawley, J.F: 1988, *Astrophys. J.* **332**, 659
Gull, S. F.: 1973, *Mon. Not. R. astr. Soc.* **161**, 47
Jones, E.M., Smith, B.W. & Straka, W.C.: 1981, *Astrophys. J.* **249**, 185
Jun, B., Norman, M.L. & Stone, J.M.: 1995 *Astrophys. J.*, submitted
Jun, B. & Norman, M.L.: 1995a, *Astrophys. J.*, submitted
Jun, B. & Norman, M.L.: 1995b, in preparation
Milne, D. K.: 1987, *Aust. J. Phys.* **40**, 771
Stone, J. M. & Norman, M. L.: 1992a, *Astrophys. J. Suppl* **80**, 753
Stone, J. M. & Norman, M. L.: 1992b, *Astrophys. J. Suppl* **80**, 791

A DISTRIBUTION FUNCTION CALCULATION OF THE Hα PROFILES OF SNR FILAMENTS

A. J. LIM
Department of Mathematics, U.M.I.S.T.
P.O Box 88, Manchester, M60 1QD, U.K.

Abstract. In shocked media of high preshock ionisation, the lack of thermal contact between atoms in the neutral component of the gas can prevent the formation of a thermal equilibrium independent of the ionic component. The behaviour of the neutral gas in such a shock is dominated by the atomic processes driven by the postshock ionic component.

A transport equation for the velocity distribution of the neutral gas is explicitly solved under the physical conditions of a 1D ionic shock transition. The resulting distributions are used to calculate predicted Hα line emission from such a system.

Key words: Supernova Remnants – Shock Waves

1. Introduction

Multifluid simulations of flows in the interstellar medium are often characterised by formation of separate shockfronts in the ionised and neutral components of the gas, of which the ion shock propagates with a higher velocity. Previous numerical studies of multifluid flows (for example, Draine 1980, Tóth 1994) have usually concentrated on the parameter range necessary the formation of the magnetic precursor caused by ion streaming. The shock velocities considered here are of the order of those found in the Balmer dominated filaments of non-radiative shocks occuring in Supernova blast waves (Raymond 1991), and the line profiles calculated are intended to be qualitatively compared with the Hα emission from this type of object.

The level of preshock ionisation encountered by a shock transition plays an important part in determining the shock structure. Upon the arrival of the ion shock, the neutrals are unaffected by bulk processes and are not thermalised, leading to the formation of an 'intershock region' where cold neutrals moving at the preshock speed are in contact with the hot, slow-moving postshock ions. The relative densities of ions and neutrals will therefore have a defining influence upon the immediate postshock region, with energy and momentum being transferred to the cold neutral gas in this region. The interaction between the neutral and ionic components in this region takes place via atomic processes, such as collisional ionisation and resonant charge transfer. If the neutral fraction is low enough, then ion neutral collisions may be the dominant effect upon the state of the neutral gas, and in this case the neutrals will not be able to establish a thermal equilibrium with themselves, in this regime fluidic behaviour cannot be assumed. The modelling of this

Astrophysics and Space Science **233**: 273–280, 1995.

component cannot therefore be accurately performed with the use of the usual hydrodynamic continuity equations (1-3), and the approach used here is to numerically solve a transport equation to obtain the velocity distribution of the neutral gas under the physical conditions imposed by a steady state ionic shock .

2. Hydrodynamic Calculation

The hydrodynamic simulation yields the steady state temperature structure of the ions, under the assumption that the neutral gas has no significant effect upon the behaviour of the ionic shock. A preionisation fraction of 0.99 is assumed. These conditions enable a two-fold method to be used; a steady state solution for the ionic shock structure is found using the Euler equations (1-3), in the usual way.

$$\frac{\partial \rho_i}{\partial t} + \frac{\partial}{\partial x}(\rho_i u_i) = 0, \tag{1}$$

$$\frac{\partial \rho_i u_i}{\partial t} + \frac{\partial}{\partial x}\left(\rho_i {u_i}^2 + P_i\right) = 0, \tag{2}$$

$$\frac{\partial}{\partial t}\left(\frac{\rho_i {u_i}^2}{2} + \frac{P_i}{\gamma - 1}\right) + \frac{\partial}{\partial x}\left(u_i\left(\frac{\rho_i {u_i}^2}{2} + \frac{\gamma P_i}{\gamma - 1}\right)\right) = -G_e, \tag{3}$$

where G_e is the Coulomb energy exchange term with the electron gas, as given in Draine (1980).

The equations are integrated using the Flux-Vector Splitting method due to Van Leer (1972), the calculation is to second order accuracy in space and time. Resolution is increased near the shockfront and near postshock regions with the use of regridding routines based on the gradients of the conserved variables between grid cells.

The electron gas is assumed to be tied to the ions by electrostatic interactions and an adiabatic condition is imposed via an electron entropy continuity equation;

$$\frac{\partial S_e}{\partial t} + v_e \frac{\partial S_e}{\partial x} = \frac{{\rho_e}^{-\frac{5}{3}}}{3}\{2G_e\}, \tag{4}$$

where,

$$S_e = P_e {\rho_e}^{\frac{-5}{3}}. \tag{5}$$

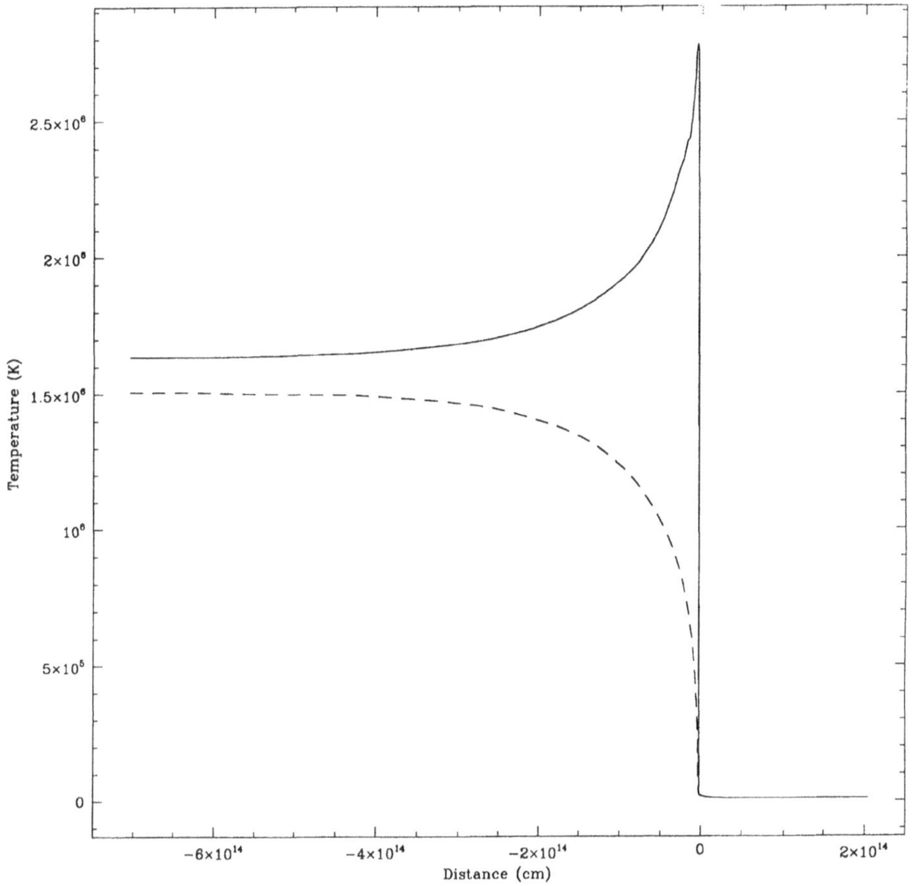

Fig. 1. Temperature structure of a 313 km s^{-1} shock. The solid line represents the ion temperature and the dotted line, the electron temperature.

The shock is produced by placing a reflection boundary condition at the extreme left of the grid upon which the preshock medium, moving right to left, impacts. Equations (1-4) are marched forward until a steady state is reached. The above method is used to calculate ionic shock structures for shockspeeds of 313 km s^{-1}, 625 km s^{-1}, 938 km s^{-1}, 1250 km s^{-1}, 1875 km s^{-1}, and 2500 km s^{-1}, and the temperature structure of the slowest model is shown in Figure 1.

This calculation gives the steady state structure of the ionic component in the approximation of dynamically negligible neutral gas. The temperature and velocity structures of these shocks are then used to solve a transport equation for the neutral velocity distribution function.

3. Solution of the Transport Equation

A neutral distribution is set up in the preshock region to be in equilibrium with the ionic component, which is then followed downstream through the shockfront. The effect of the atomic interactions with the ions in each computational cell is a function of the ion temperature and velocity in that cell. In order to remove any explicit time dependence from the transport equation, the calculation is performed in the frame of reference comoving with the shockfront.

Two atomic process are considered in the current model; these, and their effects upon the distribution profile, are as follows:

i) electronic collisional ionisation; serves to remove mass from the distribution. This process imposes a limit upon the penetration of the neutral gas into the postshock region, and therefore upon the total Hα emission from the shock. (Data from: Cox 1970, and Lotz 1967.)

ii) resonant charge transfer; the exchange of an electron between a neutral and an ion is more probable than ionisation by electron impact and for a preshock neutral in contact with the postshock ions the reaction can take the form;

$$\text{fast neutral} + \text{slow ion} \longrightarrow \text{slow neutral} + \text{fast ion}.$$

The neutral fraction is such that transferred ions are rapidly absorbed by the Maxwellian ionic distribution, but transferred neutrals form a significant part of the total neutral density. These neutrals will have a velocity distribution determined by the ionic post shock conditions, i.e. large thermal speeds and a mean velocity of the order of $0.25v_{shock}$. It is this transferred distribution which forms the broad component of the Hα line profiles calculated here. (Data from: Peek 1966, and McClure 1966.)

Other processes which modify the transport equation include momentum transfer by elastic collisions and, in very strong shocks, ionisation by proton impact; these processes will be included in future models.

An interesting consequence of charge transfer is that a significant fraction of neutrals are transferred to upstream velocities comparable to the preshock flow speed. Some of these particles can overtake the shockfront and deposit energy in the preshock gas. A complete model of this process must involve a time dependent calculation allowing for the effect of this energy upon the

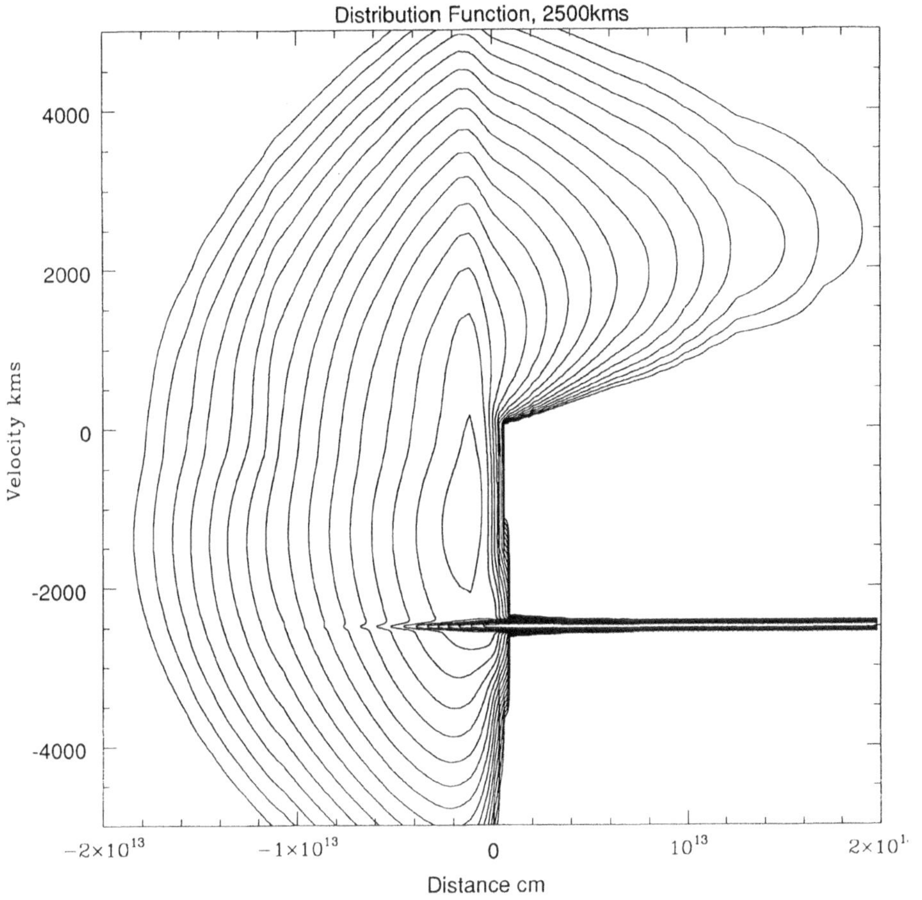

Fig. 2. Spatial dependence of the distribution function for a 2500 km s^{-1} shock.

preshock ions. For this steady state calculation, however, it is sufficient to follow the upstream-moving gas in a separate iteration to the downstream-moving gas. Charge transfer between this broad component precursor and the preshock ions will increase the preshock neutral density, and this is accounted for by performing several downstream-upstream iterations to obtain a steady state. The deposition of energy in the preshock region will, if the assumption of dynamically insignificant neutrals is relaxed, raise the preshock temperature and it has been observed (Hester, Raymond and Blair 1994) that the width of the narrow component of the line profiles of many SNR filaments

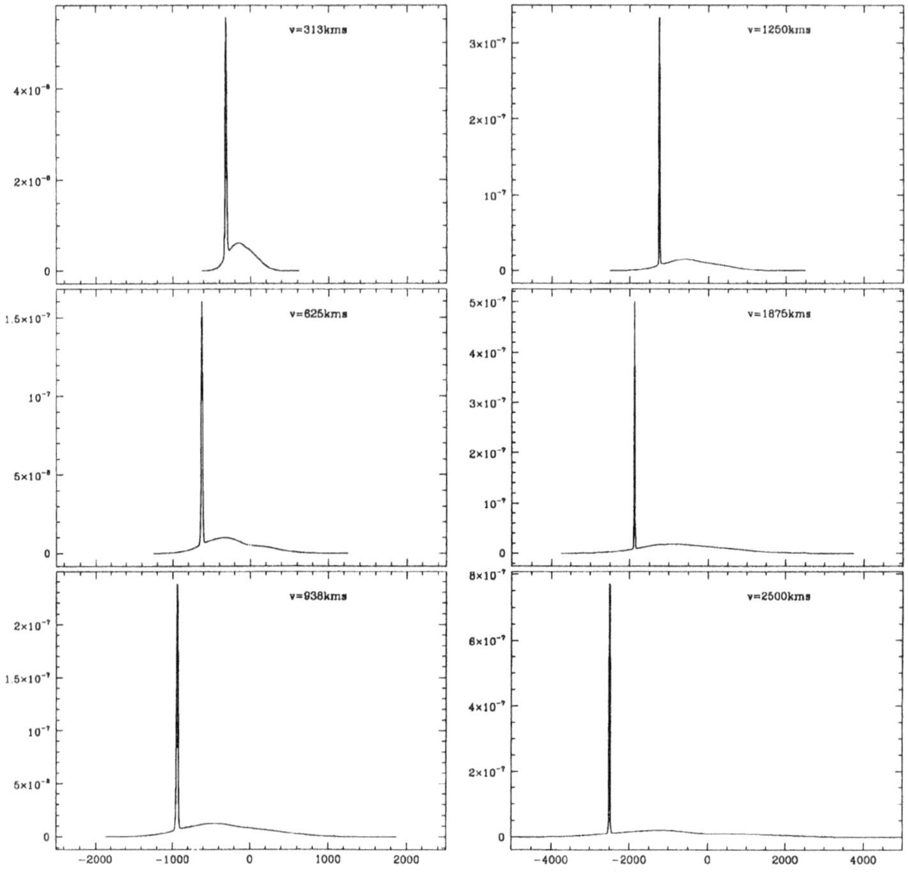

Fig. 3. Integrated line profiles for models of varying shock velocity.

are significantly wider than expected. Detailed time dependent calculations of this effect are in progress, but for the steady-state calculation, a contour plot of the spatial dependence of the distribution function in a 2500 km s^{-1} shock is shown in Figure 2, where the upstream-moving neutrals can be clearly seen.

4. Hα Line Profiles

The Hα profiles calculated from the steady state shock models above are shown in Figure 3, where a clear trend of decreasing broad-to-narrow component ratio, with increasing shock velocity, is seen. This can be explained in terms of the more extreme postshock conditions in the stronger shocks and the competing processes of charge transfer and collisonal ionisation. The ratio of the electron to ion temperatures determines the relative rates of these processes. The electrons undergo an adiabatic compression of a factor ≈4 at the shockfront while the ions are completely thermalised, Coulomb equlibration then takes place between these two fluids. A higher shock speed implies a higher postshock ion temperature suggesting relatively faster charge transfer. However, the ionisation rate is very sensitive to temperature rises above ≈15000K, and this effect, combined with the rapid Coulomb equilibration in the immediate postshock region increases the relative rate of ionisation. Consequently, in stronger shocks, a larger fraction of the neutral gas will be is ionised before transfer can take place and the narrow component of line profile becomes more significant. The cross-section for the resonance resulting in charge transfer also decreases rapidly for very energetic collsions, with the result that the trend is reinforced for very strong shocks.

Current research is concerned with the time dependent calculation of these line profiles which allow the modelling of the broad component neutral precursor. A study of the energy deposition into the preshock gas by these fast neutrals will shed some light upon the feasibility of this mechanism to explain the unexpected width of the observed narrow components. Preliminary results from this type of model, for 50% preionisation, indicate a precursor temperature in the region of 30000K.

Finally, another serious limitation of the steady-state 1-D model is that it only allows the calculation of the line profiles which would be seen by an observer with a line of sight perpendicular to the shockfront. To calculate the profiles which would be seen from any given line of sight angle, it is necessary to consider the velocity structure of the neutrals in the plane of the shock. A two- or three-dimensional transport equation is therefore necessary, even though the model will still only yield results for plane parallel shocks. Preliminary results from such a calculation are currently in preparation for publication.

References

Cox, D. P.: 1970, *PhD Thesis* , University of California, San Diego

Draine, B. T.: 1980, *Astrophys. J.* **241**, 1021

Hester, J. J., Raymond, J. C., Blair, W. P.: 1994, *Astrophys. J.* **420**, 721

Lotz, W.: 1967, *Astrophys. J. Suppl.* **14**, 207

McClure, G. W.: 1966, *Phys. Rev.* **148**, 47

Peek, J. M.: 1966, *Phys. Rev.* **143**, 33

Raymond, J. C.: 1991, *Pub. Ast. Soc. Pac.* **103**, 781

Tóth, G.: 1994, *Astrophys. J.* **425**, 171

SUPERNOVA REMNANTS IN THE STARBURST GALAXY M82

A. PEDLAR & T.W.B. MUXLOW
University of Manchester, NRAL, Jodrell Bank, Nr Macclesfield, Cheshire SK11 9DL, U. K.

Abstract. We have used the enhanced MERLIN at 1.5 and 5 GHz to image the central 700pc of the nearby starburst galaxy M82. Of order 40 discrete sources are detected and it appears that most of these sources are supernova remnants. Not only do many show shell structure and have a non-thermal radio spectrum, but they also follow a surface brightness/diameter relation consistent with that found in the LMC and Galactic supernova remnants. The detected M82 remnants are more compact and brighter than Galactic remnants which implies that they must be less than a few hundred years old and hence supernova rates are of order 0.05 per year. The 1.5 GHz measurements have shown that many of the remnants have low–frequency spectral turnovers which are probably due to free–free absorption in localised ionised gas with emission measures $> 10^6$ pc cm^{-6}.

Key words: Starbursts – Supernova Remnants – Radio Synchrotron

1. Introduction

The strong 60-100μm emission observed from the nuclear regions of many spiral galaxies is generally interpreted as thermal emission from dust grains heated by a population of early-type stars created in a burst of star formation (Rieke et al. 1980). The high star formation rate naturally results in a correspondingly high supernova rate, in some cases estimated to be as high as one supernova per year. Unfortunately the visual extinction to the majority of starburst regions is between 10 and 20 magnitudes and hence most of the supernovae in these regions will not be detected optically.

Although free-free absorption can be present at long (ie $\gtrsim$ 20 cm) wavelengths (see section 3), the radio emission at short cm wavelengths is unaffected by dust or ionised gas and arrives essentially unattenuated from starburst regions. Many starburst galaxies show significant non-thermal radio emission over extended areas which is inferred to be consequence of acceleration of relativistic particles via supernovae. However the detection of discrete, compact radio sources in several starburst nuclei provides direct evidence of supernova activity in these regions.

The best studied starburst galaxy is in the nearby (3.2 Mpc) galaxy M82. MERLIN and VLA observations have revealed ~40 compact objects (Unger et al. 1984, Kronberg, Biermann and Schwab 1985) which were tentatively identified as either radio supernovae or young supernova remnants (SNR). Although the strongest source (41.9+58) has a radio luminosity several hundred times that of the nearby galactic SNR Cassiopeia A, its compact shell-

Astrophysics and Space Science **233**: 281–286, 1995.

like structure (Wilkinson and deBruyn 1990) strongly suggest that it is an unusual SNR about 50 years old. In the decade since their discovery it has become clear that most of the compact objects in M82 do not show the rapid variations expected of radio supernovae (Kronberg and Sramek 1992) and indeed a number have been partially resolved by the 0.17" (2.5 pc) beam of VLA at 15 GHz (Kronberg et al. 1985) consistent with their being supernova remnants. VLBI observations (Bartel al. 1987) resolve out the majority of the compact sources giving lower limits to their sizes. If the compact sources in M82 are young supernova remnants then an intermediate resolution is required to establish their properties. The extended MERLIN at 1.5 and 5 GHz, with angular resolutions of 120 and 50 mas (0.75 pc) and an rms sensitivity of a few tens of μJy is ideally suited to the detailed study of these discrete sources.

2. The Observations and Results

2.1. 5 GHz MERLIN

The MERLIN array has been used to study the nuclear region of M82 at high angular resolution, complementing the well established VLA studies (eg Kronberg et al. 1985). Our study with the enhanced MERLIN at 5 GHz (Muxlow et al. 1994), with an angular resolution of 50 mas, resolved all of the discrete sources. Many showed clear shell structure (e.g. Figure 1) which, combined with their steep spectral index, confirmed them to be supernova remnants. Furthermore, it was found that these sources followed a surface brightness/diameter relation consistant with SNR found in the LMC and Galactic remnants (Figure 2). Approximately 20 of the remnants are brighter and more compact than Cassiopea A suggesting that they are younger than $\sim$ 300 years old. The cumulative number as a function of diameter appears to increase linearly to diameters of at least 3 pc. This is consistent with the SNR being in the free expansion phase, rather than the adiabatic (Sedov) phase, implying ambient densities of < 10 atoms cm^{-3}. The cumulative number-diameter plot indicates a supernova rate of $\sim$ 0.05 per year if the shells are expanding at 5000 km s^{-1}. The flux density of the components decreases approximately as (diameter)$^{-1.2}$, which appears to extrapolate to the flux densities measured for both LMC and Galactic remnants.

2.2. 1.5 GHz MERLIN

This study is an attempt to fully image the whole of the nuclear region of M82 with $\sim$ 0.1 arcsec resolution at L–Band. In addition to the original 1.42 GHz data (April 1993), a second run was obtained (May 1993). Furthermore, in order to enhance the density of the spatial frequency coverage an additional

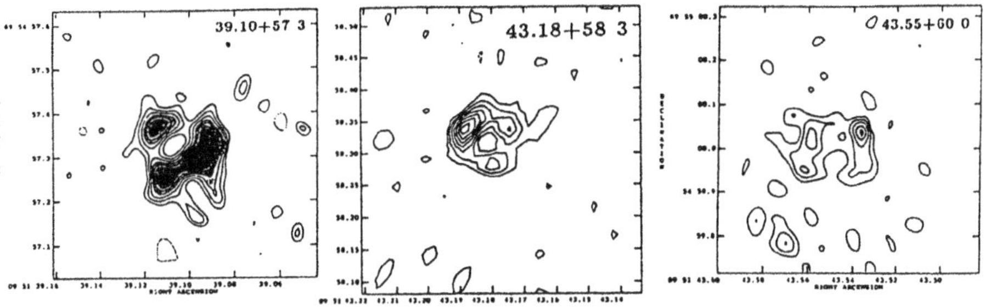

Fig. 1. A selection of the compact sources in M82 which show shell structures. These are MERLIN observations at 5 GHz with an angular resolution of 50 mas. See Muxlow et al. (1994) for details.

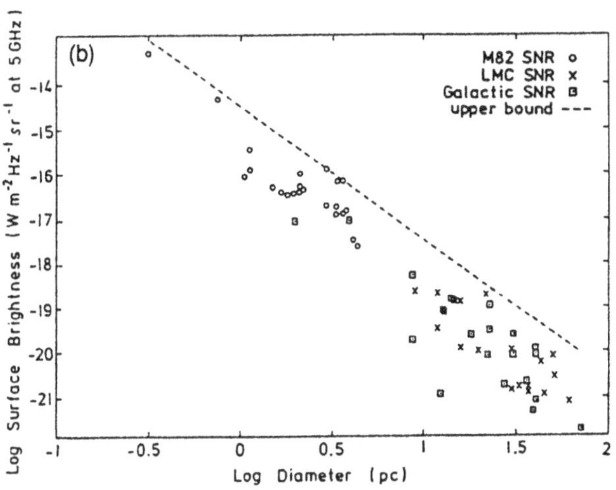

Fig. 2. The surface brightness – diameter relation for the 23 brightest SNR in M82, plotted together with data from the LMC and the Galaxy.

run was obtained during May 1993, cycling every few minutes between 1.612, 1.658, and 1.720 GHz. This technique is known as Multi–Frequency Synthesis (MFS). In total around 50 hours of data were collected. The remaining low spatial–frequency coverage was obtained by adding a 4 hour VLA A–array observation (May 1994).

The analysis of the L–Band data is still continuing; however, a preliminary image at 0.3 arcsec resolution is shown in Figure 3. The dynamic range of this image is close to 2000:1 and residual image errors are close to this level. Although further improvements are expected, the present image is substantially correct. The flux densities and sizes of the more compact SNRs in

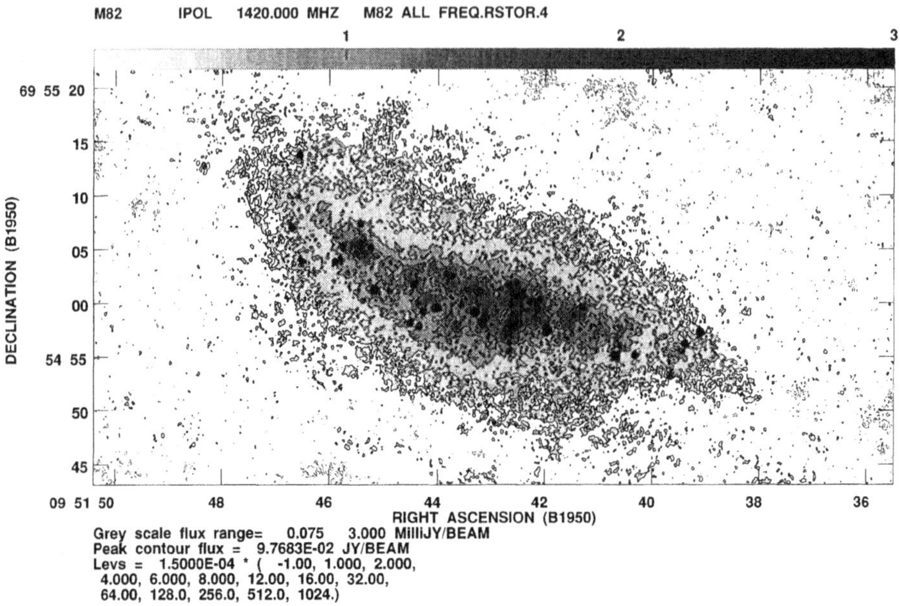

Fig. 3. The combined MERLIN/VLA 1.5GHz image of the central kpc of M82. The resolution is 0.3 arcsec

The analysis of the L–Band data is still continuing, however a preliminary image at 0.3 arcsec resolution is shown in Figure 3. The dynamic range of this image is close to 2000:1 and residual image errors are close to this level. Although further improvements are expected, the present image is substantially correct. The flux densities and sizes of the more compact SNRs in Figure 3 agree well with the single frequency analysis of Sanders (1994), however, for the first time the extended emission at L–Band is fully imaged.

3. Discussion

A comparison of Figure 3 with a VLA 5GHz A-Array image of M82 (data kindly supplied by Kronberg and Sramek) observed in December 1992 with a angular resolution of 0.33 arcsec show both striking similarities and interesting detailed differences. Figure 3 is more sensitive to the extended emission around the inner 1 kpc of the nucleus of M82 than the 5GHz image (1 kpc

= 66 arcsec). However even at higher flux density levels there are significant differences in the extended emission between L–Band and 5GHz which cannot be assigned to imaging errors.

Several SNR show significant spectral flattening, the most extreme case being the remnant 44.01+596 which has strongly rising spectrum between 1.4GHz and 5GHz. By including flux density measurements from VLA 8.4GHz observations (Huang it et al. 1993) it is clear that the spectra for the SNRs are steep between 5GHz and 8.4GHz. Thus we are detecting low–frequency turnovers. Around 40% of the more compact SNR show evidence for such low–frequency turnovers in their spectra. As the 5GHz MERLIN observations resolve all these compact SNRs (see above) the resulting sizes are too large to allow sychrotron self–absorption to explain the spectra. We suggest that free–free absorption in the gas surrounding such SNRs can explain the low–frequency turnovers seen. Using the gradient of the spectrum between 5GHz and 8.4GHz (where we assume there is no absorption) we can extrapolate back to the value of the SNR flux density that we would expect at 1.5GHz. By comparing this with the measured value we can calculate the optical depth and hence the emission measure (E) for a number of SNRs. Assuming an electron temperature of 10^4K we get values of E of around a few x 10^6 pc cm^{-6} which is consistant with the emission measure in giant Galactic HII regions. It is also comparable with the well documented emission measure of the brightest SNR 41.95+575 which shows a frequency turnover below 1GHz (Wilkinson & deBruyn 1984). This strongly suggests that free–free absorption is the cause of the low–frequency turnover in many of these SNR. The random distribution of the remnants exhibiting this behaviour implies that the absorption is being caused by local clumps of gas rather than one single cloud. It may also explain the differences between the 1.4GHz and 5GHz images of the extended structure in the nuclear region. A further point to note is that the central 12 arcsec (180 pc) in general shows an overall steeper spectral index than the surrounding regions. The reason for this is not as yet fully understood, but it is interesting to note that starformation in M82 is interpreted as moving outwards from the centre of the galaxy, and that CO observations by Nakai *et al.* (1986) show a ring of molecular gas with a radius centered on 200pc. This is the region of current starformation and the radio SNRs appear to be situated on its inner rim. The innermost 12 arsec is thus that region lying inside the molecular gas ring.

References

Bartel N., Ratner M.I., Rogers A.E.E., Shapiro I.I., Bonometti R.J., Cohen N.L., Gorenstein M.V., Marcaide J.M., Preston R.A.,: 1987, *ApJ* **323**, 505

Kronberg, P.P., Biermann, P., & Schwab, F.R.,: 1985, *ApJ.* **291**, 693

Kronberg P.P., Sramek R.A.,1992, In MPE Conference 235 'X-Ray emission from Active Galactic Nuclei and the Cosmic X-ray background'

Huang, Z.P., Thuan, T.X., Chavalier, R.A., Condon, J.J., & Yin, Q.F., 1993. ApJ.: submitted.
Muxlow, T.W.B., Pedlar, A., Wilkinson, P.N., Axon, D.J., Sanders, E.M & deBruyn, A.G.: 1994, *MNRAS* **266**, 455
Nakai, N., Hayashi, M., Hasegawa, T., Sofue, Y., Handa, T., & Sasaki, M.: 1986, *IAU Symp.* **115**, 614
Rieke G.H., Lebofsky M.J.,Thompson R.I.,Low F.J., Tokunaga A.T.: 1980, *ApJ* **238**, 24
Sanders, E.M., 1994. *M. Sc. Thesis,* University of Manchester.
Unger S.W.,Pedlar A.,Axon D.J.,Wilkinson P.N., Appleton P.N.: 1984, *MNRAS* **211**, 783
Wilkinson, P.N., & deBruyn, A.G.: 1990, *MNRAS* **242**, 529

SELF-CONSISTENT THEORY OF WHITE DWARF BURNING IN SUPERNOVA IA EVENTS

VITALIY V. BYCHKOV and MICHAEL A. LIBERMAN
Department of Physics, Uppsala University, Box 530, S-751 21 Uppsala, Sweden

Abstract. A self-consistent model of white dwarf burning in the Supernovae Ia events is proposed which includes the consequent stages of the flame, the spontaneous explosion and the detonation. The flame is ignited locally at several points near the center of the star, but soon it is pushed out of the center due to the Rayleigh-Taylor instability. Most of the hot fuel of the central part of the star remains unburnt by the flame. The unburnt fuel explodes spontaneously after the time delay determined by the average temperature near the star center. Until the time of the explosion the flame consumes more than 0.1 of the white dwarf mass out of the star center, which causes pre-expansion of the unburnt fuel in the outer layers of the star. The spontaneous explosion triggers the detonation which incinerates the rest of the star and produces the intermediate mass elements in the pre-expanded layers. The detonation overcomes gravitational binding and causes mass ejection. The proposed theory provides the physical basis for the explanation of the observed spectrum of Supernovae Ia.

Key words: Supernovae: general – Hydrodynamics – Instabilities

1. Introduction

According to the observed spectrum supernova events are divided into two large groups: supernovae of Type I and Type II, and smaller subgroups (Woosley & Weaver, 1986). The supernovae Ia (SNIa) include about 80% of the supernovae of Type I and, besides, it is the subgroup with the most regular properties.

It is a general belief that different supernovae are caused by absolutely different physical mechanisms of the explosions. Particularly, the SNIa are explained by the thermonuclear explosions of carbon-oxygen (CO) white dwarfs accreting mass. The CO white dwarf is a star which has already burnt hydrogen and helium. The gravitational pressure inside the CO white dwarf is balanced by the pressure of degenerate electrons, which is almost independent of the star temperature. If the white dwarf accretes mass, then the temperature at the white dwarf center increases and the carbon reaction may be ignited. However there are some cooling processes preventing from the self-accelerating reaction (thermal run-away), among which convection is the fastest. Convection becomes unimportant when the central temperature of the white dwarf is higher than $6.6 \cdot 10^8 K$ (Woosley & Weaver 1986, Khokhlov 1989). At the moment of the thermal run-away the reaction time at the white dwarf center is about $3 - 4s$, the typical central density is $\rho_c \approx 3 \cdot 10^9 g/cm^3$ and the mass of the white dwarf is close to the Chandrasekhar limit $M_{WD} \approx 1.4 M_\odot \approx 2.8 \cdot 10^{33} g$.

Astrophysics and Space Science **233**: 287–292, 1995.

Observations show that the most abundant nuclei in the SNIa ejecta are the Fe-peak elements. Still the intermediate mass elements like Si are also produced by SNIa in considerable amount, which caused most of the problems in constructing models of white dwarf burning. The thermal run-away in the white dwarf may lead to a flame or to a detonation. Both regimes were employed in construction of the models of white dwarf burning, however both the detonation model and the deflagration (flame) model failed. The model of white dwarf burning in the pure detonation regime (Arnett, 1969) was unable to explain the presence of the intermediate mass elements. The deflagration model reproduced the observed spectrum rather well (Nomoto et al., 1984), but it was unable to explain the mass ejection. Imshennik & Khokhlov (1984) pointed out that the detonation in white dwarfs may also produce sufficient amount of the intermediate mass elements, if a considerable part of the star is consumed by the detonation at low densities $< (3-4)\cdot 10^7 g/cm^3$. Therefore the observed SNIa spectrum may be explained if the white dwarf is pre-expanded by some reason prior to the detonation ignition. The desirable pre-expansion happens if some part of the star is consumed by the flame before the detonation is triggered. That is why the numerical model was developed where the white dwarf burning started as a flame and after a certain time the detonation was artificially "turned on" (Khokhlov 1991a, Arnett & Livne 1994). The best agreement of the numerical model with the observations of SNIa events was achieved if the detonation is "turned on" when the flame has consumed more than 10% of the white dwarf mass. The numerical model of the delayed detonation might be a platform for the theory of SNIa events, but unknown mechanism of the transition from the flame to the detonation remained a "stumbling block" for the theory.

To justify the developed numerical model of white dwarf burning Khokhlov (1991a) assumed that the detonation is triggered by the flame accelerating due to instabilities. This assumption failed because the flame velocity u_f is considerably smaller than the sound speed a_s in white dwarfs (Bychkov & Liberman 1995a, b). The possibility of the detonation ignition on the compression stage of white dwarf pulsations was discussed also by Khokhlov (1991b), but numerical simulations (Arnett & Livne, 1994) show that the pulsations can not trigger the detonation. Besides if the detonation is ignited on the compression stage of the star pulsations, then the detonation consumes the compressed white dwarf at large densities, which contradicts the main idea of the detonation propagating through the pre-expanded star.

In the present paper we consider the flame and detonation ignition, the flame propagation and stability in white dwarfs. We show that the correct physical approach to these problems explains the delayed initiation of the detonation in SNIa events. A self-consistent model of white dwarf burning in the consequent stages of flame–explosion–detonation is proposed, which provides the explanation of the observed SNIa spectrum.

2. The flame–explosion–detonation model

The thermonuclear reaction time depends strongly upon the initial temperature (Fowler et al., 1975)

$$\tau_c = \tau_0 exp(\sqrt[3]{E/T_c}) \tag{1}$$

where τ_0 is a weak function of temperature, the activation energy is equal $E = 84^3 \cdot 10^9 K$ for the carbon reactions. As it follows from Eq.(1) even slight increase of the temperature due to energy release in the reaction shortens the reaction time considerably. That is why the reaction time τ_c is determined at most by the time of the slowest initial induction stage of the reaction, for which the temperature increases by a negligible value $\delta T < 3T_c\sqrt[3]{T_c/E}$ (Zeldovich et al., 1985). For the same reason almost all energy of the reaction is released in a very short time, much shorter than $3\sqrt[3]{T_c/E}\tau_c$. The self-accelerating reaction may be stopped if the induction time is longer than the time of any cooling process. When the average induction time at the star center is shorter than the typical convection time ($\sim 3-4s$), then the self-accelerating reaction may proceed as a spontaneous explosion on the length scales comparable to the white dwarf sizes ($\sim 100 - 600km$).

The thermal run-away may also happen locally in a small volume of hot fuel with temperature higher than the average. The local thermal run-away may ignite the flame even if the length scale of the volume is negligible $\sim \sqrt{\chi\tau_c} < 1m$ compared to the white dwarf sizes (χ is the thermal diffusivity). The real structure of the white dwarf is multi-dimensional because of the preceding convection stage of the star evolution. Therefore the flame starts first at several points near the center of the white dwarf where the local temperature is slightly higher than the average one and the local induction delay is much shorter than the average induction time at the center of the white dwarf. The detonation may be ignited also, but to trigger the detonation the reaction must happen on a large length scale comparable to the white dwarf scales $\sim 100-600km$ (rough dimensional considerations give the length scale $a_s\tau_c \sim 10^4km$, which is even larger than white dwarf sizes). That is why the detonation in white dwarf may be triggered only by the spontaneous explosion, which happens after the time delay equal to the average induction time near the star center. This time is available for the flame to propagate and to consume some part of the star. If the flame is able to burn the whole hot central part of the white dwarf during this time then the spontaneous explosion does not happen and the detonation is not triggered. Thus we come to the question whether the flame can prevent from the spontaneous explosion of the hot fuel at the white dwarf center or not.

Let us consider one of the sources of flame ignition. The flame propagates and forms a bubble of burnt matter. The density of the burnt matter

ρ_b is lower than the density of the surrounding fuel ρ_c ($\Theta = \rho_c/\rho_b > 1$), therefore the position of the bubble near the center of the white dwarf is unstable against the instability similar to the Rayleigh-Taylor one. The typical time scale of the instability may be obtained from the equation of the bubble motion near the white dwarf center (Bychkov & Liberman, 1995 c):

$$\tau_i = \frac{\Theta}{\sqrt{4.19\Theta(\Theta - 1)\Gamma\rho_c}} \tag{2}$$

where Γ is the gravitational constant. In a short time determined by Eq.(2) the bubble of the burnt matter together with the flame front is pushed out of the center to the layers of lower density. The bubble is pushed out faster than the flame propagates: during the time τ_i the flame front may consume the fuel inside the sphere of the radius only about $u_f \tau_i \approx 10km$. The fuel mass inside this sphere is negligible in comparison with the total mass of the central part of the white dwarf, that is why almost all the hot fuel in the central part of the white dwarf remains unburnt by the flame.

The unburnt fuel at the center of the white dwarf is able to explode. Most of the fuel explodes after a time delay equal to the average induction time at the white dwarf center. This time is available for flame to burn a considerable part of the white dwarf fuel in the outer layers of the star. Estimates (Bychkov & Liberman, 1995 c) show that during this time the flame may consume as much as 10% of the white dwarf mass:

$$M_f \approx 4\pi/3N\rho_b(u_f\tau_c)^3 \approx (0.06 - 0.1)NM_{WD} \tag{3}$$

where N is the number of points of flame ignition. It follows from Eq.(3) that even one or two points of flame ignition ($N = 1, 2$) provide the necessary mass burnt by the flame to explain the observed spectrum of the SNIa.

Starting from the spontaneous explosion the following burning of the white dwarf is in the supersonic regime. The spontaneous explosion of the fuel with the initial temperature gradient $T = T(r)$ can be treated as a wave front propagating with the velocity, determined by the gradient (Zeldovich, 1980):

$$u_{exp} = (d\tau_c/dT)^{-1}\frac{\nabla T}{(\nabla T)^2}. \tag{4}$$

At the center of the white dwarf the velocity is infinite which corresponds to the constant volume explosion. The spontaneous explosion may trigger the detonation at the point where the explosion velocity is equal to the detonation velocity $u_{exp} = D$. It was checked numerically for the white dwarf conditions interesting in scope of SNIa theory (Khokhlov, 1989) that the spontaneous explosion does trigger the detonation.

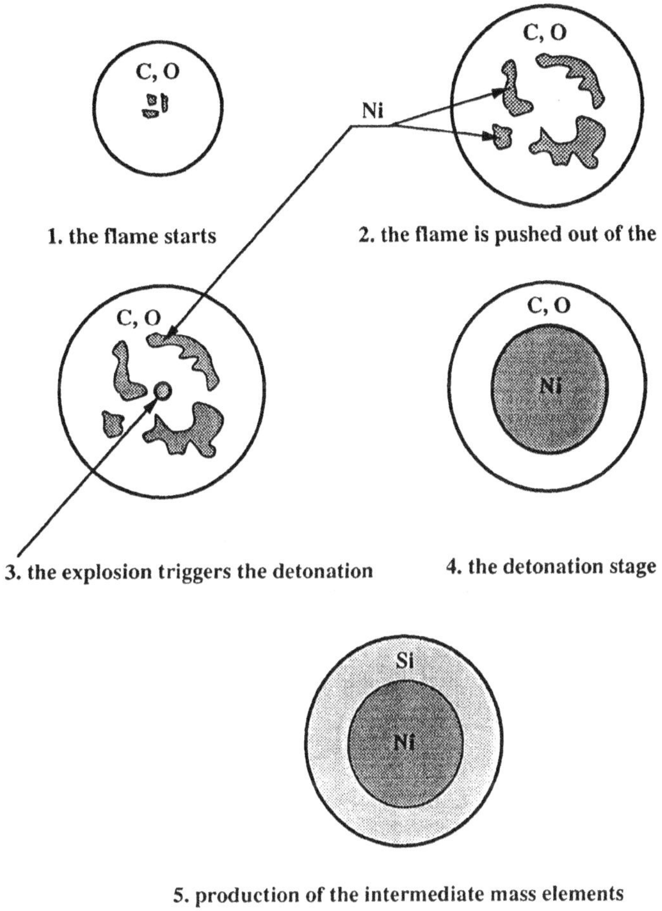

Fig. 1. The schematic diagram of the consequent stages of white dwarf burning.

3. Conclusion

The consequent stages of flame–explosion–detonation according to the physical model of white dwarf burning in SNIa events proposed in this paper are shown on Fig.1. The process of burning is the following:

1. The local thermal run-away ignites the flame at several points near the white dwarf center, the flame propagates and forms bubbles of burnt matter.

2. The bubbles of the burnt matter together with the flame front are pushed out of the white dwarf center because of the instability. The bubbles move out faster than the flame propagates, so that the hot center of the white dwarf remains unburnt.

3. After a time delay equal to the average induction time at the white dwarf center the abrupt explosion happens on a length scales comparable to the size of the white dwarf. Until that time the flame propagating in the outer regions consumes more than 10% of the white dwarf mass, which leads to the pre-expansion of the white dwarf.

4. The explosion propagates like a wave front due to the initial temperature gradient. The detonation is triggered by the spontaneous explosion at the point where the explosion velocity is equal to the detonation velocity. After that time the reaction propagates in the detonation regime.

5. When the detonation propagates through the layers pre-expanded by the flame, it produces the intermediate mass elements. The detonation overcomes the gravitational binding and causes mass ejection.

The proposed model provides the physical explanation of the observed SNIa spectrum.

4. Acknowledgments

This work was supported in part by The Royal Swedish Academy of Science, grant Nr. 1482. One of us (V.B.) is grateful to Alex Raga and Tom Millar for their hospitality at UMIST.

References

Arnett, W.D.: 1969, *Astrophys. Sp. Sci.* **5**, 180

Arnett, W.D. & Livne, E.: 1994, *Astrophys. J.* **427**, 315

Bychkov, V. & Liberman, M.: 1995a, 'Hydrodynamic Instabilities of the Flame Front in White Dwarfs', *Astron. Astrophys.* , accepted

Bychkov, V. & Liberman, M.: 1995b, 'Thermal Instability and Pulsations of the Flame Front in White Dwarfs', *Astrophys. J.* , submitted

Bychkov, V. & Liberman, M.: 1995c, 'On the theory of Type Ia Supernova events', *Astron. Astrophys.* , submitted

Fowler, W., Caughlan, G., Zimmerman, B.: 1975, *ARAstron. Astrophys.* **13**, 69

Imshennik, V.S. & Khokhlov, A.M.: 1984, *Sov.Astron.Lett* **10**, 262

Khokhlov, A.M.: 1989, *Sov.Sci.Rev.* **E8**, 1

Khokhlov, A.M.: 1991a, *Astron. Astrophys.* **245**, 114

Khokhlov, A.M.: 1991b, *Astron. Astrophys.* **245**, L25

Landau, L.D. & Lifshitz, E.M.: 1989, *Hydrodynamics*, Pergamon Press, Oxford

Nomoto, K., Thielemann, F.K., Yokoi, K.: 1984, *Astrophys. J.* **286**, 644

Timmes, F.X. & Woosley, S.E.: 1992, *Astrophys. J.* **396**, 649

Woosley, S.E. & Weaver, T.A.: 1986, *Ann. Rev. Astron. Astrophys.* **24**, 205

Zel'dovich, Ya.B.: 1980, *Comb.Flame* **39**, 211

Zel'dovich, Ya.B., Barenblatt, G.I., Librovich, V.B., Makhviladze, G.M.: 1985, *The Mathematical Theory of Combustion and Explosion*, Consultants Bureau, NY.

SHOCKS AND SHELLS IN HOT STAR WINDS

A. FELDMEIER, J. PULS, C. REILE, A.W.A. PAULDRACH and R.P. KUDRITZKI*
Universitäts-Sternwarte, Scheinerstr. 1, 81679 München, Germany

and

S.P. OWOCKI
Bartol Research Institute, Newark, Delaware 19716, USA

Abstract. Radiation-driven winds of hot, massive stars show *variability* in UV and optical line profiles on time scales of hours to days. *Shock heating* of wind material is indicated by the observed X-ray emission. We present time-dependent hydrodynamical models of these winds, where flow *structures* originate from a strong instability of the radiative driving. Recent calculations (Owocki 1992) of the unstable growth of perturbations were restricted by the assumptions of 1-D spherical symmetry and isothermality of the wind. We drop the latter assumption and include the energy transfer in the wind. This leads to a severe numerical shortcoming, whereby all radiative cooling zones collapse and the shocks become isothermal again. We propose a method to hinder this collapse. Calculations for dense supergiant winds then show: (1) The wind consists of a sequence of narrow and dense shells, which are enclosed by strong reverse shocks (with temperatures of 10^6 to 10^7 K) on their starward facing side. (2) Collisions of shells are frequent up to 6 to 7 stellar radii. (3) Radiative cooling is efficient only up to 4 to 6 R_*. Beyond these radii, cooling zones behind shocks become broad and alter the wind structure drastically: all reverse shocks disappear, leaving regions of *previously* heated gas.

Key words: Stars: early type – Stars: winds – Hydrodynamics – Instabilities

1. Introduction

The driving mechanism of OB star winds is the momentum transfer from the stellar radiation field to heavy metal ions, accomplished in numerous UV line transitions. Building upon the work by Lucy & Solomon (1970), Castor et al. (1975), and Abbott (1980, 1982), current wind models solve the stationary 1-D hydrodynamics (Pauldrach et al. 1986), including a full non-LTE treatment (Pauldrach 1987), the radiative transfer allowing for multiple scattering (Puls 1987), and an empirical shock emission in the EUV (Pauldrach et al. 1994). These models are able to predict the correct mass loss rates and terminal velocities, and to reproduce the complete UV line spectrum (Pauldrach et al. 1994).

A major restriction of these models is their assumption that the wind is *smooth and stationary*, whereas a wealth of observational evidence exists for wind structure and variability (the latter on time scales of hours to days). Among these are: "Discrete absorption components" (DACs) in unsaturated

* affiliated to Max-Planck-Institut für Astrophysik, 85740 Garching, Germany

Astrophysics and Space Science **233**: 293–299, 1995.

P Cygni line profiles (Prinja & Howarth 1986; Henrichs 1988; Kaper 1993; Prinja & Fullerton 1994); broad and black absorption troughs (Lucy 1982; Puls et al. 1993; Prinja & Howarth 1986) and variability of the blue absorption edge (Henrichs 1991) in saturated UV P Cygni profiles. The X-ray emission from OB stars is discussed, e.g., by Hillier et al. (1993) and Cassinelli et al. (1994).

In the following we describe some recent efforts (Feldmeier 1993, 1995) in numerically modeling the non-linear evolution of initially small wind perturbations, which grow to mean flow scales due to the strong radiation-hydrodynamical instability of the wind (Lucy & Solomon 1970; Carlberg 1980; Abbott 1980; Lucy 1984; Owocki & Rybicki 1984, 1985). The correspondence of such wind structures to observed variability features is discussed in Puls et al. (1993, 1995). Our main interest here is the inclusion of the energy transfer in the wind, (1) to calculate the X-ray emission, and (2) to *test* the assumption of isothermality which underlies all supergiant wind models up to now (Owocki et al. 1988; Owocki 1991, 1995). Technical details of our approach are described in the latter papers and in Feldmeier (1993, 1995).

2. The wind model

Our present simulations are performed for a typical O supergiant with stellar parameters close to the ones of ζ Pup (O4If): $M = 42 M_\odot$, $R_* = 19 R_\odot$, $T_{\rm eff} = 42,000$ K, $N_{\rm He}/N_{\rm H} = 0.16$ (Kudritzki et al. 1983). The wind parameters are $v_\infty = 2000$ km/s and $\dot{M} = 3 \cdot 10^{-6}\ M_\odot$/yr. The source function controlling the diffuse radiation force is assumed to be the one for purely geometric dilution of the photospheric radiation field, i.e., for an optically thin wind (Owocki 1991). For the radiative cooling function, we use the simple power law form $\Lambda = A N_{\rm e} N_{\rm H} T^\alpha$, with $\alpha = -1/2$ and $A = 1.64 \cdot 10^{-19}$ erg cm^3 K$^{\frac{1}{2}}$ s^{-1}. This is a reasonable approximation to the Raymond et al. (1976) cooling function from 10^5 to some 10^7 K. Finally, 3000 logarithmically spaced grid points from the photosphere up to $10 R_*$ are used. The perturbation of the wind is due to a sound wave originating in the photosphere, with period 5000 s and amplitude 1%.

Wind model with collapsed cooling zones. Fig. 1 shows a snapshot of the wind structure 10 days after model start. The initially stationary wind (dashed line) is transformed into a sequence of very narrow and dense shells. The shells are enclosed by a strong reverse shock on their starward facing side, and by a compression wave or a forward shock on their outer side. (Which of both alternatives is actually realized cannot be decided conclusively from the still rather poor grid resolution). From about 4 to $5 R_*$ on, the reverse shocks become weaker with growing distance. The (main) reason is that the instability vanishes with vanishing velocity gradient. Beyond about $5 R_*$ the

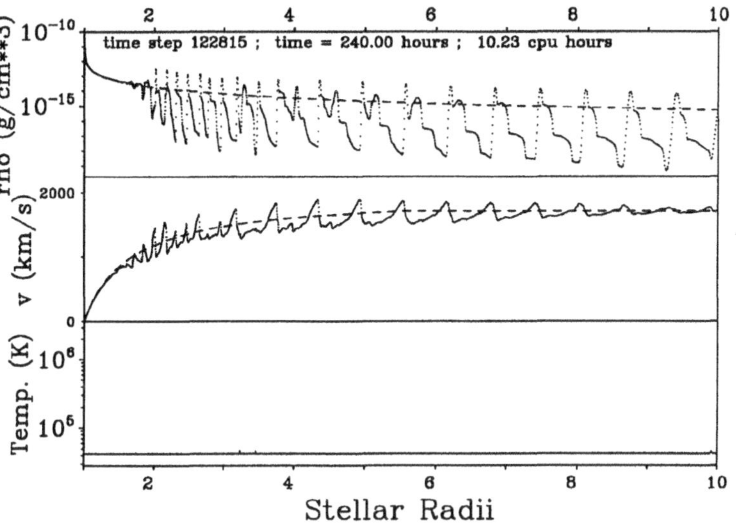

Fig. 1. Snapshot of the wind 10 days after model start. The values of ρ, v, and T at each grid point are plotted. Dashed line: stationary start model.

wind structure is periodic, whereas at lower radii it appears to be chaotic. However, the latter behavior results only from the creation of 3 or 4 shells per cycle of the sound wave, where these shells collide and merge with each other. To demonstrate the periodicity of the wind at all radii, Fig. 2 shows a grey-scale rendition of the temporal power spectrum of the velocity, plotted vs. frequency for the fixed locations $r = n\,R_*, n = 1, \ldots, 10$. In addition to the frequency f of the sound wave, the first 40 to 50 harmonic overtones of f appear at $3\,R_*$. They are excited by the nonlinear steepening of the sound wave due to the wind instability. The overtones die out with growing distance from the star. Besides these overtones, no other signal is evident in Fig. 2.

This behaviour is different in models with a large amplitude of the sound wave, and also in models without any explicit perturbation of the wind, i.e., with self-excited structure. In both of these cases, continuous intervals of frequencies are excited (the reasons for this differing behaviour are still under debate, cf. Feldmeier 1995; Owocki & Feldmeier, in preparation), which *may* indicate *turbulent* behaviour of the wind (Burgers 1950; Tatsumi & Tokunaga 1974). Observational evidence for compressible turbulence in Wolf-Rayet star winds is found from the spectral size distribution of wind blobs (Moffat 1994).

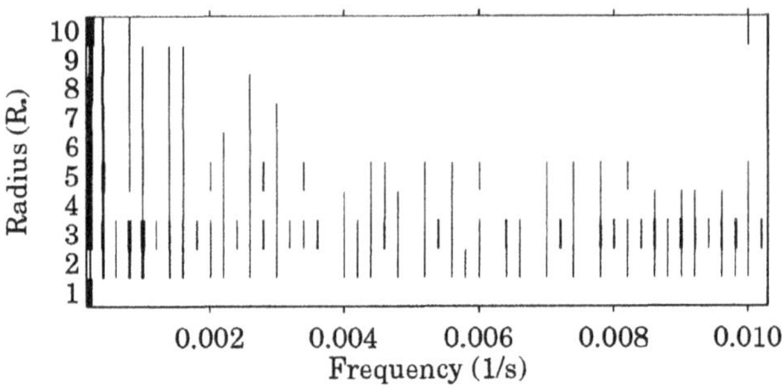

Fig. 2. Grey scale rendition of the temporal power spectrum of the velocity $v(t)$, sampled over 48 cycles of the sound wave, at several fixed locations in the wind. In addition to the frequency of the sound wave (broad line at the left edge of the figure), its first ≈ 40 to 50 harmonic overtones are excited.

Finally, from the velocity jumps in Fig. 1 one expects post-shock temperatures of order 10^6 K. *Contrary to this, no hot material is seen in Fig. 1.* Let us now consider the reason for this discrepancy.

The collapse of radiative cooling zones. Due to the global thermal instability (Langer et al. 1981; Chevalier & Imamura 1982; Falle, this proceedings), the radiative cooling zones behind strong shocks contract and expand periodically. We claim, that *during the contractional phase the cooling zone may fall below the grid resolution*, whereby the numerical scheme "forgets" about the existence of the cooling zone, and an *isothermal shock* is left. The reversible oscillation of the cooling zone has turned into an irreversible collapse. This is demonstrated in Fig. 3 for a test calculation with gas streaming supersonically against a wall (located at $x = 0$). The left panel shows a calculation on a grid fine enough to resolve the thermal oscillation, and the middle panel the same calculation on a coarse grid, where the cooling zone collapses at the wall.

To hinder this collapse, we propose to modify the cooling function at low temperatures. More specifically, we chose a stable cooling exponent $\alpha = +2$ below a temperature T_{swi}. The stabilizing influence of the low temperature regime can then hinder the collapse altogether. This is seen in the right panel of Fig. 3 for a calculation on the same coarse grid as in the mid panel. Since (1) only the *hot* gas emits X-rays, and (2) again the *hot* gas, by its very inefficient radiative cooling, fixes almost alone the total extent of the cooling layer, this modification of Λ at *low* temperatures should have an only minor influence on the interesting wind physics.

There is a second, purely numerical cause for the collapse of cooling zones, termed "advective diffusion". For a discussion of this defect (which is also hin-

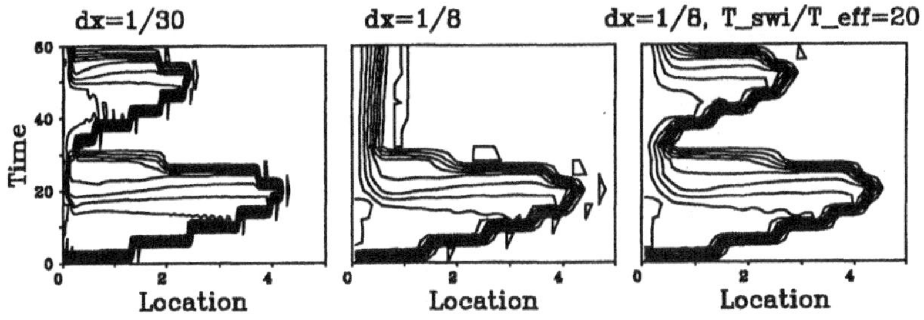

Fig. 3. Global thermal instability oscillation (in terms of iso-velocity contours) on a fine grid (resolved; left panel), a coarse grid (collapsed; mid panel), and on the coarse grid using our modified cooling function (right panel, see text).

dered by the above modification of the cooling function), we refer to Cooper (1994) and Feldmeier (1995).

Wind model with resolved cooling zones. Fig. 4 shows a wind snapshot, where now the cooling function is modified in the described manner below $T_{swi} = 10\ T_{eff}$. Hot gas of order 10^6 K is present, and cooling zones behind strong shocks are resolved. Up to about $4\,R_*$, the structure is very similar to the one displayed in Fig. 1: the cooling zones are short compared to shell-shell distances, *confirming the assumption of dynamical isothermality in this region.*

However, beyond $4\,R_*$ this model shows striking differences to Fig. 1. Within the sequence marked with (a) to (d) in Fig. 4, *all reverse shocks are destroyed*: The intershell material (being fed into the shells) is depleted, which leads to a broadening of the cooling layers. Since the material at the cold edge of the cooling zone is very dense and therefore very inert, this broadening occurs by driving the reverse shock into the preshock gas (i.e. to the left). Hereby, the shock encounters material of always lower velocity, and the shock jump becomes smaller until it finally vanishes. Thus, the hot material at large radii ($> 5\,R_*$) is the left-over from *previous* shock heating at smaller radii. This gas cools away steadily due to radiative energy losses and spherical expansion on its further advection through the wind.

3. Conclusions

The observational consequences of these results for the X-ray emission from hot stars have now to be analyzed. This work is presently under way in our group. Since only the hard X-ray component ($E \gtrsim 1$ keV) of ζ Pup stems from radii smaller than $10\,R_*$ (Hillier et al. 1993), we can conclude temptatively

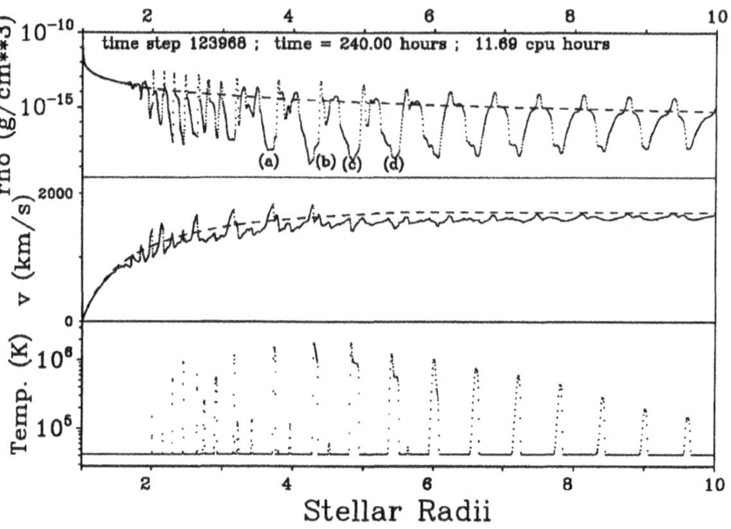

Fig. 4. Snapshot of the wind using a modified radiative cooling function below $T_{\rm swi}/T_{\rm eff} = 10$.

that only this component should (partially) originate from radiative shocks, whereas the soft X-rays far out in the wind should originate from "old" hot gas. However, it is not clear at present if the latter material is not too rarefied in our models to produce an observable X-ray flux. Finally, the maximum temperatures of order 10^6 K in Fig. 4 are too small to account for the ROSAT observations of ζ Pup, which instead indicate temperatures ranging up to 3 to $4 \cdot 10^6$ K. However, hotter material is more readily produced in models with larger photospheric wind perturbations. The principal shock destruction occurs in the same way as described above, but is shifted now to somewhat larger radii ($\approx 6\,R_*$). Interestingly, the volume filling factor of hot material at large radii can reach $\approx 60\%$ then (Feldmeier 1995). Assuming a density depletion factor of 100 of this hot material compared with stationary wind densities, this results in a filling factor (relative to the stationary wind) of $\approx 0.6\%$, not too different from the few % derived by Hillier et al. (1993).

Acknowledgements

We thank A. Fullerton and S. Haser for interesting discussions. A.F. and A.P. were supported by the DFG in the "Gerhard-Hess-Programm" under contracts Pa 477/1-1 and 1-2. C.R. was supported by the BMFT under contract 05-2MU114(7). S.P.O. acknowledges partial support from NSF grant AST 91-15136 and from NASA grant NAGW-2624. The calculations were performed on a Cray Y-MP at the Leibniz-Rechenzentrum in München.

References

Abbott, D.C.: 1980, *Astrophys. J.*, **242**, 1183.
Abbott, D.C.: 1982, *Astrophys. J.*, **259**, 282.
Burgers, J.M.: 1950, *Proc. Acad. Sci. Amst.*, **53**, 247, 393, 718, 732.
Carlberg, R.G.: 1980, *Astrophys. J.*, **241**, 1131.
Cassinelli, J.P., Cohen, D.H., MacFarlane, J.J., Sanders, W.T., Welsh, B.Y.: 1994, *Astrophys. J.*, **421**, 705.
Castor, J.I., Abbott, D.C., Klein, R.I.: 1975, *Astrophys. J.*, **195**, 157.
Chevalier, R.A., Imamura, J.N.: 1982, *Astrophys. J.*, **261**, 543.
Cooper, R. G. 1994, PhD thesis, University of Delaware.
Feldmeier, A.: 1993, PhD thesis, Universität München.
Feldmeier, A.: 1995, *Astron. Astrophys.*, in press.
Henrichs, H.F.: 1988, in Conti, P.S., Underhill A.B., ed(s)., *O Stars and Wolf-Rayet Stars*, NASA SP-497: Washington, D.C., 199.
Henrichs, H.F.: 1991, in Baade, D., ed., *Rapid Variability of OB-Stars: Nature and Diagnostic Value*, ESO Conf. Proc. No. 36: Garching, 199.
Hillier, D.J., et al.: 1993, *Astron. Astrophys.*, **276**, 117.
Kaper, L.: 1993, PhD thesis, Amsterdam.
Kudritzki, R.P., Simon, K.P., Hamann, W.R.: 1983, *Astron. Astrophys.*, **118**, 245.
Langer, S.H., Chanmugam, G., Shaviv, G.: 1981, *Astrophys. J.*, **245**, L23.
Lucy, L.B., 1982: *Astrophys. J.*, **255**, 278.
Lucy, L.B., 1984: *Astrophys. J.*, **284**, 351.
Lucy, L.B., Solomon, P.M.: 1970, *Astrophys. J.*, **159**, 879.
Moffat, A.F.J.: 1994, in Klare, G., ed., *Reviews in Modern Astronomy 7*, Astronomische Gesellschaft: Hamburg, 51.
Owocki, S.P.: 1991, in Crivellari, L., Hubeny, I., Hummer, D.G., ed(s)., *Stellar Atmospheres: Beyond Classical Models*, Kluwer: Dordrecht, 235.
Owocki, S.P.: 1992, in Heber, U., Jeffery, S., ed(s)., *Atmospheres of Early-Type Stars*, Springer: Heidelberg, 393.
Owocki, S.P.: 1995, in Moffat, A.F.J., et al., ed(s)., *Instability and Variability of Hot Star Winds*, Kluwer: Dordrecht, in press.
Owocki, S.P., Rybicki, G.B.: 1984, *Astrophys. J.*, **284**, 337.
Owocki, S.P., Rybicki, G.B.: 1985, *Astrophys. J.*, **299**, 265.
Owocki, S.P., Castor, J.I., Rybicki, G.B.: 1988, *Astrophys. J.*, **335**, 914.
Pauldrach, A.: 1987, *Astron. Astrophys.*, **183**, 295.
Pauldrach, A., Puls, J., Kudritzki, R.P.: 1986, *Astron. Astrophys.*, **164**, 86.
Pauldrach, A.W.A., Kudritzki, R.P., Puls, J., Butler, K., Hunsinger, J.: 1994, *Astron. Astrophys.*, **283**, 525.
Prinja, R.K., Fullerton, A.W.: 1994, *Astrophys. J.*, **426**, 345.
Prinja, R.K., Howarth, I.D.: 1986, *Astrophys. J. Suppl.*, **61**, 357.
Puls, J.: 1987, *Astron. Astrophys.*, **184**, 227.
Puls, J., Owocki, S.P., Fullerton, A.W.: 1993, *Astron. Astrophys.*, **279**, 457.
Puls, J., Feldmeier, A., Springmann, U., Owocki, S.P., Fullerton, A.W.: 1995, in Moffat, A.F.J., et al., ed(s)., *Instability and Variability of Hot Star Winds*, Kluwer: Dordrecht, in press.
Raymond, J.C., Cox, D.P., Smith, B.W.: 1976, *Astrophys. J.*, **204**, 290.
Tatsumi, T., Tokunaga, H.: 1974, *J. Fluid Mech.*, **65**, 581.

SIMULATIONS OF THE WARDLE INSTABILITY OF C-TYPE SHOCK WAVES

G. TÓTH*
Dept. of Astrophysical Sciences, Princeton University, Princeton, USA
and
Dept. of Atomic Physics, Eötvös University, Budapest, Hungary

Abstract. C-type shocks in the partially ionized ISM are modelled by numerical simulations. Under certain conditions the shocks are subject to the Wardle instability, which initially makes the shock front rippled, then in the non-linear stage can produce density variations in both the ion and neutral fluids. A systematic search in the numerically accessible parameter space is done to determine the wave vector $\mathbf{k}_{max}$ and the growth rate s_{max} of the fastest growing modes. The neutral Alfvén number, and the angle θ_s between the shock normal and the upstream magnetic field determine the strength and obliqueness of the shock, as well as the dimensionless parameters of the fastest growing mode. The results confirm and extend Wardle's linear analysis.

The non-linear evolution shows saturation of the instability and the formation of high density regions that detach from the shock front with the downstream flow. Numerical difficulties are partially solved by an implicit treatment of the ion-neutral friction terms, but strong shocks still can not be modelled efficiently. A fully implicit method for the ions and the magnetic field is used to model C-type shocks with low fractional ionization and high ion Alfvén speed.

Key words: ISM – Shock Waves – Instabilities – MHD – Methods: numerical

1. Introduction

C-type shock waves occur in a partially ionized medium with magnetic field present, when the ion Alfvén speed $v_A^{(i)} \equiv B/(4\pi\rho^{(i)})^{1/2}$ exceeds the shock velocity v_s, while the neutrals flow at a supersonic speed throughout the shock. $\rho^{(i)}$ and $\rho^{(n)}$ are the densities of the ion and neutral fluids, respectively, B is the magnetic field. The density, velocity, and temperature of both fluids change in a smooth fashion, but the overall jump conditions are the same as for the discontinuous ideal magnetohydrodynamical (MHD) shock wave of a single fluid. The slip between the ion and neutral velocities is also referred to as ambipolar diffusion. Supernova explosions, strong stellar winds, and collisions between molecular clouds can all drive C-type shocks into the ISM at typical speeds of $v_s \approx 10\,\mathrm{km\,s^{-1}}$. The smoothness of the shock plays an important role in various processes, including the chemical reactions, charged particle acceleration, and emission and absorption line features (see references

* Present address: Astronomical Institute, Utrecht University, Postbus 80000, 3508 TA, Utrecht, The Netherlands

Astrophysics and Space Science **233**: 301–305, 1995.

in Draine and McKee 1993, and in Hartquist, Flower and Pineau des Forêts 1990).

Under certain circumstances C-type shocks are subject to the Wardle instability (Wardle 1990, 1991a,b), which is a rippling perturbation along the shock-front with an exponentially growing amplitude. The mechanism is the following: if the magnetic field lines bend slightly, the ions are pushed along the field lines into the "troughs" by the ion-neutral drag force, which increases the local ion density. The higher ion density increases the drag force in the troughs and the field lines are bent further. Linear analysis of this instability by Wardle showed that the critical parameter is the "neutral Alfvén number" $A^{(n)} \equiv v_s/v_A^{(n)}$ where $v_A^{(n)} \equiv B/(4\pi\rho^{(n)})^{1/2}$. Only very weak shocks with $A^{(n)} \lesssim 5$ show stable behavior; stronger shocks can produce several e-foldings of the initial perturbation during one dynamical time, $t_{\rm flow} \sim 10^3\,{\rm yr}$. The fastest growing mode has a wave length $\lambda_{\rm max} \sim L_{\rm flow} \sim 0.01\,{\rm pc}$, the width of the shock. The numerical values, roughly appropriate for a $10\,{\rm km\,s^{-1}}$ shock propagating into a diffuse molecular cloud, are quoted only to give the reader a sense of scales.

Many problems, available only by means of fully dynamical numerical modelling, are left open. Most importantly, the non-linear effects might suppress the instability at a low amplitude, thus leaving the steady state models essentially identical to the dynamical solution. The linear regime of the growth of the perturbation may hold surprises as well. Wardle neglected ion pressure and inertia in his calculations, which is perfectly valid for low fractional ionization $x_e \equiv n^{(i)}/n_H \lesssim 10^{-6}$, typical in dense molecular clouds, with hydrogen number density $n_H \gtrsim 10^4\,{\rm cm^{-3}}$ and magnetic field $B \gtrsim 10^2\,\mu{\rm G}$. In diffuse clouds with $x_e \approx 10^{-4}$, $n_H \approx 10^2\,{\rm cm^{-3}}$ and $B \approx 10\,\mu{\rm G}$, however, the ion Alfvén speed $v_A^{(i)} \approx 40\,{\rm km\,s^{-1}}$ is comparable to the shock speed $v_s \sim 10\,{\rm km\,s^{-1}}$, thus the thermal and ram pressure of ions will affect the dynamics to some extent. In perpendicular shocks $\mathbf{k}_{\rm max}$ is always parallel to $\mathbf{B}$, but oblique shocks can have $\mathbf{k}_{\rm max}$ off the plane of $\mathbf{B}$ and the shock normal, and this possibility was not treated by the linear analysis. Finally, the intermediate magnetosonic waves of the ion fluid may play an interesting role in a special family of oblique shocks, which were left out from Wardle's investigations.

For a detailed study on this subject the reader is referred to the full paper by Tóth (1995).

2. Numerical methods

In a previous paper (Tóth 1994) I described in detail the numerical methods for modelling C-type shocks. The code is a slightly modified version of the explicitly 2 dimensional Eulerian difference scheme, the Flux Corrected

Transport method (Book, Boris and Zalesak 1988). Following DeVore (1991) the magnetic field is included according to the Constrained Transport method (Evans and Hawley 1988), which keeps $\nabla \cdot B = 0$ to the accuracy of round off errors. The two fluids interact via the drag force and energy transfer, which appear as source terms in the momentum and energy equations. A semi-implicit treatment of the source terms (Tóth 1995) is necessary for the efficient simulation of the non-linear stage. When the Alfvén speed of the ions is very high and thus the Courant condition becomes very restrictive for the explicit or semi-implicit FCT scheme, a fully implicit treatment of all the equations for the ions and the magnetic field is used.

For studying the Wardle instability the initial conditions are taken to be high accuracy numerical solutions of the steady state equations plus a small sine wave perturbation, parallel to the shock front, in the transverse velocity components. The boundary conditions are continuous at the upstream and downstream boundaries, and periodic in the other direction.

3. Results

A systematic search is done for determining the fastest growing modes of the Wardle instability in C-type shocks of various strength and obliqueness. The numerical simulations of perpendicular shocks fully confirm the results of Wardle's linear analysis. Even for relatively high fractional ionization, where the ion pressure and inertia are not completely negligible, there is a good agreement between the analytical and numerical growth rates and wave numbers for the fastest growing modes. In the case of oblique shocks the $k_{||}$ component parallel to the symmetry plane of the magnetic field and the shock normal is found to be the most important parameter, while the ϕ_k angle between k and the plane plays a secondary role. The growth rates $s_{\mathbf{max}}$ and $k_{||\mathbf{max}}$ of the fastest growing modes are found to be in good agreement with Wardle's analytical results which were restricted to modes with $k = k_{||}$ and $\phi_k = 0$. A simulation with an "intermediate shock" in the ion fluid did not reveal any special behavior, with the possible exception of a somewhat smaller growth rate and wave number than expected for a shock with the same strength and obliqueness but a sub-intermediate shock speed.

Simulation of a weak perpendicular shock shows saturation of the instability in the non-linear regime. A new quasi-stationary state is reached with the exception of periodic formation of high and low density regions behind the shock front (see Figure 1). This result has been confirmed by different numerical methods, and by varying the grid resolution.

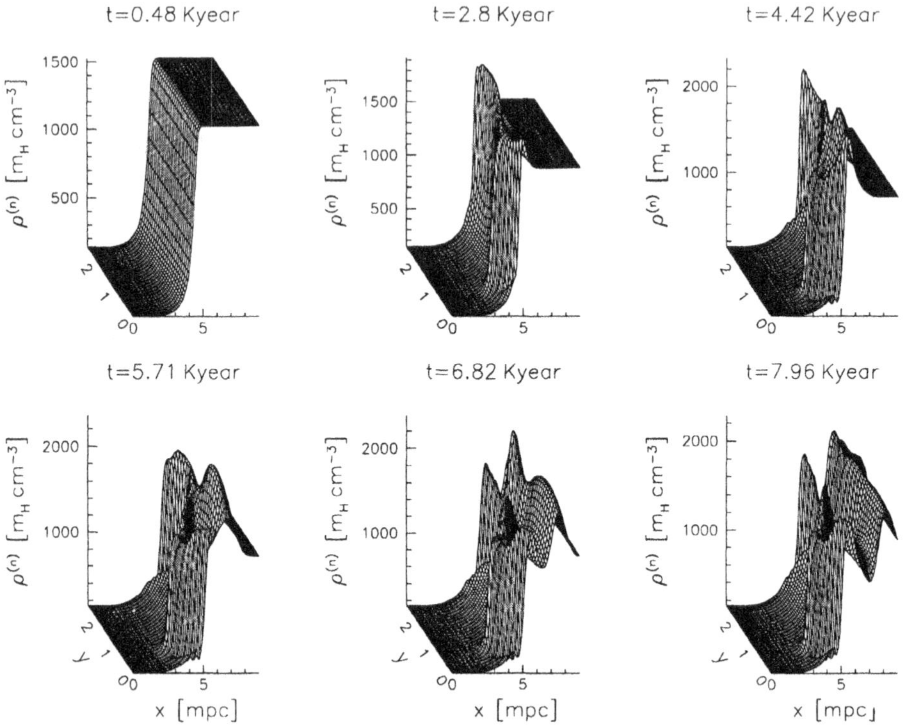

Fig. 1. The evolution of the neutral density in a weak ($A^{(n)} = 8$) perpendicular shock. Note how the first density "clump" detaches while a new one is forming behind the shock.

Acknowledgements

This research was supported in part by NSF grant AST-9017082 and OTKA, the Hungarian Science Foundation, grant F-4491.

References

Book, D.L., Boris, J.P. and Zalesak, S.T.: 1981, 'Finite-Difference Techniques for Vectorized Fluid Dynamics Calculations' in D.L. Book, ed(s)., *Springer-Verlag*, New York, p.29

DeVore, R.C.: 1991, *J. Comput. Phys.* **92**, 142

Draine, B.T. and McKee, C.F.: 1993, *Ann. Rev. Astr. Ap.* **31**, 373

Evans, C.R. and Hawley, J.F.: 1988, *Astrophys. J.* **332**, 659

Hartquist, T.W., Flower, D.R. and Pineau des Forêts, G.: 1990, 'Molecular Astrophysics – A Volume Honouring Alexander Dalgarno' in T.W. Hartquist, ed(s)., *Cambridge University Press*, Cambridge, p.99

Tóth, G.: 1994, *Astrophys. J.* **425**, 171

Tóth, G.: 1995, *Mon. Not. R. astr. Soc.* **274**, 1002

Wardle, M.: 1990, *Mon. Not. R. astr. Soc.* **246**, 98

Wardle, M.: 1991a, *Mon. Not. R. astr. Soc.* **250**, 523

Wardle, M.: 1991b, *Mon. Not. R. astr. Soc.* **251**, 119

A NUMERICAL METHOD FOR PARTICLE ACCELERATION IN SHOCKS

R. BANDIERA and M. SALVATI
Osservatorio Astrofisico di Arcetri
Largo E.Fermi 5, I-50125 Firenze, Italy

Abstract. We have developed a reasonably fast and robust code to solve the standard Diffusion-Convection Equation, in plane parallel geometry, with synchrotron losses included. Standard iterative algorithms are rather inefficient for the solution of this equation: 1. they are time consuming, and convergence is not guaranteed in all cases; 2. the distribution is not adequately determined at levels much lower than its maximum value. We devised a completely different kind of approach for solving the finite difference equation. It does not require iterations, but is based on the solution of a linear system of equations. The number of operations required is of order of the cube of the grid size. We present the description of the numerical code, and some examples with different velocity profiles and levels of synchrotron losses.

Key words: Acceleration Mechanisms – Shock Waves – Synchrotron Losses

1. Introduction

The acceleration of cosmic rays in a converging flow is described by the Diffusion-Convection Equation (DCE). The steady-state equation, in plane parallel geometry and with synchrotron losses included, reads:

$$u\frac{\partial f}{\partial x} - \frac{1}{3}\frac{\partial u}{\partial x}p\frac{\partial f}{\partial p} = \frac{\partial}{\partial x}\left(\kappa\frac{\partial f}{\partial x}\right) + \frac{1}{p^2}\frac{\partial}{\partial p}\left(sp^4 f\right),$$

where the distribution function $f(x,p)$ is defined by $dN = 4\pi p^2 f(x,p)dp$. Moreover let the diffusion coefficient be $\kappa(x)$, and the spatial distribution of synchrotron losses $s(x)$. We found it convenient to define new variables, as follows:

$$y \text{ so that } \frac{\partial}{\partial y} = \kappa\frac{\partial}{\partial x}; \qquad q = \ln(p); \qquad w = p^4 f; \qquad h = \kappa s.$$

The DCE then changes into:

$$u\frac{\partial w}{\partial y} - \frac{\partial}{\partial y}\frac{\partial w}{\partial y} + \frac{4}{3}\frac{\partial w}{\partial y} - \frac{1}{3}\frac{\partial u}{\partial y}\frac{\partial w}{\partial q} - h\exp(q)\frac{\partial w}{\partial q} = 0.$$

The boundary conditions at $y_{\rm min}$ (upstream) and $y_{\rm max}$ (downstream) are derived from the analytic solution, in the case of constant velocity and no losses:

$$\begin{aligned} w(y,q) &= w_-(q) + w_\circ(q)\exp(u_- y), \text{ near } y = y_{\rm min}; \\ w(y,q) &= w_+(q), \text{ near } y = y_{\rm max}; \end{aligned}$$

Astrophysics and Space Science **233**: 307–310, 1995.

w_- is the distribution at $y = -\infty$, while w_+ is that at $y = +\infty$; $w_\circ$ is determined by the DCE.

As a reference, in the case of a shock with compression ratio $r = u_-/u_+$, without losses, the high energy tail of w_+ will be a power law of momentum, with index $(r-4)/(r-1)$.

2. The Basics of this Numerical Method

We found that generic iterative algorithms are rather inefficient for the solution of this equation: 1. they are time consuming, and convergence is not guaranteed in all cases; 2. the distribution is not adequately determined at levels much lower than its maximum value.

We then devised a completely different kind of approach for solving the finite difference equation. It does not require iterations, but is based on the solution of a linear system of equations. If the numerical grid has N_y points along y and N_q along q, the system has a size N_q, and therefore $\sim N_q^3$ operations are required. Beyond that, $\sim N_y N_q^2$ operations are needed for obtaining the coefficients of the linear system. The goal is, given an input distribution of particles at $y = -\infty$ (w_-), to derive that at $y = +\infty$ (w_+), as well as the distribution at all intermediate y.

At first, let us reverse the problem, and seek that w_- which evolves into a given, arbitrary w_+. At $y = y_{\rm max}$:

$$w(y_{\rm max}, q) = w_+(q); \qquad \left.\frac{\partial w(y,q)}{\partial y}\right|_{y_{\rm max}} = 0.$$

A finite difference form of the DCE is used to evolve w upstream from the above initial conditions. At $y = y_{\rm min}$:

$$w_-(q) = w(y_{\rm min}, q) - w_\circ(q)\exp(u\, y_{\rm min}); \; w_\circ(q)\exp(u\, y_{\rm min}) = \left.\frac{\partial w(y,q)}{\partial y}\right|_{y_{\rm min}},$$

and w_- can be deduced. If this is performed for a complete set of functions $w_+[i]$, and then we find which linear combination of the corresponding $w_-[i]$ ($\sum c[i]\, w_-[i]$) gives the true w_-, the same combination ($\sum c[i]\, w_+[i]$) is the true w_+. In our code, the functions $w_+[i]$ are equal to unity at the i-th q, and vanish elsewhere.

Finally, by evolving upstream the true w_+, we can evaluate the distribution $w(y,q)$ in all points of the grid (this requires only $\sim N_y N_q$ operations).

3. Details of the Numerical Method

Let us consider a grid equally spaced both in y (stepsize Δy) and in q (stepsize Δq); let j and l be the respective counters along the two axes.

The derivatives along y are evaluated using standard second-order finite differences formulae. In order to quench instability, an upwind differencing scheme is required for the derivative along q. We use a second-order scheme in order to reproduce well the energy profile with a reasonably low N_q. The code allows also a variable stepsize in y, so that the steps could be adapted to the velocity profile. The memory required is $\simeq 5N_q^2$, and is almost independent of the value of N_y. In fact it is not needed to keep the distribution function along the whole grid in memory during the computations.

4. Numerical Results

Here we present, for the sake of illustration, a few numerical results. We take the case of a compression ratio $r = 4$, which for a sharp shock gives a flat w_+ distribution of the accelerated particles. We instead use a ramp with half-width Δy_w. Seed particles come from $y = -\infty$, with a gaussian-like profile centered on $p = 1$ (the scaling of p is arbitrary). The distribution function is represented in two ways: a) by a contour plot on the plane $\log_{10}(p) - y$; b) by a series of spectral profiles at various values of y. The plots span 15 decades in p and more than 10 orders of magnitude in the level of the distribution. Synchrotron losses are present only in the last model, where they have been confined between $y = -1$ and $y = 1$.

The figures show that a smoother velocity change is less efficient in accelerating particles, and generates a steeper profile. The main effect of synchrotron losses is to produce a cutoff in p, while at lower momenta the distribution closely resembles that without losses. A nice feature of the code is that it gives solutions which are smooth over a large dynamic range.

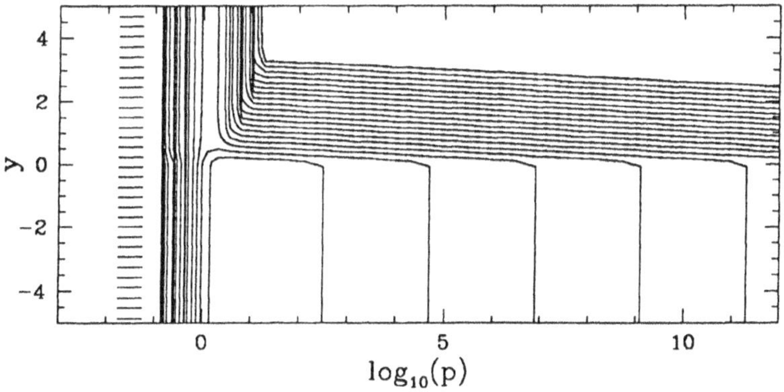

Fig. 1. Model with $\Delta y_w = 0.1$, without losses. Contour plot. The horizontal ticks on the left side indicate the values of y for which profiles are drawn in the following figures.

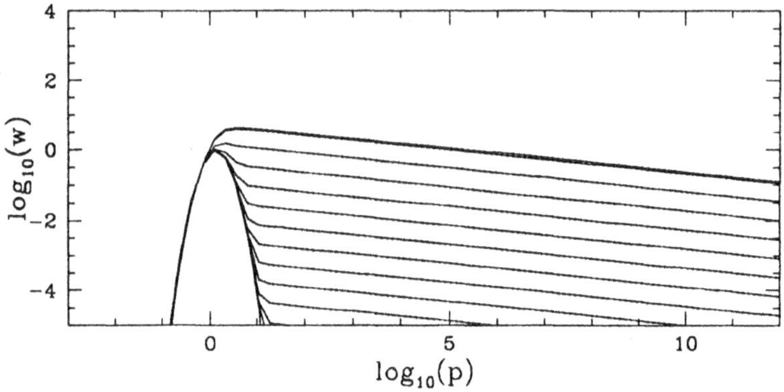

Fig. 2. Same model as in Fig. 1. Spectral profiles.

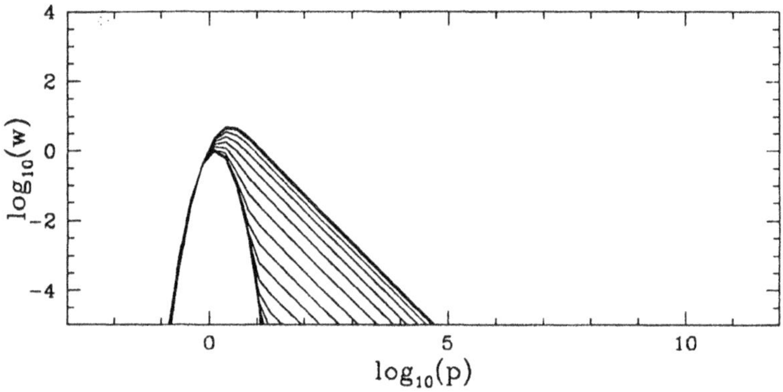

Fig. 3. Model with $\Delta y_w = 1$, without losses. Spectral profiles.

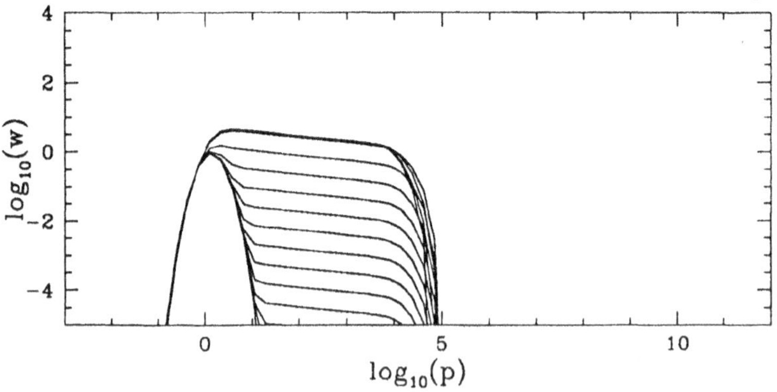

Fig. 4. Model with $\Delta y_w = 0.1$, and synchrotron losses. Spectral profiles. This figure should be compared with Fig. 2.

THE DEFLECTION OF JETS BY CLOUDS

STEVE HIGGINS,* TIM O'BRIEN and JAMES DUNLOP
Astrophysics Group, Liverpool John Moores University,
Byrom St., Liverpool, L3 3AF, UK

Abstract. The role of collisions between extragalactic jets and dense clouds in determining the appearance of high-redshift radio galaxies is discussed and investigated through numerical hydrodynamic simulations in three dimensions. The code has the facility to track jet material separately from ambient material. This allows us to use simplifying assumptions to calculate synthetic radio images. The results indicate that the most powerful radio sources are likely to be observed during or shortly after an interaction, and that such interactions can explain both the radio structures and the spatial association between optical and radio light found in powerful radio galaxies. In some cases such a scenario may provide an alternative explanation of jet properties to mechanisms based on variations in the source or fluid-dynamical instabilities.

Key words: Extragalactic Jets – Shock Waves

1. Introduction

Many examples of complex or disrupted structure in extragalactic jets have been attributed to the effects of an inhomogeneous environment: for instance small scale structure in high-power sources (e.g., Lonsdale & Barthel 1986; Barthel & Miley 1987), wide- and narrow-angle tail sources (Burns 1986) and compact steep spectrum sources (Fanti et al. 1986). Since then the discovery that the optical and radio structures of many high-redshift radio galaxies are closely aligned (McCarthy et al. 1987; Chambers, Miley & van Breugel 1987), along with of a number of important correlations, have provided further evidence to support this suggestion (Dunlop & Peacock 1993; McCarthy, van Breugel & Kapahi 1991; Liu & Pooley 1991).

2. Numerical Simulations

Previous three-dimensional simulations used schemes with only 1st-order accuracy (De Young 1991) or concentrated only on the impact region (Norman 1992). We have advanced these simulations by using a second-order accurate scheme, on a grid covering a larger region, and by tracking jet material and estimating synchrotron emission, which allows us to plot crude synthetic radio images.

* This author is supported by a PPARC research studentship

Astrophysics and Space Science **233**: 311–315, 1995.

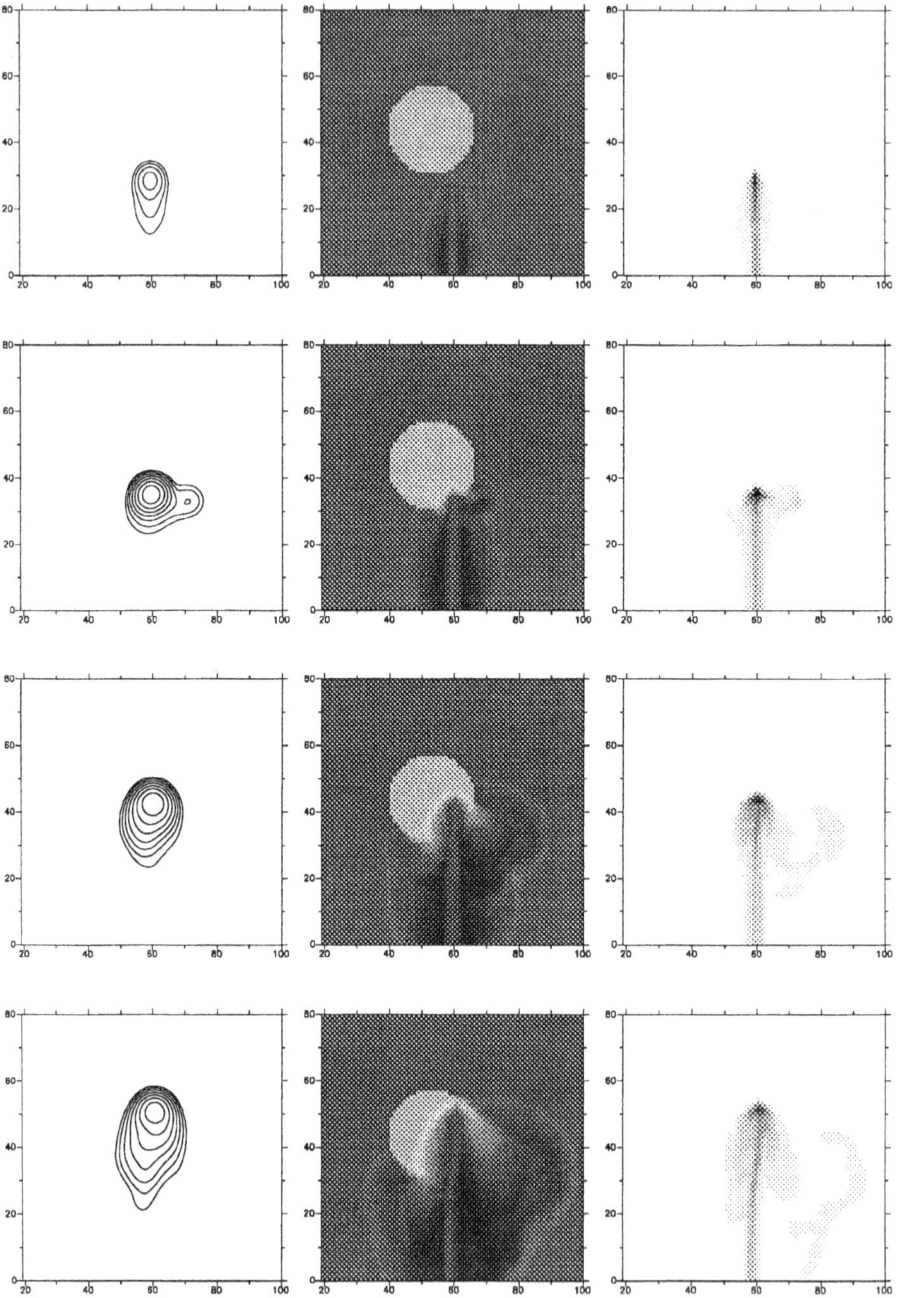
0
20
40
60
80
100

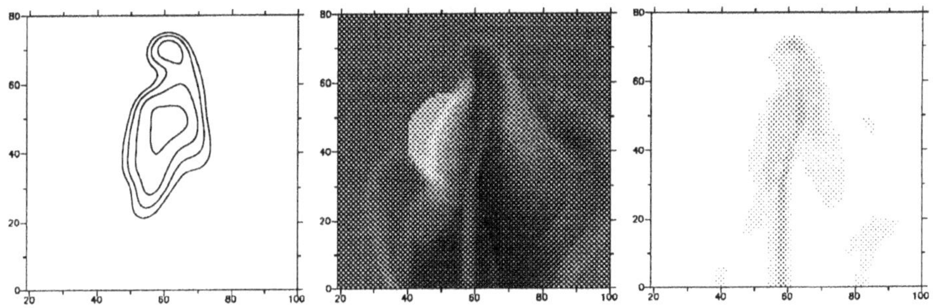

Fig. 1. Five epochs of a jet-cloud collision. Synthetic radio maps are shown to the left, logarithmic total gas density in the middle, and jet gas density on the right.

The simulations are carried out with a hydrodynamics code based on the second-order Godunov method of Falle (1991). We tested the program using Sod's shock tube problem (Sod 1978) and Nittman's shock-discontinuity problem (Nittman, Falle & Gaskell 1982). To follow the interaction for a significant time the computational grid ($50 \times 100 \times 100$ cells) had to be mapped onto a large region. To check for effects due to the low resolution, we repeated the simulation in a smaller region around the point of impact. We found no significant differences between the simulations. Example results are shown in the accompanying figure.

We make two basic assumptions to estimate the synchrotron emission: that the magnetic field is frozen into the jet plasma, so that the field energy density is related directly to the density of the jet material; and that the energy density of particles accelerated to relativistic energies in shocks is simply proportional to the thermal energy density in the gas. We then calculate the intensity of synchrotron emission from a combination of the density and pressure in the jet material. We discuss this scheme in more detail in a paper in preparation.

The figure shows a sequence of synthetic radio map contour plots, after convolution with a gaussian representing the observing beam, together with corresponding greyscale plots of logarithmic gas density and density of jet material in the symmetry plane of the interaction. The results are shown before, and at four different epochs during the collision of a jet with a gas cloud ($t = 0.08, 0.16, 0.32, 0.48$ and 0.64 million years). In this case the jet has a velocity of 0.1c, and a radius of 0.05 kpc. It is in pressure balance with the ambient medium and has the same density. The cloud is also in pressure balance, and has a density of 100 times the ambient density, and a radius of 0.7 kpc.

3. Implications

This simple off-axis collision with a single cloud can produce a complex radio structure which evolves through the simulation. As a result of the interaction, the radio source becomes much more luminous. There are three main structural features (a hotspot at the deflection point, a hotspot at the head of the jet and a secondary hotspot) but the relative brightness of these features changes rapidly with time. After collision, the gas cloud becomes elongated and closely aligned to the radio structure.

The increased radio luminosity persists for a considerable time after the collision. This leads us to suggest that samples of ultra-luminous radio sources (particularly when detected at low frequency) could be biased towards sources which have recently been involved in a jet-cloud interaction.

These simulations also predict that the jets in powerful/steep-spectrum radio sources should be bordered by the elongated remnants of disrupted gas clouds. If such fragments are the sites of line emission and scattering of continuum from a hidden quasar, then this would appear to provide a natural explanation for the radio-optical alignment effect.

The variation of the apparent spatial correspondence between optical and radio structures with the epoch of observation can explain features observed in various sources, as discussed in Dunlop (1995). Close spatial correspondence between the radio and optical structures, such as is observed in 4C41.17 (Miley et al. 1992) can be produced shortly after the collision. In the fifth epoch of the figure there are multiple hotspots within 1 kpc of the aligned optical light, as in 4C41.17 (Carilli, Owen & Harris 1993). At later times the spatial correspondence is less exact, but there is still a clear alignment (within $\simeq 20°$) as observed in 3C324 (Rigler et al. 1992; Dickinson 1995). Alignments of this accuracy are in fact typical of what is seen in most powerful high-redshift radio galaxies.

References

Barthel, P.D. & Miley, G.K.: 1987, *Nature* **333**, 319
Carilli, C.L., Owen, F. & Harris, D.E.: 1993, *Astron. J.* **107**, 480
Chambers, K.C., Miley, G.K. & van Breugel, W.J.M.: 1987, *Nature* **329**, 624
De Young, D.S.: 1991, *Astrophys. J.* **371**, 69
Dickinson, M.: 1995, in *'Galaxies in the Young Universe'*, Ringberg Workshop, in press.
Dunlop, J.S. & Peacock, J.A.: 1993, *Mon. Not. R. astr. Soc.* **263**, 936
Dunlop, J.S.: 1995, in *'Galaxies in the Young Universe'*, Ringberg Workshop, in press.
Eales, S.A.: 1992, *Astrophys. J.* **397**, 49
Falle, S.A.E.G.: 1991, *Mon. Not. R. astr. Soc.* **250**, 581
Hill, G.J. & Lilly, S.J.: 1990, *Astrophys. J.* **367**, 1
Liu, R. & Pooley, G.: 1991, *Mon. Not. R. astr. Soc.* **249**, 343
Lonsdale, C.J. & Barthel, P.D.: 1986, *Astrophys. J.* **303**, 617
McCarthy, P.J. et al.: 1987, *Astrophys. J.* **321**, L29
McCarthy, P.J., van Breugel, W. & Kapahi, V.K.: 1991, *Astrophys. J.* **371**, 478

Miley, G.K. et al.: 1992, *Astrophys. J.* **401**, L69
Nittman, J., Falle, S.A.E.G. & Gaskell, P.H.: 1982, *Mon. Not. R. astr. Soc.* **201**, 833
Rigler, M.A. et al.: 1992, *Astrophys. J.* **385**, 61
Sod, G. A.: 1978, *J.Comp.Phys.* **27**, 1

INTERACTING WINDS IN CLASSICAL NOVA OUTBURSTS

H. M. LLOYD,* T. J. O'BRIEN and M. F. BODE
Astrophysics Group, Liverpool John Moores University,
Byrom St., Liverpool L3 3AF, UK

Abstract. Classical nova outbursts occur in binary systems containing a white dwarf accretor and a Roche-lobe-filling main-sequence star. The outburst is due to a thermonuclear runaway in the accreted material on the surface of the white dwarf, and results in the ejection of up to $10^{-4} M_{\odot}$ of material at velocities of several hundred to a few thousand kilometres per second. There is now strong evidence that the mass ejection takes place via a wind with secularly increasing velocity. The fast ejecta catches up with slower moving material ejected earlier in the outburst, forming a layer of shock-heated gas which gives rise to a short burst of soft X-ray emission. This emission was observed in V838 Her (Nova Herculis 1991), and was succesfully accounted for by the 'interacting winds' model. In this paper, we present 2.5-D numerical hydrodynamics calculations of interacting winds in novae which consider the effects of the binary system on shaping the mass-loss, and show that many of the features seen in the optical shells of novae many years after outburst can be accounted for.

Key words: Novae – Numerical Hydrodynamics

1. Background

Several years after outburst, extended optical emission can be seen around many novae. This usually takes the form of a shell, although in most cases this is far from spherically symmetric. Common features include equatorial rings, polar caps and at least in the case of DQ Herculis, pronounced tropical rings above and below the equatorial band (see e.g. Slavin, O'Brien & Dunlop 1995). An attractive possibility for explaining the departure from spherical symmetry is the effect of the binary companion. The photospheric radius of the nova ejecta is greater than the binary separation for several weeks or months after the outburst and there is therefore a lengthy common-envelope phase in which the ejected matter will interact with the secondary star (e.g. Livio et al. 1990). The secondary star experiences a gravitational drag force in the nova envelope, and transfers orbital energy and angular momentum to the ejecta, causing the material to be ejected in a highly anisotropic flow. We have used a '2.5-D' hydrodynamics code to investigate the effects of orbital energy and angular momentum deposition on nova mass-loss.

* This author is supported by a PPARC research assistantship

Astrophysics and Space Science **233**: 317–321, 1995.

		Slow nova	Moderately-fast nova
Mass-loss parameters	$\dot{M}_0$	$2.5 \times 10^{-4} M_\odot \mathrm{yr}^{-1}$	$2.5 \times 10^{-4} M_\odot \mathrm{yr}^{-1}$
	$\dot{M}_1$	$3 \times 10^{-5} M_\odot \mathrm{yr}^{-1}$	$3 \times 10^{-5} M_\odot \mathrm{yr}^{-1}$
	v_0	100 km s^{-1}	300 km s^{-1}
	v_1	1000 km s^{-1}	2000 km s^{1}
Binary parameters	M_1	0.6 $M_\odot$	1.0 $M_\odot$
	M_2	0.33 $M_\odot$	0.5 $M_\odot$
	P	3.0hr	4.5hr

TABLE I

2. The Model

The code, and the source terms due to the drag force on the secondary, are described by Lloyd, O'Brien & Kahn (1995). The mass ejection is assumed to take place via a wind. The mass ejection in novae can be well modelled by an optically thick, transonic wind (e.g. Bath & Harkness 1989). The wind velocity does not remain constant, however. There is strong evidence to point to the fact that the ejection velocity in novae increases with time (see O'Brien, Lloyd & Bode 1994). In the model, we assume that the mass-loss rate and velocity in the wind take initial values $\dot{M}_0, v_0$ for some time before varying linearly with time to reach final values $\dot{M}_1, v_1$.

Two runs of the code were carried out, with parameters appropriate for a moderately-fast nova and a slow nova respectively (see Table 1). Due to computational constraints, it is impossible to evolve the models over the timescales required for moderately-fast and slow novae (tens to hundreds of days) and the slow mass-loss and subsequent acceleration are carried out over a timescale of order one day.

3. Results

Figures 1(a) and 1(b) show contours of the logarithm of density in the meridional plane for the moderately-fast nova at two evolutionary times. Figure 1(a) shows the situation at an early time. Acceleration of the wind is yet to commence and the slow wind has clearly been shaped into a 'double cone' structure by its interaction with the secondary. Material which flows into the interaction region is heated by drag on the secondary and subsequently drives a shock wave into the quiescent wind, giving rise to the double cone structure. Figure 1(b) shows the solution after the acceleration of the wind is complete. The fast wind has now caught up with the slow wind, sweeping

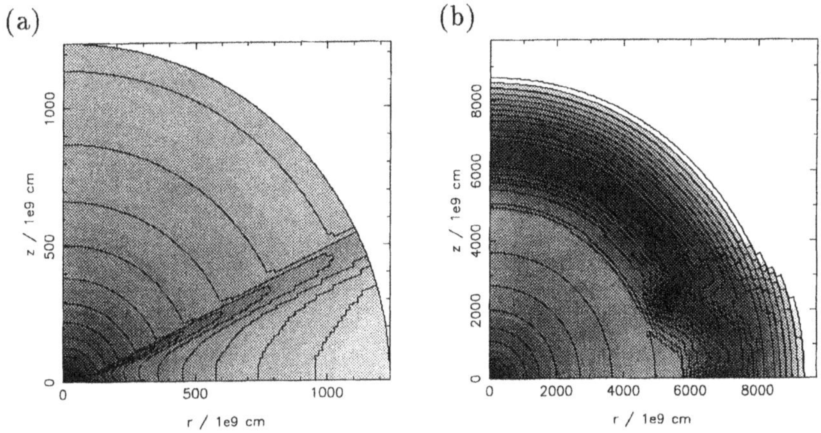

Fig. 1. Logarithmic contours and greyscale of density for the moderately-fast nova model at two evolutionary times.

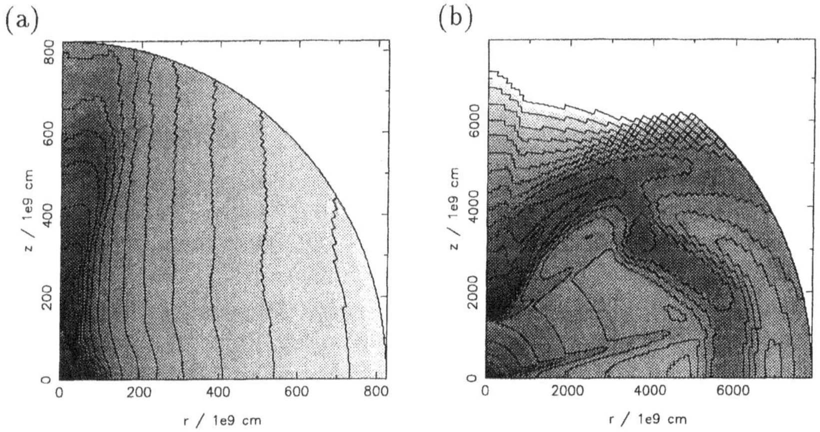

Fig. 2. Logarithmic contours and greyscale of density for the slow nova model at two evolutionary times.

it up into a shell. The density in the shell is enhanced at the points where it has swept up the double cone, giving rise to two ring structures above and below the equatorial plane.

Figures 2(a) and 2(b) show the results for the slow nova calculation. Figure 2(a) shows the situation prior to acceleration of the wind. Due to the lower wind velocity, the drag luminosity is increased, and the ratio of drag luminosity to mechanical luminosity of the wind is higher by a factor of ~ 100 than in the moderately-fast case. The material heated by the wind now expands as far as the symmetry axis, where a dense pillar of shocked

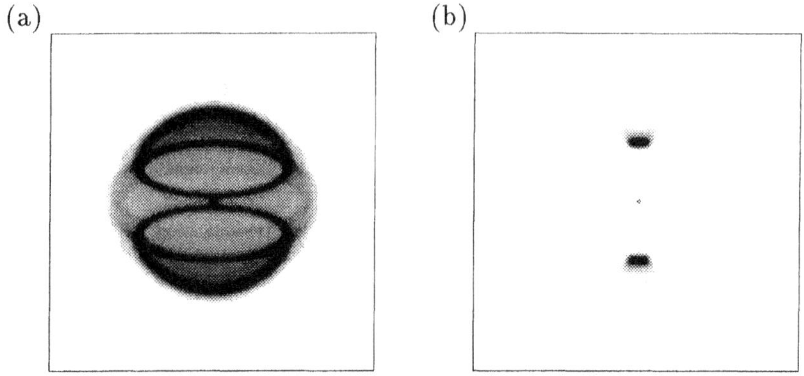

Fig. 3. Synthetic images calculated from the results of the hydrodynamic simulations. The images are formed by integrating the square of the gas density along several lines of sight through the hydrodynamic grid

gas is formed. Figure 2(b) shows the solution after the acceleration of the wind is complete. In this case, sweeping up the slow wind produces a pair of dense polar blobs.

Figure 3 shows synthetic images derived from the hydrodynamic solutions shown in Figures 1(b) and 2(b). The tropical rings are clearly seen in the image of the moderately-fast nova, along with a weaker equatorial ring. The whole is bounded by a thin, approximately spherical shell, which is cusped at the tropics with a bulge at the equator. These features are all present in the remnant of DQ Her, although the observed shell is markedly prolate. The slow nova image is dominated by two polar blobs. Polar blobs are a common feature in the optical remnants of slow novae, such as HR Del and RR Pic.

It is clear that many of the features present in nova shells can be formed in the hydrodynamic interaction of the binary companion with a wind of increasing velocity, and that the types of features seen correlate with the speed class (via the relationship between wind speed and drag luminosity) in the observed manner. However, there remain some problems in fully reconciling these results with the observations. In particular, the shells of novae are typically prolate whereas the mechanism outlined here appears to produce spherical, or oblate, shells. In the moderately-fast nova calculation, the equatorial ring is weaker than the tropical rings whereas this tendency is reversed in DQ Her. It is likely that both these problems would be resolved if the density in the wind from the nova outburst was equatorially enhanced. It is known that the white dwarf in the DQ Her binary is a fast rotator, which may provide the required equatorial enhancement in the mass-loss. Equatorial enhancement of the slow wind could also produce an equatorial ring in addition to the polar blobs in the slow nova, a feature which is observed in, for example, RR Pic (and perhaps HR Del, Slavin, O'Brien & Dunlop 1995). Future calculations will attempt to quantify this effect.

References

Bath, G.T. and Harkness, R.P.: 1989, '' in Classical Novae, ed(s)., *Bode, M. F. and Evans, A.*, Wiley: Chichester, p. 61

Livio, M., Shankar, A., Burkert, A. and Truran, J.W.: 1990, *Astrophys. J.* **356**, 250

Lloyd, H. M., O'Brien, T. J. and Kahn, F. D.: 1995, *Mon. Not. R. astr. Soc.* **273**, L19

O'Brien, T. J., Lloyd, H. M. and Bode, M. F.: 1994, *Mon. Not. R. astr. Soc.* **271**, 155

Slavin, A. J., O'Brien, T. J. and Dunlop, J. S.: 1995, *Mon. Not. R. astr. Soc.* , in press

THE LOCATION AND X-RAY EMISSION OF SHOCKS AROUND ISOLATED BE STARS

JOHN M. PORTER
Astrophysics Group, Chemical and Physical Sciences,
Liverpool John Moores University, Byrom St.,
Liverpool L3 3AF

Abstract. All hot stars are observed to have X-ray emission: O stars have $L_X/L_{\rm bol} \sim 10^{-7}$, whilst B stars' emission drops off with spectral subtype. Dynamical instability of OB star radiatively driven winds generates shocked regions which may be responsible for the bulk of the X-rays observed. The wind-compressed disc model of Bjorkman & Cassinelli (1993) presents another site for X-ray emission. The disc formed in the equatorial plane of a fast rotating Be star from equatorward drift of wind streamlines is confined on both sides by a shock which may also generate X-rays. As the X-ray emission originating from the wind shocking is ubiquitous amongst B and Be stars then the wind-compressed disc model näively predicts that Be stars should generate more X-rays than B stars of equivalent spectral subtype.

The X-ray emission from the shocks confining compression discs has been calculated and compared to a limited set of observations. The excess X-ray emission from the Be star disc shocks is found to be undetectable over the inherent wind shocking emission.

1. Introduction

Studies of hot OB stars reveal that they all have radiatively driven winds with terminal velocities of a few$\times 10^3$ km s^{-1} (Snow 1981), which have been shown to be dynamically unstable (Owocki, Castor & Rybicki 1986). In the case of Be stars, in addition to the fast wind, there is observational evidence of a disc-like structure in the equatorial plane (Pockert & Marlborough 1978, Waters, Cotè & Lamers 1987, Quirrenbach et al. 1994). Be stars are also found to rotate at a significant fraction of their break-up velocity (Slettebak 1982).

Bjorkman & Cassinelli (1993) have put forward a mechanism which may generate the disc structure implied by Be stars observations: for many B stars, high rotation causes the wind streamlines to drift towards the stars' equatorial plane. As this occurs from both hemispheres, then a shock will form near the equatorial plane where the two opposing flows collide generating a "wind compressed disc" (WCD).

The opening angle of the disc generated by numerical hydrodynamic simulations of WCDs has been found to be small, $\theta \sim 1$–$4°$ (Owocki, Cranmer & Blondin 1994). In contrast Waters et al. (1987) and Cotè & Waters (1987) find that to generate enough IR emission consistent with IRAS observations of Be stars, a much thicker disc, $\theta \sim 15°$, is needed. Therefore it may be that

Astrophysics and Space Science **233**: 323–327, 1995.

although streamline drift towards the equator *will* occur, it may not be the only agent which generates the observed disc in Be stars.

Irrespective of the mechanism which generates the disc, flow streamlines *will* intersect the disc, and so Be stars will have two distinct independent regions which emit X-rays (see fig.1); (1) the strong shocks ubiquitous in radiatively driven winds, (2) the shocks bounding the disc.

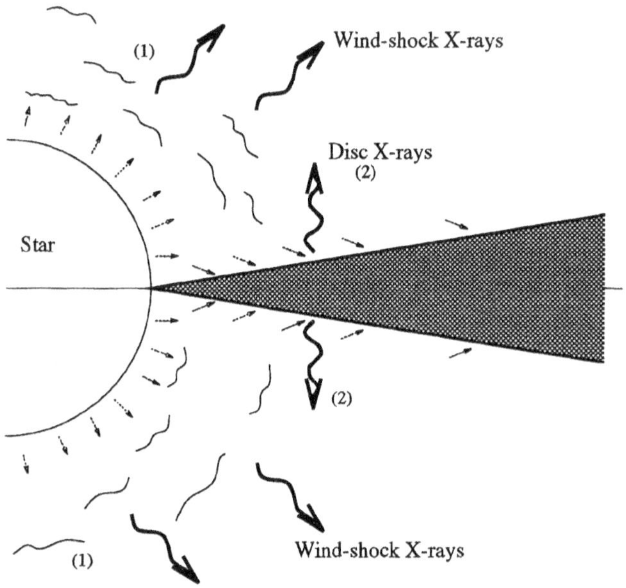

Fig. 1. Cross section through the star and disc showing the two possible sources for X-ray emission around Be stars. The small dotted arrows represent the wind flow. (1) Be stars have radiatively driven winds, which are dynamically unstable leading to strong shocks forming in the wind. (2) As a consequence of high rotation in Be stars, the wind flows into a compression disc in the equatorial plane with shocks bounding the surface of the disc. Almost all of the disc X-rays are generated close to the star ($r < 2R_*$), where the disc shock is still strong.

In contrast, normal B stars have no disc, and so will only have one source of X-rays: the shocks in the wind. Hence, there may be a detectable difference between B and Be stars if the X-ray emission from the disc shocks in Be stars make up a significant fraction of the total X-ray output of the Be star.

2. The Model

The flow streamlines were calculated according to Bjorkman & Cassinelli's (1993) prescription. Although radiatively driven winds are unstable, a time independent model was used here to isolate the emission solely due to the shocks confining the disc.

A set of model stars was used - the spectral type sequence of stellar parameters was taken from Schmidt-Kaler (1982) and the calculation was repeated for different rotational velocities of the B star: 0.5, 0.7 and 0.9 of the critical rotation. The disc was assumed to have a constant opening angle of 10°- the discrepancy in the opening angle of the disc between the observations and numerical simulations has little effect here, as the change in velocity normal to the disc from 1-10° is small.

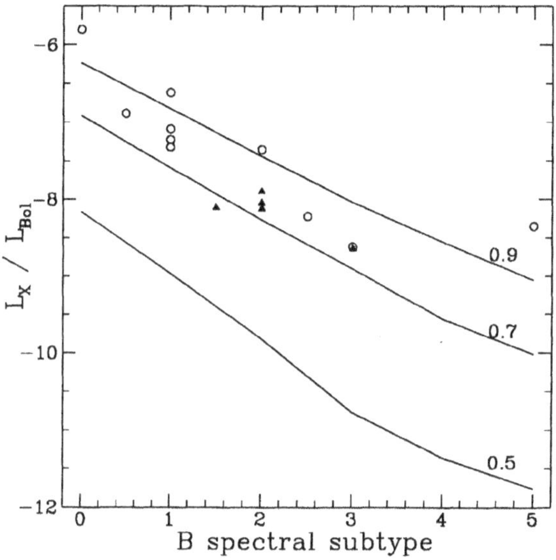

Fig. 2. Total X-ray emission scaled to the bolometric luminosity of the star at the disc confining shocks as a function of spectral type. Stellar parameters for the spectral sequence are from Schmidt-Kaler (1982). The three lines represent the emission from the stars with rotation velocities of 0.5, 0.7 and 0.9 of the critical rotation velocity. The points are the ROSAT observations of B and Be stars from Cassinelli et al. (1994) and Cassinelli & Cohen (1994); the filled triangles are Be stars.

The X-ray luminosity for each streamline which intersects the disc was calculated from the mass loss rate along the streamline given from radiation driven wind theory, and the shock temperature (which is proportional to velocity normal to the shock squared). The mass loss rates of the stars are extrapolated from the fit to O star mass loss rates of Garmany et al. (1981). The total emission of the disc is then the sum of X-ray luminosity over all streamlines which intersect the disc. Almost all of the emission arises close to the star ($r < 2R_*$), as the shocks bounding the disc are strongest in this region - for larger radii, the shocks become weaker leading to very little or no emission. The total X-ray emission was then compared to the bolometric luminosity of the model star to determine whether it may be detectable.

3. Results and Discussion

The X-ray emission from the disc declines with spectral subtype in a fashion similar to B star observations. In fig.2, the symbols are from Cassinelli et al.'s (1994) and Cassinelli & Cohen's (1994) ROSAT observations. This sample (spectral types B0 to B5 with greater than 3 sigma detections) contains 6 classical B stars, 5 Be stars and 4 β Cephei variables. The filled triangles are the Be stars. The lines in fig.2 corresponding to the three rotation velocities are *not* attempted fits to the data. The motivation of this study was to examine whether the disc X-ray emission could be detected above the emission from the wind shocks.

Fig. 2 indicates that for the rotation velocities very close to break up, the disc shock emission is possibly as high as the wind shock output. However, the Be stars have been estimated to rotate at ~ 0.7 of their break up velocity (Chen & Huang 1987), in which case the disc X-ray output is lower than the wind shock emission. With the inherent scatter in the observational data, the effect of any extra disc emission will be lost, making X-ray samples of B and Be stars indistinguishable. This conclusion does not change significantly if different mass loss rates for the model stars are assumed - indeed Garmany et al.'s (1981) law used here may be too high by a factor of a few, especially at early spectral types (see Snow 1982), leading to at least a similar decrease in the total X-ray emission. This decrease supresses the disc emission further, strengthening the conclusion.

References

Bjorkman, J. E., Cassinelli, J.P.: 1993, *Astrophys J.* **409**, 429

Cassinelli, J. P., Cohen, D. H., MacFarlane, J. J., Sanders, W. T., Welsh, B. Y.: 1994, *Astrophys J.* **421**, 705

Cassinelli, J. P., Cohen, D. H.: 1994, in *IAU 162 "Pulsation, Rotation and Mass Loss in Early-Type Stars"*

Chen, H., Huang L.: 1987, *Chin. Astron. Astrophys.* **11**, 10

Cotè J., Waters, L. B. F. M.: 1987, *Astron. Astrophys.* **176**, 93

Garmany, C. D., Olson, G. L., Conti, P. S., Van Steenberg, M. E.: 198, *Astrophys J.* **250**, 660

Owocki, S.P., Castor, J.I., Rybicki G.B.: 1988, *Astrophys J..* **335**, 914

Owocki, S. P., Cranmer, S. R., Blondin, J. M.: 1994, *Astrophys J.* **424**, 887

Pockert, R., Marlborough, J.M.: 1978, *Astrophys J.* **220**, 940

Quirrenbach, A., Buscher, D. F., Mozurkewich, D., Hummel, C. A., Armstrong, J. T.: 1994, *Astron. Astrophys.* **283**, L13

Schmidt-Kaler, Th.: 1982, in *Landolt Börnstein New Series*, Vol 2b, eds. K. Schaifers & H. H. Voigt, Springer-Verlag: Berlin

Slettebak, A.: 1982, *Astrophys. J. Suppl.* **50**, 55
Snow, T. P.: 1981, *Astrophys J.* **251**, 139
Snow, T. P.: 1982, *Astrophys J.* **253**, L39
Waters, L. B. F. M., Cotè J., Lamers, H. J. G. L. M.: 1987, *Astron. Astrophys.* **185**, 206

BIPOLAR OUTFLOWS FROM STARS AND GALAXIES AS A TORNADO PHENOMENON

P.F. BROWNE
Physics Department, UMIST, PO Box 88, Manchester M60 1QD, UK

Abstract. At an early stage in the lives of stars and galaxies when they are surrounded by discs, vorticity in the disc concentrates into a central vortex, thus converting a Keplerian velocity field $u_\phi \propto r^{-\frac{1}{2}}$ into an irrotational velocity field $u_\phi \propto r^{-1}$, which implies inward transfer of angular momentum. Centrifugal forces due to spin-up of the inner region and gravity dominant in the outer region then squeeze gas at intermediate layers, increasing pressure gradient in the axial direction sufficiently to drive a wide-angle low-velocity bipolar outflow from the disc. A logarithmic singularity of vorticity at the axis implies strong centrifugal forces which expand plasma to radius R where pressure gradient balances centrifugal force density of ions; the much weaker centrifugal force density of electrons cannot balance pressure gradient, so that electrons are driven inwards relative to ions until charge separation limits the relative displacement. Now the radial gradient of u_ϕ causes ions to rotate at a different rate to electrons, generating an azimuthal current density j_ϕ which is the source of an axial magnetic field B_z in the core of the central vortex. Centrifuging carries lines of $\mathbf{B}$ to the core wall, where they are wound into helical force-free configuration with $\mathbf{B} \propto \mathbf{j}$. An annular channel of radius R and thickness ΔR into which parallel helical lines of $\mathbf{j}$ and $\mathbf{B}$ are compressed constitutes a "magnetic vortex tube"(MVT). An MVT separates an inner high-velocity highly collimated outflow from the outer low-velocity wide-angle outflow, and is responsible for jets. Magnetic pinches in the MVT may constrict the core flow at HH objects.

Key words: Outflows – Jets

1. Bipolar Outflows Driven by Pressure Gradients

The equations of motion for a centrifugally supported spherical gas are

$$u_\phi^2/r = g_r(r,z), \qquad u_\phi r = l_z, \qquad \ddot{z} = -g_z(r,z) \tag{1}$$

where the constant l_z is angular momentum per unit mass and where $\mathbf{g} = -\nabla\phi_g$, where ϕ_g is gravitational potential.

In the special case when

$$g_r(r,z)/g_z(r,z) = r/z \tag{2}$$

we obtain, with $\dot{\phi} = u_\phi/r$,

$$\ddot{z} = -\dot{\phi}^2(r,z)z \tag{3}$$

If one assumes, further, that $\dot{\phi}(r,z) = \dot{\phi}(r)$, then

$$z = z_0 \cos(\dot{\phi}t + \alpha) \tag{4}$$

Astrophysics and Space Science **233**: 329–336, 1995.

where z_0 and α are integration constants. Thus particles perform simple harmonic motion about the equatorial plane with different amplitudes z_0 and different phases α. The period of the oscillation is that of rotation, $P = 2\pi/\dot{\phi}$. The cloud preserves its spherical shape because the phases are random, but collisions between oppositely moving particles dissipate energy, reducing slowly amplitudes of the simple harmonic motions, so that the sphere evolves slowly into an increasingly oblate spheroid and then a disc.

The rotational velocity field of the disc at this stage will be approximately Keplerian, for which $u_\phi \propto r^{-\frac{1}{2}}$ corresponding to angular momentum distribution and $l_z = ru_\phi \propto r^{\frac{1}{2}}$. On the basis of the discussion in Section 2 we argue now that enstrophy is concentrated into a single axial vortex, after which the rotational velocity field has become $u_\phi \propto r^{-1}$ corresponding to angular momentum distribution $l_z = ru_\phi =$ constant. Thus angular momentum is transferred inwards. Conservation of angular momentum then requires spin-up of the inflowing gas. For unit mass we have $l_z = ru_\phi = r^2\dot{\phi}$, so that $\dot{\phi} \propto r^{-2}$ and centrifugal force increases as $\dot{\phi}^2 r \propto r^{-3}$.

Dominance of outward centrifugal force at small values of r compared to dominance of inward gravitational force at large values of r squeezes gas at intermediate values of r, say $r \simeq r^*$, causing an increase of pressure gradient in the axial directions. If axial pressure gradient $|\delta p(r,z)/\delta z|$ exceeds the axial component of gravity $|\rho g_z|$, a wide-angle low-velocity bipolar outflow is initiated.

In the core of the central vortex there develops an outward pressure gradient for another reason - the lowering of pressure due to centrifuging as a function of z. Axial pressure gradient in the vortex core drives a well-collimated high-velocity bipolar outflow. There are therefore two outflows, a high-velocity well-collimated outflow within a low-velocity wide-angle outflow Fig.1a). The high degree of collimation often is delineated by emission-line jets. Outflow velocity acquired between z and ∞ is

$$\dot{z} = [2p(r,z)/\rho - 2\phi_g(r,z)]^{\frac{1}{2}} \tag{5}$$

Tornado-like bipolar outflows enable the system to shed angular momentum, which is required for further contraction.

2. Evolution of Vorticity in a Rotating Disc

The equation of fluid motion usually is written in the form

$$\rho D\mathbf{u}/Dt = -\rho\nabla\phi_g - \nabla p + \rho\nu\,\nabla^2\mathbf{u} \tag{6}$$

where $\mathbf{u}(\mathbf{r},\mathbf{t})$ is the velocity field, ρ is mass density, ϕ_g is gravitational potential, $p(\mathbf{r},t)$ is pressure and ν is kinematic viscosity. Noting that

$$D\mathbf{u}/Dt = \delta\mathbf{u}/\delta t + (\mathbf{u}.\nabla)\mathbf{u} = \delta\mathbf{u}/\delta t + \frac{1}{2}\nabla u^2 - \mathbf{u}\times(\nabla\times\mathbf{u}) \tag{7}$$

the curl of (6) yields the well known form

$$\delta \mathbf{w}/\delta t = \nabla \times (\mathbf{u} \times \mathbf{w}) + \nu \nabla^2 \mathbf{w} = -\mathbf{u}.\nabla \mathbf{w} + \mathbf{w}.\nabla \mathbf{u} + \nu \nabla^2 \mathbf{w} \qquad (8)$$

where $\mathbf{w} = \nabla \times \mathbf{u}$ so that $\nabla.\mathbf{w} = 0$, and where $\nabla.\mathbf{u} = 0$ because the fluid is assumed incompressible ($D\rho/Dt = 0$). An alternative form of (8) is

$$D\mathbf{w}/Dt = \mathbf{w}.\nabla \mathbf{u} + \nu \nabla^2 \mathbf{w} \qquad (9)$$

A measure of the amount of vorticity in the fluid in volume V is provided by the enstrophy defined as $\int \frac{1}{2} w^2 \, dV$. Following Batchelor (1967),

$$d/dt\left(\int \frac{1}{2} w^2 dV\right) = \int \mathbf{w}.(D\mathbf{w}/Dt)dV = \int \mathbf{w}.[(\mathbf{w}.\nabla)\mathbf{u} + \nu \nabla^2 \mathbf{w}]dV \qquad (10)$$

The interesting term in (10) is $\mathbf{w}.\nabla \mathbf{u}$ which is the gradient of $\mathbf{u}$ along a vortex line. If $\mathbf{u}$ increases with distance the vortex line must stretch, and if $\mathbf{u}$ decreases with distance the vortex line must contract. The term $\mathbf{w}.\nabla \mathbf{u}$ vanishes for two dimensional motion because $\mathbf{w}$ is normal to the plane of the motion and it vanishes also for unidirectional motion. In commenting on the result (10) Batchelor (1967) writes: "There are many flow fields in which the total amount of vorticity increases, and continues to increase in a remarkable way, until through change of the vorticity distribution the loss by viscosity is able to balance the gain by extension of vortex-lines. This is one of the most striking properties of turbulent flow, in which the integral of w^2 over unit volume of fluid reaches a large value (proportional to a positive power of the Reynolds number of the flow) before the loss by viscous action equals or exceeds the gain by extension of vortex-lines."

We are interested in the evolution of vorticity within a rotating disc under conditions of high Reynolds number. Vorticity cannot be created or destroyed in the interior of a homogeneous fluid, and is produced only at boundaries, but it may be redistributed. The problem is of primary interest in atmospheric physics, in which connection a numerical study led McWilliams (1984) to conclude that "under a broad range of circumstances, the flow structure has its vorticity concentrated in a small fraction of the spatial domain, and these concentrations typically have lifetimes long compared with the characteristic time for non-linear interactions in turbulent flow (i.e. an eddy turnaround time)".

Batchelor (1967) gives the following simple example of the flow field associated with a tornado. It is assumed that $\mathbf{w}$ is unidirectional so that $(\mathbf{w}.\nabla)\mathbf{u}$ is a vector parallel to $\mathbf{w}$ (see 9) and axisymmetry is assumed. Then, introducing cylindrical polar coordinates (r, ϕ, t), $\mathbf{u} = [u_r(r,t), u_\phi(r,t), u_z(r, z, t)]$ and $\mathbf{w} = [0, 0, w(r,t)]$. Consider a meridional plane $\phi =$ constant, in which the velocity components are (u_r, u_z). Because $\mathbf{w}$ is independent of z, motion in this plane is irrotational, and the origin 0 is a stagnation point. Then the flow is $\mathbf{u} = \nabla\phi$ where $\nabla^2\phi = 0$, specifically

$$\begin{aligned}\phi &= \frac{1}{2}\alpha(z^2 - \frac{1}{2}r^2)\\ u_r &= -\alpha r/2, \qquad u_z = \alpha z\end{aligned} \tag{11}$$

where α is a positive constant. The equation of motion (8) then yields

$$\delta w/\delta t = \frac{1}{2}\alpha r^{-1}\delta(wr^2)/\delta r + \nu r^{-1}\delta(rw')/\delta r \tag{12}$$

where $w' = \delta w/\delta r$. Under steady-state conditions $\delta w/\delta t = 0$. Then, since $\delta w/\delta r = dw/dr$, (12) has the integral

$$rdw/dr + \alpha wr^2/2 = A \tag{13}$$

where A is a constant. Batchelor chooses $A = 0$, in which case (13) yields

$$w(r) = w_1 \exp(-\alpha r^2/4\nu) \tag{14}$$

If A does not vanish, we may integrate (13) by putting $w = uv$ and choosing v to satisfy $dv/dr + \alpha vr/2 = 0$, obtaining

$$w(r) = A\nu^{-1}\exp(-kr^2)\int r^{-1}\exp(kr^2)dr \tag{15}$$

where $k = \alpha/4\nu$. For $r \simeq 0$, we may approximate to (15) by taking the exponential factor out of the integral, so that

$$w(r) = A\nu^{-1}\ln(r/a) \tag{16}$$

where $a \to 0$.

Batchelor restricts his attention to the solution (14). He writes: "It can be shown that (14) is in fact the distribution to which w tends, as $t \to \infty$, for an arbitrary initial distribution of w with respect to r subject only to the conditions that $w \to 0$ faster than r^{-2} as $r \to \infty$ and that $\int_r^\infty w2\pi r dr$ is finite and nonzero".

3. The Magnetic Field

The logarithmic divergence of $w(r)$ as $a \to 0$ is of considerable importance in regard to the origin of magnetic fields. The steep increase of centrifugal force as $a \to 0$ causes plasma to expand outward until limited by pressure gradient. The centrifugal force on an ion of mass m_i greatly exceeds tht on an electron of mass m_e, with the consequence that pressure gradient forces electrons inwards until the charge separation is arrested by electrostatic forces. There develops a radial polarization layer of thickness ΔR in which is a radial electrostatic field of strength E_r, where

$$eE_r = m_i u_\phi^2/R, \qquad E_r = 4\pi e n_i \Delta R \tag{17}$$

Current density j_ϕ and magnetic field B_z are given by

$$j_\phi = en_i(\delta u_\phi/\delta r)\Delta R, \qquad B_z = 4\pi j_\phi \Delta R/c \tag{18}$$

The field B_z approximates that in a solenoid with current per unit length equal to $j_\phi \Delta R$. Equilibrium between centrifugal force and pressure gradient yields

$$n_i m_i u_\phi^2/R = -\,\delta p/\delta r \tag{19}$$

Eliminating E_r, j_ϕ, n_i and u_ϕ from the above five equations leads to

$$B_z^2/4\pi = (\delta p/\delta r)(\delta u_\phi/c\delta r)^2 \Delta R^3 \simeq p(\Delta R/\Delta r)^3 (u_\phi/c)^2 \tag{20}$$

where we write $\delta p/\delta r = p/\Delta r$ and $\delta u_\phi/\delta r = u_\phi/\Delta r$. Between the positive and negative charge layers the magnetic fields of the opposite currents reinforce each other, so that there is a field B_z without differential rotation.

Winding of field lines into helical form can cause a magnetic pinch. Initially current density $(0, j_\phi, 0)$ is the source of a field $(0, 0, B_z)$. Winding creates helical lines of magnetic field $(0, B_\phi, B_z)$ whose source is $(0, j_\phi, j_z)$. Then

$$\mathbf{j} \times \mathbf{B} = (j_\phi B_z \;-\; j_z B_\phi)\hat{\mathbf{r}} \tag{21}$$

where $\hat{\mathbf{r}}$ is a unit radial vector. Whether $\mathbf{j} \times \mathbf{B}$ is directed outwards or inwards depends on helicity of the $\mathbf{B}$ lines. Defining $h = j_z B_\phi / j_\phi B_z$, we see from (20) that expansion occurs for $h < 1$ and pinching for $h > 1$. For $h = 1$ we have the force-free situation in which helical lines of $\mathbf{j}$ and $\mathbf{B}$ are parallel.

The annular channel into which parallel helical lines of $\mathbf{j}, \mathbf{B}$ and w are concentrated in a force-free configuration is termed "magnetic vortex tube" (MVT). An MVT evolves from a singularity of vorticity (17). Essentially the MVT has a core $(r < R - \Delta R)$ with axial magnetic field $(0, 0, B_z)$ and an external region $(r > R)$ with an azimuthal magnetic field $(0, B_\phi, 0)$, the current density being zero in both these regions. The transition for the axial to the azimuthal field occurs in the core wall $(R - \Delta R < r < R)$ where lines of field $\mathbf{B}$ and current density $\mathbf{j}$ are helical and parallel. The current is therefore confined to the annular channel of cross sectional area $2\pi R \Delta R$. The current is the superposition of (i) a circular current $j_\phi \Delta R$ per unit length which is the source of axial core magnetic field $B_z = 4\pi R j_\phi / c$, and (ii) an axial current $2\pi R \Delta R j_z (\equiv J)$, which is the source of azimuthal external magnetic field $B_\phi = 2J/cr$.

Parallel MVTs of like polarity attract if $h > 1$ and repel if $h < 1$, behaving in the former case like parallel line currents and in the latter case like parallel solenoids. Such attraction and repulsion is seen during growth and then decay of sunspots, which are made up of discrete flux tubes intersecting the surface of the Sun at "magnetic knots". Magnetic forces cause magnetic knots to aggregate when vorticity is strong enough for $h > 1$, and then cause them to disperse when $h < 1$. Thus vorticity controls both growth and decay of the sunspot.

The MVT phenomenon is seen in galaxies as well as stars. Jets and counterjets delineated by synchrotron emission which emanate from opposite poles of the nucleus of many galaxies can be identified with MVTs. Trailing and bending of the jets is expected because lines of $\mathbf{w}$ and $\mathbf{B}$ comove with the fluid (Browne, 1985; 1993).

4. Outflow Phenomena

1. The strong magnetic field in the MVT wall channels the high velocity outflow (50 - 400 km s^{-1}) within the vortex core. Outside the MVT the outflow has low velocity (1 - 10 km s^{-1}) and extends over a much larger solid angle. Jets in emission lines have been seen to extend from the star in several bipolar outflows (Mundt et al., 1984), indicating a high degree of collimation which is expected for an outflow confined to the core of an MVT.
2. A bipolar outflow carries away angular momentum, first the angular momentum of inner disc gas and then angular momentum of outer disc gas. A change in the direction of angular momentum from inner to outer layers of the disc will induce precession of the MVT, as happens for the vortex at the exit of a bath tub.
3. The rate of outflow of gas is controlled by the rate of influx of gas in the disc. If the latter varies for any reason, the variation will feed through to the outflow at a later time. "Knots" in the outflow may therefore be attributed to variation in the gas feed at the base of the outflow. For example, radial pulsation of the disc might generate a periodic sequence of knots in the outflow.
4. Double and occasionally multiple vortices, such as are associated with some tornados, may occur also for bipolar outflows (Walker et al., 1988; Meyers-Rice and Lada, 1991; Lopez et al., 1991; Liljestrom, 1991).
5. Far from the star, where plasma density is low, the magnetic field in the MVT wall may be strong enough to pinch the gas flow. The condition for constriction is $\rho \dot{z}^2 < B_\phi^2/8\pi$, where ρ is mass density in the flow. Assuming that most flow mass is in the form of H_2 molecules present in density $n = 3000$ cm^{-3} (Lada, 1985), and adopting the observational value $\dot{z} = 400$ km s^{-1}, the condition for constriction is satisfied by $B_\phi >$ 0.26 mG. Magnetic field as large as 3mG have been inferred from Zeeman

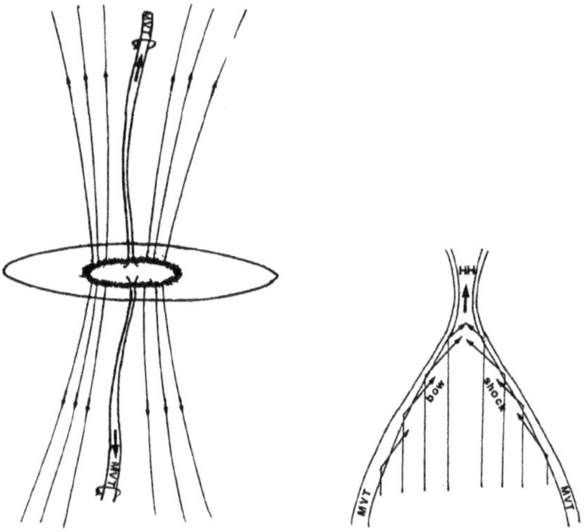

Fig. 1. (a) Two-component bipolar outflow, the inner outflow confined by an MVT. (b) Herbig-Haro source (HH) where a pinched MVT constricts outflow. A bow-shaped shock is produced by reflection of the outflow by the converging MVT wall.

splitting of OH maser lines in the bipolar outflow in Cepheus (Cohen et al., 1984). At the constriction pressure gradient will build up so as to increase the flow velocity through the constriction. Herbig Haro objects may represent such magnetic pinches. Gas excited at the constriction carries the excitation with it, causing HH sources to appear to move if the flow is clumpy. One could understand why some HH sources resolve into a string of knots if the aperture at the constriction were to oscillate.

6. A high resolution study of the outflow from young stellar object IRS 5 using the 2.6 mm ^{12}CO and ^{13}CO lines have led Frilund et al. (1989) to conclude as follows, "The high-velocity gas lies in the middle of the blue outflow region while the low velocity gas is found at the edges. This is strongly suggestive of a cone-like shell structure. In the middle of the outflow we have found apparently accelerated components, the velocity gradient being very steep, about 70 km s^{-1} pc^{-1} for one component which may be interacting with the Herbig Haro object HH 29. Another component which can be traced to 6‘ (0.25 pc) from IRS 5 may have a velocity gradient of about 30 km $s^{-1}pc^{-1}$. The observations also indicate that at least part of the bipolar molecular outflow emanates from the surface of an extended disc. High velocity gas is also seen at the position of IRS itself."
7. "Bow shocks" are simulated by the encounter between the supersonic outflow and the regular paraboloid shape of the converging MVT wall (Fig.1b).

5. MVTs in the Laboratory

There exists strong evidence for MVTs in the laboratory. In the homopolar plasma rotator a discharge is struck between concentric cylindrical electrodes in the presence of an axial magnetic field B_z. The radial electric field E_r due to the interelectrode voltage drives a radial current density j_r which decreases as the plasma takes up the rotational drift velocity cE_r/B_z. If no forces oppose the electric drift, j_r decreases to zero, but after inertial forces cease there is always viscous forces at the walls. In experiments reported by Kunkel et al. (1963), j_r was concentrated into rotating "spokes"; some 10-13 "spokes" passed over small probe electrodes with 10 μs intervals, implying rotational velocity $\sim 3 \times 10^6$ cm s^{-1}.

The spokes have been photographed (Bostock 1986; Fuegas and von Pamel 1989) in another device, the "plasma focus". Current from a capacitor bank enters coaxial cylindrical electrodes at one end, flowing a distance z along the anode, then crossing the interelectrode gap in "spokes" confined to a plane, and finally returning distance z to the cathode. This current is the source of azimuthal magnetic field B_ϕ, so that the radial electric field E_r creates an axial drift with velocity cE_r/B_ϕ. Now "spokes" move axially, and appear stationary from an axial perspective, permitting photographs.

MVTs permit current to flow transversely to a magnetic field **B**. Lines of **B** are stretched and wound into pairs of opposite coils which elongate into a pair of helices in the direction normal to **B**, forming a pair of current channels between the electrodes.

References

Batchelor, G.K.:1967, *An introduction to Fluid Dynamics* Cambridge : Cambridge Univ. Press, pp. 266-273, 105-106

Bostick, H.W.: 1986, *IEEE Trans. Plasma Sci.* **PS-14**, 703

Browne, P.F.: 1985, *Astron. Astrophys.* **144**, 298

Browne, P.F.: 1993, in H.C. Arp et al., ed(s)., *Progress in New Cosmologies: Beyond the Big Bang,* Plenum Press: New York), p. 205

Cohen, R.J., Rowland, P.R. and Blair, M.M.: 1984, *Mon. Not. R. Astron. Soc.* **210**, 425

Frilund, C.V.M. et al.: 1989, *Astron. Astrophys* **213**, 310

Fuegas, J., von Pamel, O.: 1989, *J. Appl. Phys.* **66**, 1080

Kunkel, W.B., Baker, W.R., Bratenahl, A. and Halbach, K.: 1963, *Phys. Fluids* **6**, 699

Liljestrom, T. 1991, in A. Vittone & L. Errico, ed(s)., *Stellar Jets and Bipolar Outflows,* Dordrecht: Kluwer

Lopez, R. et al. 1991, in A. Vittone & L. Errico, ed(s)., *Stellar Jets and Bipolar Outflows,* Dordrecht: Kluwer

McWilliams, J.C.: 1984, *J. Fluid Mech* **146**, 21

Meyers-Rice, B.A. & Lada, C.J.: 1991, *Astrophys. J* **368**, 445

Mundt, R. et al.: 1984, *Astron. Astrophys* **140**, 17

Walker, C.K. et al.: 1988, *Astrophys. J.* **332**, 335

GPSR Compliance

The European Union's (EU) General Product Safety Regulation (GPSR) is a set of rules that requires consumer products to be safe and our obligations to ensure this.

If you have any concerns about our products, you can contact us on ProductSafety@springernature.com

In case Publisher is established outside the EU, the EU authorized representative is:

Springer Nature Customer Service Center GmbH
Europaplatz 3
69115 Heidelberg, Germany

Batch number: 10761697

Printed by Printforce, the Netherlands